BIOMECHANICS

Carl Gans

Department of Zoology
The University of Michigan
Ann Arbor, Michigan

J. B. Lippincott Company
Philadelphia • Toronto

BIOMECHANICS: AN APPROACH TO VERTEBRATE BIOLOGY

ISBN 0-397-47309-5

Library of Congress Cataloging in Publication Data

Gans, Carl
Biomechanics: an approach to vertebrate biology.

Bibliography: p.
1. Animal mechanics. 2. Morphology (Animals)
3. Vertebrates—Anatomy. I. Title. [DNLM: 1. Biomechanics. 2. Vertebrates—Anatomy and histology.
3. Vertebrates—Physiology. QL605 G199b 1974]
QP303.G33 596'.01'852 74-4338
ISBN 0-397-47309-5
ISBN 0-397-47308-7 (pbk.)

Book produced by Ken Burke and Associates
Copyeditor: Judith Fillmore
Design: Joseph Fay
Illustrations: Martha Lackey and Margaret Bruden

Cover photo:

Caecilians like this South American species *Siphonops annulatus* are an order of subterranean limbless amphibians related to frogs and salamanders. Unless one digs into their tunnel system one is unlikely to encounter them. The animals are formidable predators on various small animals, such as insects, mice, and earthworms. After grabbing the prey they retreat backward into their tunnel, twisting as they go so that the food is deformed or broken to a size that they can swallow. This provides one special strategy of feeding in a limbless animal.

PREFACE

THE STUDY OF ADAPTATION is the central theme of biology, although its generality sometimes remains unrecognized. In any environment, adaptive functioning has been the condition for survival and the basis for natural selection. Hence some degree of structural, physiological, behavioral, or ecological adaptation is obvious, as regularly confirmed by ecologists, behaviorists, cell biologists, and biochemists.

The organism's structural pattern reflects this adaptation on all levels. Structure, whether scaled in angstroms, microns, or meters, determines the functional limits; its study should permit predictions regarding selective factors. Such analysis has been called functional morphology, although some of the studies are carried out in the framework of comparative anatomy or as variants of physiology.

Characterization of the adaptive meaning of structure is important to many fields. It permits the taxonomist to recognize the functional implications of the morphological differences among species or higher categories. It allows the physiologist to note the manner and extent in which the structural limits on function may be environmentally induced or affected. It permits the ecologist and evolutionary biologist to make predictions about natural history which may then be tested, thus quantifying the process of adaptive change. Functional anatomy is the method by which the adaptive meaning of structure is characterized.

The present book gives some examples of analyses in functional anatomy. It is not intended to list adaptations, but to offer a sample of approaches to the field and of the philosophy from which approaches derive. What kinds of questions should one ask? How does one approach solutions? Such topics are often omitted in books of comparative anatomy or evolution; this short book is intended to complement these.

What follows is intensely personal. It reflects a personal view in its emphasis on biomechanics, the relation of biological structures to physical force and displacement, rather than the interaction of biological structure with other components of the environment such as light, radiation, and chemicals. It reflects a personal view in proceeding toward solutions with the tools immediately at hand, rather than with those supposedly proper to the discipline. It reflects a personal view in selecting all examples from topics on amphibians and reptiles on which I have done some work, rather than from studies on the many other organisms to which these approaches also pertain.

The examples, most of all, reflect a personal conviction that functional analysis is useful for simple as well as for complex problems. If questions are phrased clearly, they may yield important answers at each level of complexity. The examples are designed, furthermore, to emphasize the "black box" approach; certain systems are assumed to exist and their internal properties are left unplumbed. Biomechanical analysis should include the properties of the central nervous system. This level of organization is beyond the scope of this book.

Each example also describes a different level of supportive research facilities. The available equipment obviously determines the complexity of the problems that can be tackled and the level and manner in which analysis may proceed. At different times and in different places one has different options. It is important to recognize these, and even more important to see the level at which they limit the nature and reliability of the results. Thus limitations, whether I recognized them early or discovered them too late, are emphasized in the examples. They support the view that progress in science proceeds by successive approximations.

The first chapter presents a statement of how a functional anatomist views structure and function. Briefly, it presents the theory behind my views. It also includes some materials addressed to those lacking an extensive background in anatomy. The chapter should be read twice—the second time after the entire volume has been completed.

The other four chapters each deal with a distinct problem. Each chapter presents a detailed statement of the problem and the method of approach, as well as the difficulties, whether I became aware of them during study or some time after the project's completion. The four chapters represent not only distinct problems, but also different levels of complexity in analysis. Thus Chapter 5 deals with a problem attacked by electromyography and movement analysis. These techniques might well have been applied to the problem in Chapter 2 if the equipment and the skills (as well as the living animals) had then been available. Yet the first set of approaches provided answers, as did the last. The observation that significant answers may be obtained from a variety of techniques represents part of this book's message. Another part of the message is that in each case the biomechanical analysis points to conclusions that transcend mere description of the mechanisms used by an animal.

Although the problems here analyzed show some level of resolution even when the mechanical techniques utilized were inadequate, the book should not be considered an apology for ignoring modern techniques. A fundamental difference exists between work carried out with the maximum available, even though less than perfect, instruments and tools, and work carried out by an inadequate approach resulting from an unwillingness to change. When the questions asked are restricted by the limitations of the available tools, the nature of such limitation must be considered. Analysis must then proceed more cautiously, with areas of uncertainty circumscribed and identified for those who follow.

Of particular importance to me is that these biomechanical analyses often provide a connection between data from ecology and from physiology, and that they lead to correlations between physiological and ecological conclusions. They point to phylogenetic and historical questions as well, since they may help to utilize the structures preserved as fossils to furnish clues for interpreting the course of change in the niches occupied by species during their evolution.

The references following each chapter include its citations. Also listed (and starred) are references that should be consulted for discussions of general principles or expanded examples on the topic of the particular chapter. Some concepts that deserved more detailed discussion than could be offered within the text have been set off in a series of boxes. I have tried to keep the vocabulary simple and hope that all new terms have been defined.

This book is intended to supplement lectures and textbooks in courses on comparative anatomy and vertebrate biology, as well as to furnish a brief synthesis of approaches to biomechanical study. It originated in a series of lectures given during the summer of 1968 at the University of Washington as part of a refresher course in comparative anatomy. Some parts, particularly the last chapter, have undergone significant maturation both in my mind and during the process of putting them on paper. The first chapter represents an abbreviated statement of some ideas that I hope to deal with in considerably more detail in a longer book on the practice of biomechanical analysis.

Beyond the individuals who helped with this book, I should like to express my appreciation to those who have, over the years, helped me by collecting or shipping animals, often from obscure places and at considerable personal sacrifice. I am also grateful to those who contributed to my education by letting me utilize the facilities of their laboratories or instructing me in the techniques required to answer questions of current interest. The studies on sidewinding involved collaboration with H. Mendelssohn; those on amphisbaenian hearing with E. G. Wever. The experiments discussed in the fifth chapter were carried out jointly with H. J. de Jongh and W. F. Martin. I thank all of them for their pleasant collaboration and permission to deal with our joint efforts. Most of the studies were conducted with support from the U. S. National Science Foundation.

All of these topics and ideas have been discussed during the last decade with many friends and colleagues, and I acknowledge the major role they have played in my thinking. I am particularly grateful to Charles

M. Bogert, Abbot S. Gaunt, Daniel Janzen, Frank Kallen, and Paul F. A. Maderson, who took the trouble to provide me with their detailed view of a preliminary draft of the whole manuscript. D. G. Broadley, G. H. Dalrymple, F. L. de Vree, C. O. Diefenbach, S. S. Easter, B. E. Frye, C. W. Helms, E. Kochva, R. G. Northcutt, R. B. Payne, K. R. Porter, P. Regal, A. S. Romer, H. I. Rosenberg, W. F. Walker, E. G. Wever, and E. E. Williams among others provided more general comments on concept and style or commented on particular sections of the manuscript. I am delighted to express my gratitude for their support, but the responsibility for the final formulation is clearly my own.

I thank various colleagues, particularly D. Baic, J. P. Bogart, C. M. Bogert, A. K. Bruce, H. Fukada, A. Mittelholzer, W. Porter, and R. G. Zweifel, for permission to use their photographs and the New York Zoological Society for making Mr. Sam Dunton's pictures available for reproduction. I am indebted to J. P. Hailman for allowing me to use his Venn diagrams in the comparison of similarities (Box 1-1). Louis Martonyi furnished a number of photographs and converted the color slides, and Martha Lackey and Margaret Bruden drew the diverse illustrations. I thank the three of them for the effort they expended in helping me to communicate. Carl W. May and Ken Burke clearly spared no effort in helping to design a format that would express my ideas. Above all I thank my wife for counsel and support throughout my biological career.

Carl Gans
Ann Arbor, Michigan

CONTENTS

4 Analysis by Comparison: Burrowing in Amphisbaenians

5 Analysis by Quantification: Air Breathing and Vocalization in Frogs

STRUCTURE AND FUNCTION 1

The Study of Structure

Animals appear and feel solid, but their bodies actually consist largely of spaces filled with liquids and gases. Most of the solid components are flexible; almost none are rigid. The contrast of soft and hard tissues is a real one, even though the relative softness of such tissues as muscle may change with the physiological state of the organism. Some of this capacity for flexibility is externally apparent, especially when animals are active. A slowly moving elephant on a cement floor looks rigid and clumsy; when an elephant runs in the field the head and girdles move freely against the body and the body bends and twists.

The three-dimensional body of an animal represents a composite of diverse parts whose varying shape at any instant constitutes the animal's external configuration. Yet the internal structures not only shift their positions relative to each other, but also they can modify the external configuration itself. A man swinging an arm induces a certain internal shift in the position of the pectoral muscles of his chest. When he chins himself, the contraction of the biceps muscle and the concomitant shifting of the other muscles and of the skin can be clearly seen from the outside. Such external indications provide invaluable clues during the first stages of functional analysis; they also complicate the problem of description.

Descriptions of structures provide a basis for comparison. Such descriptions for comparison may be conceived of as proceeding on three levels, each of which tends to relate to the other two. These levels are:

1. *The identification of a number of discrete elements in the system.* Here one might ask: How many discrete muscles bridge a joint? Which nerves transmit information to or from a muscle?
2. *The determination of the nature of the multiple components.* Here the questions might be: Are the muscles parallel or pinnate? Is the bone compact or spongy? Is the nerve somatic or visceral?
3. *The establishment of the spatial configuration of the components and their change with time.* How does the relative position of the muscles along the tibia of a stork standing on one leg change when it stands on two legs? How does the pattern change during growth? What changes occur during walking?

The identification of the structures ordinarily involves dissection. The determination of the nature of components may involve histology, histochemistry, and biochemistry, all of which are dissection on a finer level. The establishment of spatial interrelation and particularly of the extent to which this changes in time requires special approaches, especially if only a limited number of animals is available for study.

Most kinds of dissection are destructive. Hence there is some need to address the specimen in a repeatable fashion. Some understanding of the postures that characterize the organism is essential for establishing the direction of incision, for organizing the structures observed, and for facilitating future comparisons. Such understanding is critical even when there is no interest in function per se, as when the primary concern involves comparing the structures observed in various specimens. Furthermore, the study of structure, whether called anatomy and restricted to the adult condition of the larger components of the animal, or termed morphology and including ontogenetic shifts and cellular conditions, requires some understanding of the range of movements that the animal can perform.

Some concept of a "normal" position may be obtained by observing or handling a living animal or by manipulating an anesthetized or freshly killed specimen. Living animals may not be available, however, and freshly killed specimens pose difficulties, such as the risk of infection by microorganisms or other parasites. A "fresh" cadaver, furthermore, represents a very transient phenomenon, and one's eyes and nose will soon indicate that some major changes are taking place. The processes of desiccation and decay quite rapidly distort the configurations existing in life.

Rather than racing to complete studies of fresh cadavers, it is preferable to use preservatives. These should inhibit postmortem changes, at least in gross shape and appearance; through chemical action some of the dissolved substances in cells and tissues are precipitated and liquids are changed into gels or solids. Inhibition of decay results from the introduction of various toxic substances (which also kill surviving parasites), and desiccation is controlled by moistening agents. Unfortunately, preservatives also have obvious disadvantages; they are irreversible and tend to be toxic, to smell, and to change the natural color of structures. Most deleterious is the production of major and differential shrinkage in the several kinds of soft tissues, the flexibility of which simultaneously undergoes drastic change. Preservation may be necessary and useful, but the process may affect the information that can be obtained from the specimen. Freezing, particularly of specimens injected with moistening agents, may provide an interim approach, but it is an expensive method.

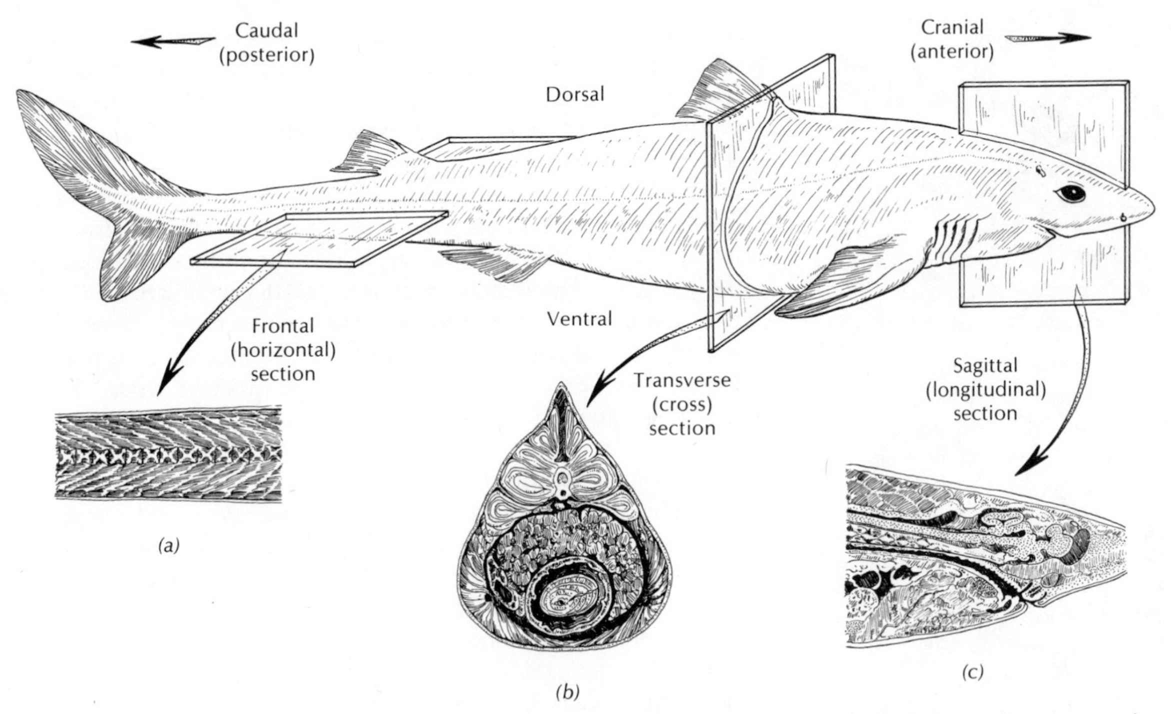

Figure 1-1 Sketch of a dogfish shark to illustrate anatomical directions and the three standard planes of sections used to visualize different parts of organisms.

The description of the animal's structures may proceed in any sequence (top to bottom, left to right, anterior to posterior, proximal to distal). The histological approaches involved in characterizing particular tissues may be determined by the shape of the organ, and such techniques are defined in many books (Romeis, 1948; Weibel and Elias, 1967). The sequence becomes critical when dissection is to serve topographic comparison–in other words, when the geometric relation of structures is to be compared. Three concepts should be kept in mind: (1) The standard "posture" of the organism may not be the obvious one. (2) Not all views may be obtained on the same specimen. (3) Dissection from the outside inward might not provide the most information. A few examples may clarify these concepts:

1. The pectoral muscles of the cat are ordinarily dissected on specimens fixed with the limbs spread wide apart, a posture into which one can place a live cat only when one is willing to face loud squeals and sharp claws. Especially the muscles suspending the rib cage from the scapulae (shoulder blades) are totally distorted in this posture.
2. Vertebrates have endoskeletons, which means that the bones and

their ligaments are exposed only after the supervening muscles and other soft tissues have been removed. Thus several parallel and destructive dissections may be necessary to understand fully the components of a joint and the topography of the muscles positioning it.

3. The most important aspect of dissection for topography is the use of different plans of approach. Whole animals may be frozen and sectioned on a bandsaw; dissection may thereafter proceed from the inside outward. The angle of the section has to be determined by the objective. The respiratory muscles of the turtle are best seen from the inside after the specimen has been sectioned transversely at midbody and the viscera removed; the architecture of a pinnate muscle may best be visualized if it can be dissected from a plane passing longitudinally through its main tendon.

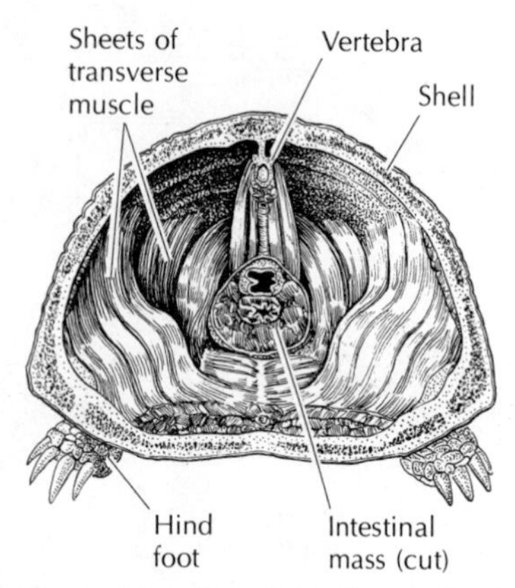

Figure 1-2 View into the shell of a tortoise to show the muscular sling that supports the viscera and changes the volume available for the lung. Only when the dissection proceeds from the "inside out" after sectioning the shell and removing internal organs does one get a good view of these thin but extensive sheets of muscle.

Certain spatial relations are so difficult to determine by dissection that special techniques are necessary (cf. Hildebrand, 1968). The animal's body may be cleared or rendered transparent and particular structures such as the bones, the circulatory system, and the semicircular canals may either be stained or injected with contrasting media that will clarify their extent and spatial position within the animal's total envelope. Injection media can also be precipitated or polymerized in place and the surrounding tissue then digested away; this maintains the appearance of a system of spaces or channels without retaining their relation to the animal as a whole. Also useful may be the preparation of x-rays from several different directions and the comparison of these images to provide three-dimensional reconstructions of internal organs in relation to the animal's surface.

When it is known that the catalog of structures obtained through dissection will serve as a basis for future biomechanical study, there is an advantage to a particular order of description. Certain structural elements are more closely associated than others; thus the head of a mammal logically subdivides into skeletal elements, including a "skull" incorporating the upper jaw, a lower jaw (this may again be divided along the symphysis), and several cervical vertebrae. Each such unit may have associated elements (e.g., brain, eyes, pinnae, and vibrissae). Other elements—such as the masticatory muscles, their nerves, and blood vessels—obviously connect mandibles to skull. Consequently it may now be most useful to organize the description by mechanical units and connecting elements (Gans, 1969a). The subdivision should serve entirely for the momentary convenience of the describer and the user of the description. No attempt should be made to imply that such mechanical units represent divisions due to particular functions. To do so confounds fact with interpretation.

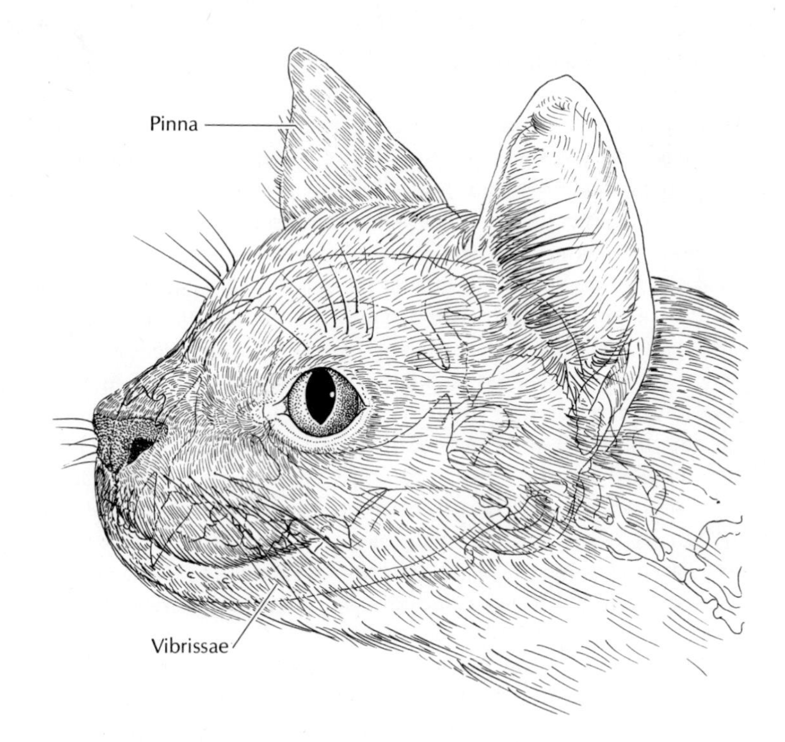

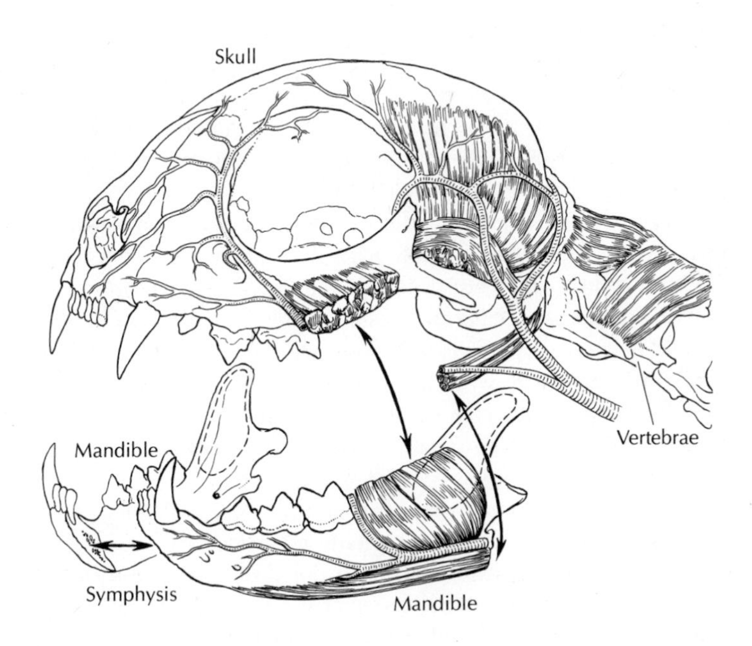

Figure 1-3 The head of a cat may be divided into such obvious mechanical units as the skull, left and right mandibles, and cervical vertebrae. The masticatory muscles, here shown cut (see arrows), serve as connecting elements. Mechanical units and connecting elements are thus convenient categories for subdividing an animal for purposes of description.

Structure and Adaptation

Adaptation represents the general or specific way (or aspect) in which an organism survives in an environment. The term adaptation also defines the process by which this state arose. The existence of a species suggests that its members can now meet the conditions of their environment, and that some minimum level of adaptation has presumably existed at all times in the past history of the species. How such adaptations arose can be explained in several ways; the hypothesis that they are due to natural selection agrees best by far with the evidence. Natural selection is here considered axiomatic, and those interested in further expositions are referred to the treatments of the "Synthetic Theory of Evolution by Natural Selection" in Grant (1963), Mayr (1963), Simpson (1953), and Williams (1966). Yet though natural selection may be accepted by itself, the implications of such acceptance are often forgotten by those dealing with biological structures; curious pre-Darwinian usages seem to hang on both in animal and plant anatomy. It thus seems useful here to restate some of the corollaries of natural selection.

It has long been known that certain characteristics of an organism are inherited; thus mice bear baby mice and elephants produce baby elephants no matter what the environment. Obviously a set of inherent instructions directs the form of the animals. At the same time, the environment does have an obvious effect on certain characteristics of an organism; thus the hair length in some mice or ear color in certain rabbits may reflect the temperature during their development, and tadpoles raised on plant material will have an intestine two or more times as long as that of members of the same egg clutch fed on meat extract (Babák, 1903). Environmental changes will influence the nature and proportions of major structures; length and surface of the frog's alimentary canal change drastically during the year (Zamachowski, 1970). This demonstrates that we are generally looking at phenotypes, the visible individuals and features that result when an organism's particular set of instructions, or genotype, has been "read" in a particular environment. The anatomist obviously deals with phenotypes and is rarely certain how much of what he sees results from the primary genetic instruction and how much results from the individual's history or, rather, the environmental circumstances during which these instructions were carried out.

The process of turning genotype into phenotype cannot be viewed in terms of one characteristic. A giraffe may not have a gene carrying the particular instruction for each spot nor does a zebra for each stripe. Almost any externally visible characteristic is produced by a sequence of biochemical reactions, with most steps being controlled by different loci (places on the chromosome) in the genetic instruction. A simple branching series demonstrating this kind of biochemical sequence appears in Figure 1-4. A more complex one may be visualized by examining one of those diagrams of the Krebs cycle ubiquitous in textbooks of introductory biology; each step there illustrated is presumably controlled by one or more loci. Interruption of any one of the steps, whether by a misreading of the genetic instructions or by such environmental effects as the absence of a necessary chemical precursor and change of temperature, may cause the end product of a series of reactions to disappear and an intermediate product to accumulate before the point of interruption. The end product, whether process or structure, may therefore disappear or be modified as the result of any one of numerous and distinct changes. Many of the instructions will, furthermore, affect the rate at which a particular compound is being produced, rather than determining its presence or absence. They may thus enhance or inhibit the ultimate size or growth rate of a structure. Slight variations in biochemical reaction sequences may consequently lead to a spectrum of terminal states (characters).

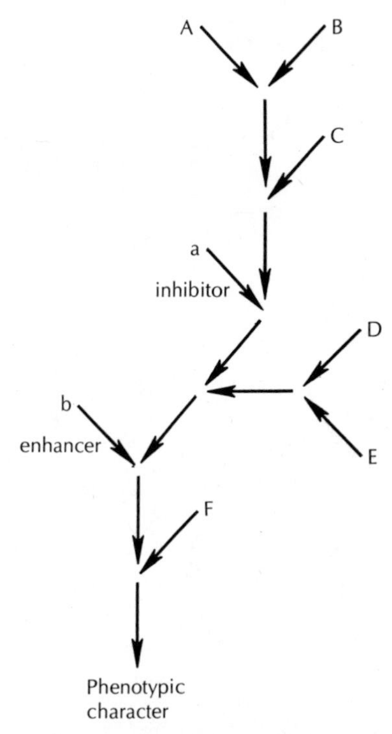

Figure 1-4 Sketch of a sample branching sequence in the transition from the genotype into a phenotypic character. Each letter refers to a genetic locus at which an allele produces a compound or controls a reaction. The loci at *a* and *b* respectively inhibit or enhance the rate of reactions; such loci affect the occurrence and magnitude of the phenotypic characteristics of the organism.

The magnitude of the phenotypic change need not reflect the size of the genetic difference. Substitution of a single allele may switch an animal's hair color from black to white; several alleles may have to be substituted to change the hair color from a light to a slightly darker gray. The probability of a change between two character states will then be seen to depend on the nature and control of the pathway by which the "character" is produced.

Genes will not participate in only a single polygene chain, but they or their products tend to interact with genes or products from various parts of the genome, the total set of genetic instructions. Each gene will have pleiotropic (multiple, diverse) effects on two or more processes and structures. Some pleiotropic effects, such as the simultaneous effect of an allele on the wing shape and eye color in fruit flies, may be seen in the adult, whereas others may be apparent at different developmental stages during the organism's life.

When a morphologist considers structure he looks at one or more individuals. Each is not only an entity in itself but a member of a population sampled from a biological species. In general, a species is defined by the concept that all members are capable of exchanging genetic material, although the exchange sometimes will proceed over several generations via geographically intermediate individuals; such genetic exchange does not ordinarily occur between members of different species. There is a further implication that within particular species members are more likely to mate with individuals from their local population than with those of geographically distant populations. This, coupled with the effect of local selection, seems to result in increased homogeneity among mem-

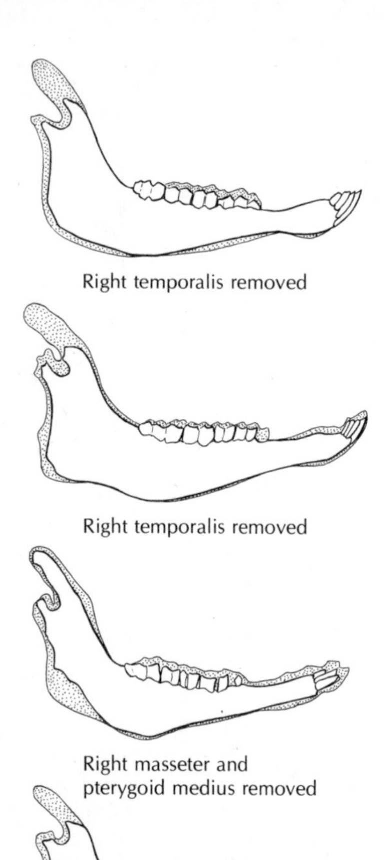

Figure 1-5 Removal of various masticatory muscles from one side of a newborn sheep produces changes in the structure of the mandibles. These sketches show the effect of such removal on the right jaw compared to that on the left (shown as a hatched background). The unilateral removal of a muscle clearly has an effect on both mandibles. Although it may be rare for animals to lose a muscle completely, many conditions result in degrees of muscular impairment. Polio, accidents, and parasites are obvious examples. (After Schumacher, 1973)

bers of a particular population and diversity between members of populations that are geographically separated.

Even members of a single local population, and of the same sex and age, have different genotypes. For any characteristics there are likely to be a few types of alleles dominating a characteristic at fairly significant frequencies, with certain rarer variant alleles persisting at very low frequencies. The expression of any of these alleles is likely to be affected by a spectrum of environmental circumstances; hence the population genotype frequencies tend to be masked in the frequencies of population phenotypes. Most characteristics will show considerable phenotypic variation, and more than one character state (or morphological condition) commonly occurs at a relatively high frequency while other variants are rarer. This means that generally no single "normal" condition exists for a characteristic such as hair color or density of bone in the ulna. Such a multimodal distribution of variables affects the kinds of statistical tests applicable to data collected about a characteristic, but, as important, it suggests that terms such as "abnormal," "throwback," or even "exception" and "variant" should be used with caution. There may indeed be specimens exceptional in one or another characteristic, but one should not argue that a character state is abnormal if it is genetically determined and occurs in some 10, 20, or 30 percent of the population.

All of the phenotypic characteristics of an organism are under the influence of natural selection. Selection may involve the death or destruction of an individual or differential reproduction whereby certain types of individuals contribute greater numbers of offspring to the next generation. Even a well-adapted form or individual may produce occasional inadaptive individuals. An individual that has suffered a nutritional deficiency during a bad season, or suffered broken bones in an accident, may continue to survive though it may never reproduce or will contribute fewer offspring to the next generation. The morphologist must constantly ask: Is the one specimen in hand typical of the species as a whole or of a variant subset? Many specimens kept, bred, or raised in zoos or laboratories develop characteristics that would militate against their survival in the wild. For some of their aspects selection has been temporarily relaxed, but only for the short range, not over evolutionary time.

This view of adaptation and the adaptive process suggests that morphologists must consider the individual specimen as a less-than-minimum sample from a population. The variability within a population becomes as much a structural characteristic as the length of a single bone or the fiber pattern in a single muscle. Thus numerous individuals must be observed for a characteristic if an approximate picture of that characteristic within a population or species is to be obtained.

Adaptation is often discussed in terms of the animal's function—that is, what a structure performs or allows the animal to perform (see Bock and Wahlert, 1965, for different uses of these terms). Each bit of calcified tissue resists deformation; it may also serve as a calcium depot. These constitute two of its functions. Each is determinable by direct or amplified observation; the term function, as used here, does not imply that the process is adaptive in the context of any particular optimization process.

Numerous studies have suggested that the function of a structure should be defined for each particular activity performed by that structure. Thus the ventral edge of the bony orbit of a vertebrate skull has a different function when facilitating vision than when it braces the teeth during mastication. Another kind of definition argues that only those imposed forces or actions that are ultimately useful to the organism—in other words, those that promote selective advantage (adaptive functions)—are truly functional. The noise produced by rubbing of adjacent parts of snake's scales is then a function only when it serves to deter predators on the snake (see p. 66), not when it inadvertently advertises the snake's presence to its prey. Such arguments rapidly involve one in a semantic morass.

A similar difficulty arises when ordering structures into so-called functional components or functional units for purposes of description. The limits of such functional units will always remain unclear, as one combination will be of use to one investigator, but another combination will be more useful to another investigator. Adjacent functional units must obviously overlap since any structure will inevitably participate in the functioning of multiple systems. Should the edge of the bony orbit be included in the unit of vision or of mastication? The definition of the systems studied will inevitably determine the kind of subdivision achieved, and the definitions will vary depending on the specialist. A neuroanatomist emphasizes sensory functional units, whereas some physiologists see the animal in terms of the flow and temperature effects of its blood. A terminology built on functional components is most useful when one wishes to understand the number and kinds of components involved in assisting the action of a particular process; even then the terminology is likely to be modified by minor changes in the selective effects of the environment (cf. Chapter 4).

A Model for Analysis

The interactions of structure, function, and selective force may be visualized on the basis of some simple diagrams. The actual conditions (or parameters) of a particular structure in an individual animal may be demonstrated by the envelope shown in Figure 1-6. The envelope

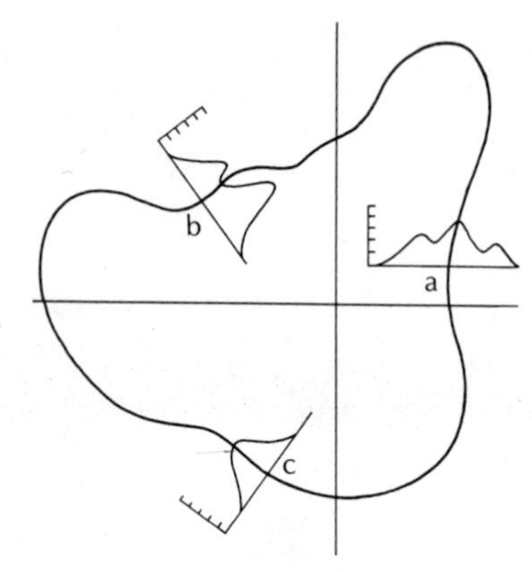

Figure 1-6 Sketch showing the outline of a bone in an individual animal traced upon a set of axes. Similar outlines for other members of the population would show a different shape. A frequency histogram of observed conditions may then be plotted for any point on the periphery. Three such histograms are here shown for points *a*, *b*, and *c*. In the same way this diagram maps the shape distribution of bones, one could map the distribution of a bone's resistance to failure in different directions, or the movement of elements at a joint. In each case one compares the condition seen in a particular individual with the condition seen in the population.

might, for instance, represent the outline of a bone or, if plotting is in polar coordinates, the excursion of which a joint is capable along various axes. Illustration in two dimensions is obviously for the sake of simplicity; the envelope could ultimately be traced in as many dimensions as there are parameters. If the same parameters of the structure are mapped for many specimens from the same population, one can obtain a frequency distribution for the position of any point of the boundary. This has been shown in Figure 1-6 for three such positions; thus the figure shows the observed phenotypic structure for the individual and its distribution for the population (at those positions). By implication the figure may also be assumed to represent the physical (or physiological) capacity of the structure in terms of maximum excursion or maximum loading (imposed forces) possible without significant likelihood of rupture.

In Figure 1-7 we see the same outline. Superimposed upon it is a frequency distribution giving the percentage of time the utilization represented by a particular point (whether in terms of stress level or excursion) actually occurs in the life of an individual. Utilization of the structure has been defined as its function. Isophenes or lines of constant conditions will then indicate conditions found 50, 60, or 90 percent of the time. The actual use pattern can be established only by monitoring the condition of the animal throughout its life cycle. One can go one step further and plot such isophenes for the population of organisms rather than for one of its members. Each line will then represent a condition in yet another frequency distribution, but such distributions have been omitted in order to keep the diagram reasonably simple.

It should be apparent that not all positions on the diagram will be equally important to the animal. The ability to withstand an extreme loading need not relate to the frequency with which this loading occurs in the life of the organism. The need to resist a particular mechanical shock or to produce sustained contraction of certain muscles may be maximal when the individual is threatened by a predator or engages in mating activity with others of its species; incapacity at that moment is likely to have a drastic effect on the probability of leaving viable offspring. Consequently one can plot another set of isophenes (Figure 1-8) to show the modified selection coefficients (representing relative reproductive success) accruing to each of these phenotypic positions. Such isophenes will shift depending on the environment within which the individual or the population is exposed to stress. As the environment changes, so do the isophenes. It is most unlikely that the isophene for selection coefficients of 0, .1, .2, or any other value, and the actual structural perimeter will regularly be parallel. Just as in structures designed by man, natural systems do not necessarily incorporate a constant "factor of safety" in all directions.

How then is the structural envelope established? What factors position it in one place rather than in another, and how is it kept there? The selective advantage for retaining each side of the structural envelope in one position rather than in another seems to be clear, since the advantage can be mapped as in Figure 1-8, but there are four complicating factors.

The *first* complication results from pleiotropic genes. Matching of the mean and minimal strength of various structures with the total selective effect has to proceed simultaneously for multiple aspects of multiple structures—for example, for bone density and excitability of nerve cells or some other seemingly unlikely combination of properties. If several such characteristics are pleiotropically determined, it becomes unlikely that all have equal fitness. The result in any given environment is a compromise resulting in an optimum fitness for the combination rather than an apparent maximum contribution to fitness by any individual aspect of the pleiotropic gene.

The *second* complication is that the diversity and fluctuation in environmental conditions will influence the spectrum of phenotypes resulting

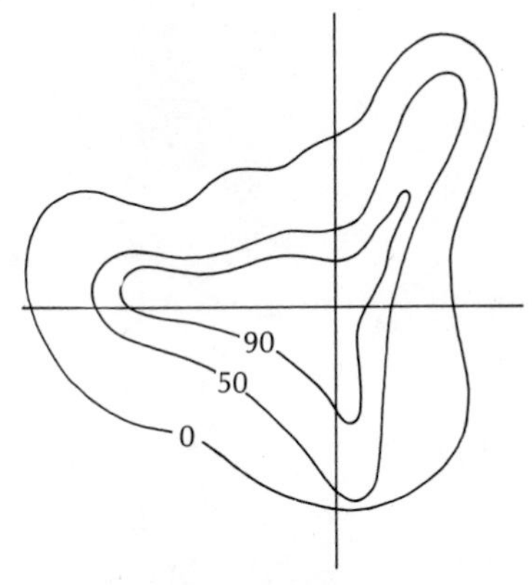

Figure 1-7 Outline of bone in Figure 1-6. The superimposed isophenes show the frequency that the material at a particular spot is used during the life of the individual. Some zones are obviously used much more frequently than others. Some peripheral material has low utilization and other peripheral regions show relatively high utilization (for instance, near the bottom of the *y*-axis). Hence there is no correlation between position and utilization frequency.

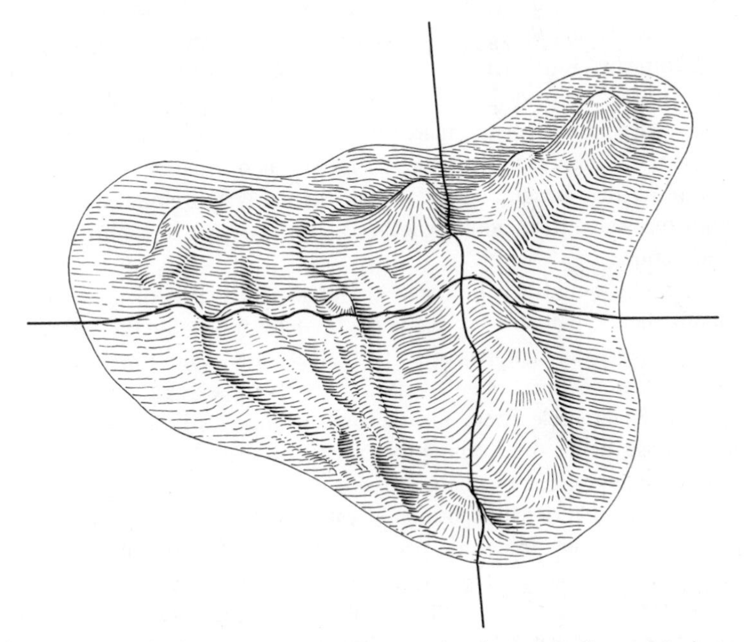

Figure 1-8 Perspective view of the outline of the element in Figure 1-7, shown this time as an adaptive surface by indicating the selective advantage for the material at each point of the cross section of the bone. Loss of material in "low" areas would not have much effect on the fitness of the individual, but loss of the zones carrying high advantage (peaks in the diagram) certainly would. The isophenes for constant selective coefficients have been omitted; they would represent lines of constant elevation in such a diagram.

from a particular genotypic instruction. The selection for maintenance of a particular structure will lead to genotypes incorporating "over-construction" for some individuals. In other words, there will be variance in the phenotypic characteristics, and many individuals will not possess some optimum phenotype but will have an even stronger structure.

The *third* complication is that the expression of this excess capacity or overdesign of the characteristics of individuals is always disadvantageous. An unnecessarily large or dense element may, for instance, be able to withstand all imposed mechanical shocks, but its size or density might interfere with movement. The energy cost of producing, maintaining, and moving a very heavy leg bone may reduce the fitness of a cursorial mammal (such as an antelope). An individual's probability of leaving offspring would hence be greater if its structures were somewhat less capable of resisting all possible mechanical failures.

A *fourth* complication is perhaps the most important. Organisms are not "designed" for their maximum adult size but must be able to grow from a small to a larger size. During growth they pass through stages that often have entirely distinct methods of making a living; consider the contrasts among egg, tadpole, and frog, or egg, caterpillar, pupa, and butterfly. Furthermore, the different sexes incur distinct environmental pressures and usually exhibit distinct phenotypes. The instructions coded in the genetic material must incorporate each contribution to the overall fitness of the various stages in such a manner that selection will result in a structural pattern that can easily be remodeled from an immature stage to an adult, whether at metamorphosis or during sexual maturation.

Mention should also be made of the so-called physiological adaptations—phenotypic variants that may be determined by the behavioral and physiological activities of an individual rather than purely environmental factors. Such phenomena may be complex. For example, the thickness of muscle fibers and the bulk of muscles reflect the amount of exercise, hence the swelling muscles on some athletes. Within limits, the tensions exerted by these muscles will also cause strengthening of tendons and additional deposition of calcium salts in the bones to which they are attached. In certain bones, such as the dentary of the rat, the shape of the bone will be influenced by muscular tension; transsection of the temporal muscle in embryos inhibits development of the coronoid process (Washburn, 1947); other disturbances during development will produce particular modifications depending on their nature and on the species involved (Figure 1-5). Warthogs feed in the kneeling position and calluses develop on the skin of their knees. Similar examples may be drawn for virtually every physiological and morphological system.

The processes of physiological adaptation are pervasive and generally

Figure 1-9 Body building is an obvious example of the application of a physiological adaptation. Regular exercise produces responses in muscles, tendons, and bones; the phenotype changes but the genotype remains constant. (Courtesy of Charles Atlas, Ltd.)

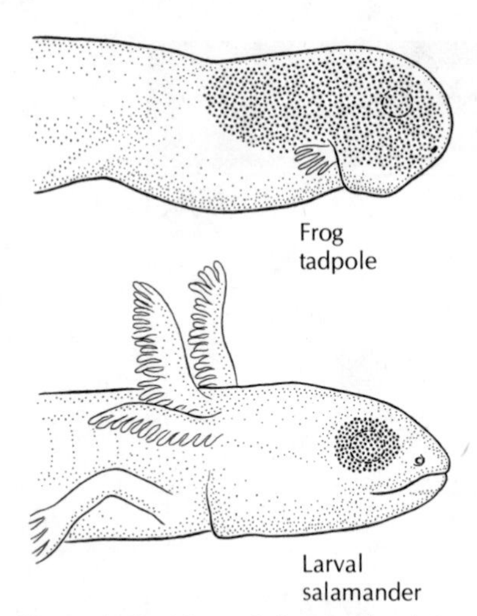

Figure 1-10 Most of the cheek of the frog (*Rana*) tadpole is covered with skin (shaded) that has the capacity to form a lens when induced by an underlying cup of nerve tissue. In the salamander (*Ambystoma*) this capacity is reduced to a small region (shaded) in the vicinity of the future eye.

increase the adaptiveness (here presumably the fitness) of the organism. Their generality has also led some people to argue that function will directly shape structure. According to this view, bone, for instance, is shaped primarily in response to stress in the connective tissues; physiological adaptation rather than genetic instruction shapes the organism. Such a simplistic view does not acknowledge that natural selection has in turn established the genetic instructions that determine the kind of physiological adaptations that can occur. The capacity of the dentary bone to form a coronoid process in response to temporal muscle pull is specific to one bony region in a limited number of mammals. In the same way, the capacity to form a lens in response to the induction of an underlying cup (Figure 1-10) is restricted to the optic region in the salamander *Ambystoma* (Brandt, 1949), but extends over the entire side of the head in some frogs (Spemann, 1936).

Physiological adaptations, then, represent mechanisms by which particular kinds of genetic instruction are transformed into a variety of phenotypes, each of which shows equivalent or higher fitness than would a phenotype formed without a feedback from behavioral and physiological input. Since the behavior and physiology will themselves be affected by environmental conditions, the physiological adaptations allow a matching of the organism to the conditions of variable environments.

These complications suggest that the pattern and range of any parameter of an animal's structure will represent a compromise of many factors and cannot be expected to be perfectly matched to the conditions of any particular moment. Each of an animal's aspects is likely to exhibit some level of "excessive construction" (while a few aspects may involve "insufficient construction"). It is not present everywhere in all specimens, but each specimen will show it somewhere. An individual may then compensate by altering its activities in ways that impose different loadings on its body, as long as the new loadings do not significantly exceed the overall capacity of the structure.

The concept of excessive construction along part of the structural envelope (Figure 1-8) makes it easier to explain the mechanism by which evolutionary changes take place. It suggests that an animal's structure is not matched so perfectly to a particular behavioral pattern or biologic role that any change would cause it to be overloaded. We may then argue that the structure is "protoadapted"[1] for letting the animals impose various new roles. As soon as any new behavior is imposed it will immediately establish a new isophene pattern for selection;

[1]The old term "preadapted" includes so many teleological uses that it seems better to use the present term.

the change in selective effect will result in a shift of the genotypic and phenotypic frequencies of the population. This shift or change in gene frequency is evolution.

Although the original organism obviously must have had some structures, we can now address ourselves to the old question: Which came first, structure or function (here equal to behavior)? The above model suggests that the *current* structure of most individual animals allows them to engage in some behavioral shifts. Variability in structure then permits a range of behavior that may facilitate the "invasion" of a new adaptive zone. A new biologic role (engendered by a new behavior) seems to precede a "new structure."

The Rationale of Comparison

The isolated description of an individual cell, a bone, or a physiological process in an individual is ordinarily of little use. Only by the end of the eighteenth century, when more and more objects were not merely described but compared, did a true biology arise from the realm of random statements. Comparison leads to generality. Which attributes are similar and which differ? What traits are shared by shark, ostrich, and emu? How do penguin and rhea differ from wood mouse and sheep? Are all the observed similarities of equal value, or do some of them have more meaning than others?

The answers depend on the aim of the comparison. One primary reason for comparison is to establish relationships. We can safely conclude that organisms have evolved and that all animals alive today and those known only from the fossil record descended from one or a few ancestral populations. This suggests that for various structures there may be different classes of similarities, superficial ones that represent a response to current (presently acting) selection and more profound ones that presumably reflect ancestral relationships. The former are often referred to as analogies and homoplasies and the latter as homologies. Basically homologous structures such as the flippers of two species of whales may simultaneously be analogous in particular details, depending on the level at which analysis is made (see Box 1-1).

Distinction between analogy, homoplasy, and homology is of obvious interest in understanding adaptation or determining affinities, and much attention has been paid to definition of the terms and to criteria for establishing homology. The earliest investigators (pre-1830) considered those structures to be "homologous" that occurred in the same place in the animal's body and bore equivalent relationships to other structures such as sense organs, nerves, and blood vessels. Some time later (about 1850), von Baer's ideas regarding embryonic origin of structure became in-

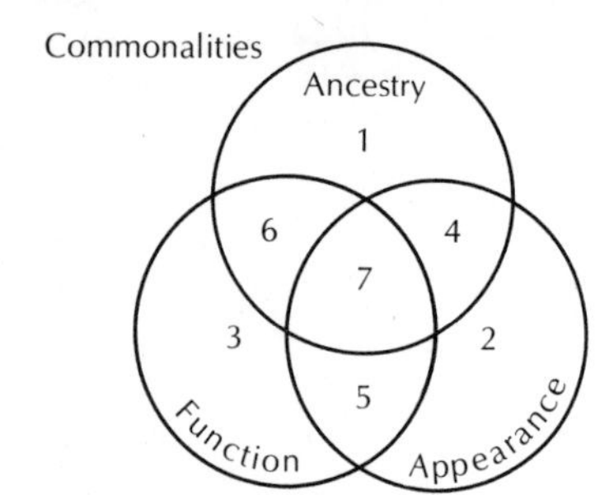

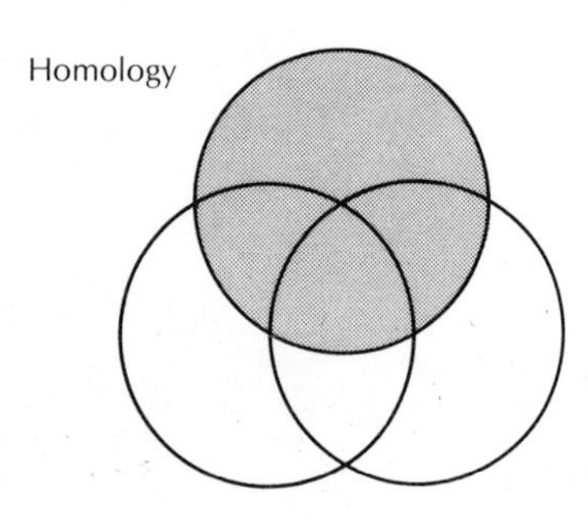

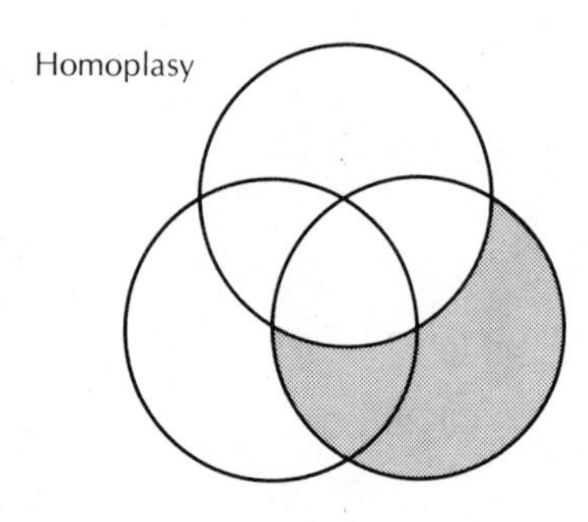

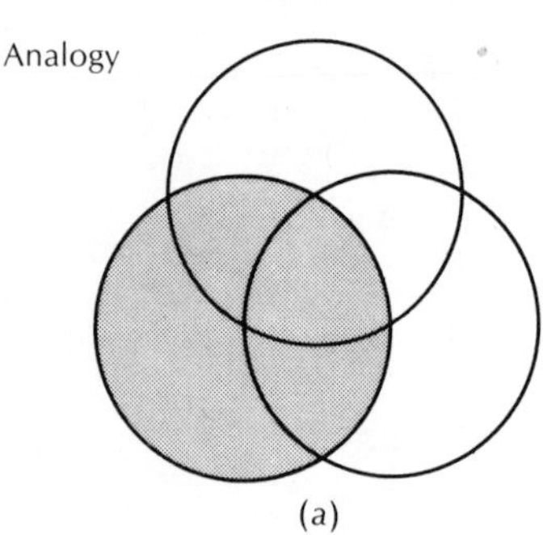

(a)

Box 1-1 Homology and Analogy

Structures may be similar in three ways, namely appearance, function, and ancestry. Although these are sometimes related, they need not be. The several similarities may be visualized by the simple Venn diagrams shown in (*a*). Thus structures may be similar because they share common ancestry or a common function; it is also possible that the similarity is accidental. In the same way we have situations in which two structures in different animals may share similar ancestry but have a distinct function and appearance. Thus the quadrate bone of the crocodile derives from the same element as the incus of a mammal. The former is massive and supports the jaw; the latter is tiny and transmits vibrations in the middle ear.

The usage adopted in this book would consider two structures *homologous* if they share common ancestry. The proof of such common ancestry is often difficult and may rest on evidence from the fossil record documenting the intermediate condition or from embryology indicating that the two structures derive from the same embryonic precursors (a muscle from a particular somite or a bone by ossification of a particular cartilage). It is always important to define the claim. For instance, two bones may be generally homologous, but a particular process on one may not be homologous to a seemingly similar bump on the other. Inspection of the humeri of three animals (*b*, a mole; *c*, a dolphin; *d*, a horse) shows how different even such clearly homologous structures may be.

Although there is widespread agreement about the meaning of the term homology, there seem to be more terms for other conditions than there are compartments. We here include only two, namely *homoplasy*, which defines those structures that are similar in appearance but not homologous, and *analogy*, which defines those structures that share a function whether or not they are similar in appearance or share a common ancestry. In view of the widespread disagreement in the literature (see footnote 2), these terms for the more superficial kinds of relationships are presented with some reservation. However, term and concept of homology are clear and useful in morphology. One can trace homologous structures throughout a series of organisms, and the occurrence of homology in several aspects of two organisms is frequently utilized to document affinities.

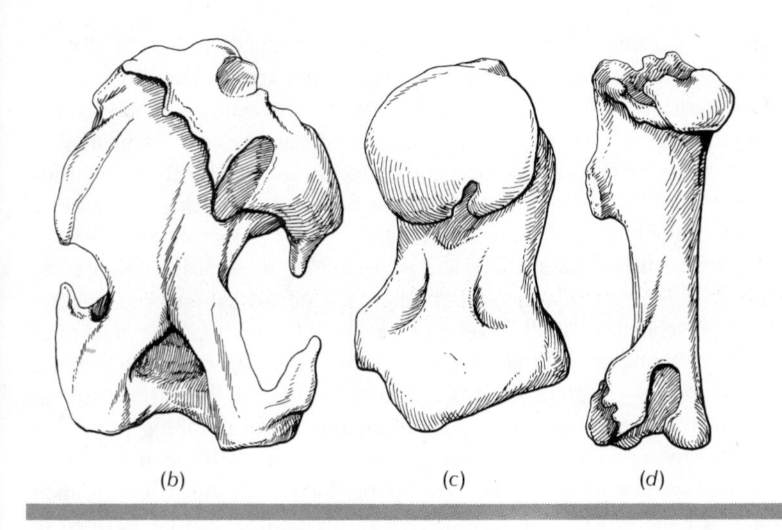

corporated in the homology concept (Russell, 1916). To be homologous, structures had to derive from the same germ layers and pass through parallel embryological sequences. Almost simultaneously, as a result of Darwin's work and its reinterpretation, von Baer's concept was used as a basis for the suggestion that structures could only be homologous when they occurred in monophyletic lines of organisms. One could then talk of the phylogeny of a structure, tracing a bone or organ's fate as its bearers diversified in multiple environments. Numerous papers, such as W. K. Gregory's classical "The Humerus from Fish to Man" (1949), reflect this view. In operational terms, demonstration of homology requires that the observed similarity be greater than random and not due primarily to contemporary selective influence.[2]

Comparison has a different but extremely important purpose in functional morphology. Functional anatomists are not only interested in the physiology and behavior of the organism, but also in the way these are reflected in the animal's structure. We have seen that an animal's structural pattern will reflect selection resulting from its current environment and behavior, but will also show phylogenic and accidental environmental influences. Ordinarily it is difficult to separate out cleanly the effects of adaptive selection on a particular structure or to extrapolate adaptive function, or even function in the broad sense, from structure.

The problem is most difficult when one attempts to separate those

[2]There have recently been numerous attempts to change the meaning of the terms analogy and homology by applying them to other conflicting or totally distinct concepts. For references and comments on some of these, as well as to other viewpoints, see Bock (1969), Gans (1969b), and Inglis (1966).

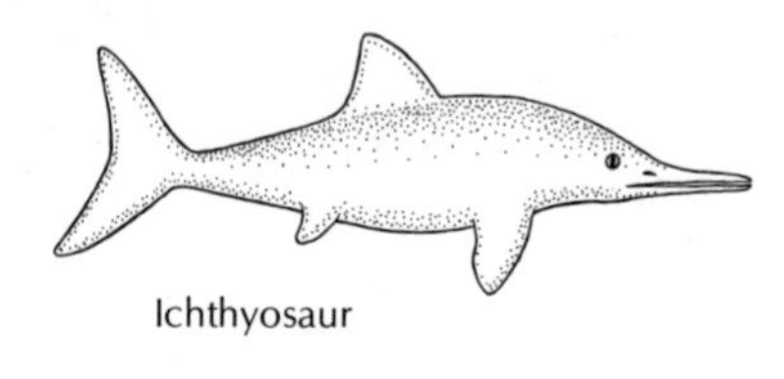

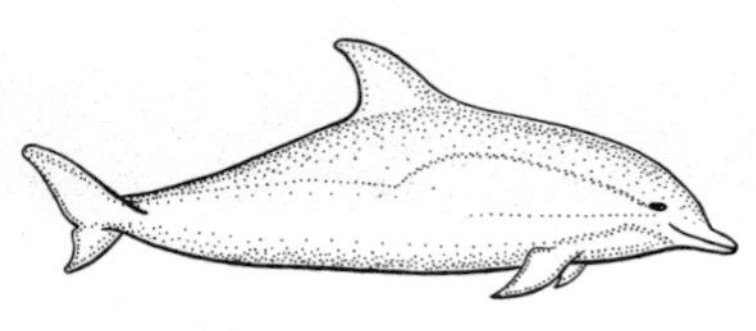

Figure 1-11 Sketches showing the superficial similarity between an ichthyosaur, an extinct reptile from the Permian (200 million years ago), and the dolphin, a Recent mammal. Both were apparently adapted for life in the ocean. Both show streamlining, as well as pectoral and dorsal fins. Other similarities are superficial only; thus the dorsal fin of the porpoise is entirely muscular, whereas that of the ichthyosaur was stiffened by bony spines, as was its vertical tail. In whales the tail lacks bony supports within the flukes, which are horizontal rather than vertical.

aspects of a single animal or species that are produced by recent adaptation from those that are purely phyletic or that are modified by chance. Comparisons among species allow one to differentiate among these three possible, but not exclusive, explanations of similarity. Similarity of an animal's features should be assumed to be due exclusively to environmental or behavior-associated selective factors (i.e., ecological or functional, rather than phyletic ones) only if (1) the particular structural pattern is always associated with a particular ecological one, (2) the structural pattern does not occur where the ecological factor is lacking, and (3) closely related forms with different habits lack the structural pattern. Similarity should be ascribed exclusively to common phyletic origin only if (1) it occurred or seems to have arisen in members ancestral to the group, (2) it occurs in all members of a grouping regardless of the environment occupied, and thus (3) it cannot be associated with adaptation to a specific ecological factor. Similarity should be assumed spurious or coincidental only when (1) the pattern can be ascribed to neither of the other two explanations and (2) a separate functional or phyletic explanation exists for each of the two similar character states observed.

Some qualifications are needed, but these general approaches to categorizing similarities among organisms seem to work well. The probability of reliable decisions in isolating adaptive from a broader range of functions can be increased in four ways: (1) using members of a single adaptive radiation (a single stock that has invaded and diversified in a series of habitats), (2) increasing the number of species being compared, (3) comparing simultaneously for multiple aspects of the organism, and (4) starting with the analysis of those species most highly specialized for the aspect studied.

Partitioning the several kinds of similarities is likely to be most successful when one is simultaneously comparing as many members as possible from what appears to be a single adaptive radiation. Such an approach involves simultaneous comparison of the results of convergent and divergent evolution on a single time level (the present)—in other words, without access to a fossil record. The results of convergent evolution will be compared by including members of different phyletic lines that seem adapted to similar environments; in contrast, the results of divergent evolution are studied in a single phyletic line, the members of which have adapted to dissimilar environments.

It is impossible, unfortunately, to find in each case groups of organisms that show a perfect match of functions. Instead one always finds slightly different solutions to similar functional problems. The closer the relation between the species included in the comparison, the easier it becomes to separate phyletic and functional influences; different solutions to functional problems are presumably contrasted most sharply

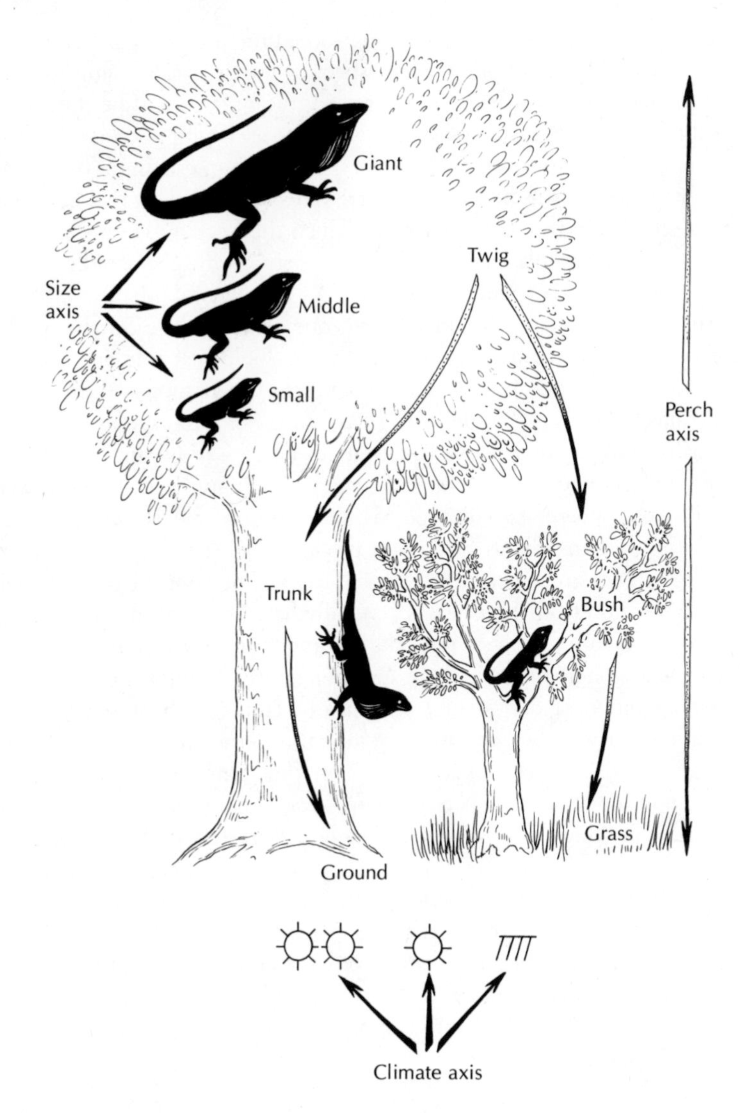

Figure 1-12 Sketch showing the diversity of the tree lizards of the genus *Anolis* in tropical environments. The animals differ in size, in the objects on which they perch, and in the amount of sunshine or moisture they tolerate or need. Several species are often found in the same general area. Not only do they differ in their preference for environmental factors, but also they commonly differ in limb proportions, claw size, toe patterns, and tail lengths. Only biomechanical analysis can separate the adaptive bases of those changes associated with the perch axis from those associated perhaps with the climate or size axes.

between organisms with similar genetic backgrounds. Widespread and diversified genera such as the tree lizards of the genera *Lygodactylus* (African geckos) and *Anolis* (American iguanids, Figure 1-12) would seem to be ideal objects for the characterization of adaptive functional influences, as might be the radiation of lungless salamanders (Plethodontidae) into the tropics or the radiation of cichlid fishes within the great lakes of central Africa. Such inclusion of multiple species then increases the probability of making correct interpretations in the same way an adequate sample size increases the reliability of any given experimental result.

Comparisons, furthermore, should proceed for as many characteristics as practical. The more limited the aspect of the organism being studied and the narrower the field of comparison, the greater will be the likelihood that the investigator may be misled into ascribing a structural condition to a single functional influence. This also suggests that one should look at the organisms in the field, there to examine their actual behavior and responses to the environment in which they occur. When multiple aspects of the organisms are studied, one may often separate the causes that have shaped particular structures. As the general adaptive pattern of the group begins to be understood, one achieves a reasonable probability that the explanations of a structure's biological role are correct.

Another useful technique for recognizing the nature and effects of functions is the examination of those members of a lineage most highly specialized for a particular adaptation—in other words, the best flyers, swimmers, jumpers. Wing specialization for gliding may be best seen in albatross, whereas hummingbirds may document extreme modifications for hovering flight. Structural modification enabling the behavior will be at a maximum in these extreme cases. The concept that one can derive function from structure is generally based on superficial examination of such forms. It may not take too much insight to recognize that bat wings support flight or shark fins facilitate swimming. Furthermore, the recognition of the general pattern of adaptation exhibited by such modified forms allows one to recognize the more subtle beginnings of tentative evolutionary "experiments" in the same direction found in seemingly unspecialized species. Many such generalists will turn out to be forms in which this adaptation has been arrested at an intermediate level.

It is hoped that this chapter and the cases detailed in the following chapters make it clear that actual observation rather than extrapolation, living specimens rather than preserved materials, manipulation rather than passive observation of the system, and comparison rather than conclusions from single forms represent optimum approaches to the study of biomechanics.

REFERENCES

The general terms used in this book may be found in any book on comparative anatomy. The starred references in this and succeeding lists should be consulted for discussions of general principles or expanded examples on the topic of the chapter.

Babák, E. (1903). Ueber den Einfluss der Nahrung auf die Länge des Darmkanals. Biol. Zentralblatt, 23:477–483.

*Beer, G. de (1962). Embryos and ancestors, 3rd ed. Oxford Univ. Press, xii + 197pp.

Bock, W. J. (1969). The concept of homology. Ann. N.Y. Acad. Sci., 167:71–73.

——————, and G. von Wahlert (1965). Adaptation and the form-function complex. Evolution, 19(3):269–299.

Brandt, W. (1949). Lehrbuch der Embryologie. S. Karger, Basel, xii + 648pp.

*Cain, A. J., ed. (1959). Function and taxonomic importance. Systematics Assoc. (London), publ. no. 3, 140pp.

*Gans, C. (1966). Some limitations of and approaches to problems in functional anatomy. In De anatomia functionali. Folia Biotheoretica (Leiden), 6:41–50.

—————— (1969a). Functional components versus mechanical units in descriptive morphology. J. Morph., 128 (3):365–368.

—————— (1969b). Some questions and problems in morphological comparison. Ann. N.Y. Acad. Sci., 167:506–513.

Grant, V. (1963). The origin of adaptations. Columbia Univ. Press, New York and London, x + 606pp.

Gregory, W. K. (1949). The humerus from fish to man. Amer. Mus. Novit. (1400):1–54.

Hildebrand, M. (1968). Anatomical preparations. Univ. of California Press, Berkeley and Los Angeles, viii + 100pp.

Inglis, W. G. (1966). The observational basis of homology. Syst. Zool., 15(3):219–228.

*Mayer, E. (1963). Introduction to dynamic morphology. Academic Press, New York and London, x + 545pp.

Mayr, E. (1963). Animal species and evolution. Harvard Univ. Press, Cambridge, xiv + 797pp.

Romeis, B. (1948). Mikroskopische Technik. Leipniz Verlag, Munich, xi + 695pp.

*Russell, E. S. (1916). Form and function. A contribution to the history of animal morphology. J. Murray, London, ix + 383pp.

Schumacher, G. H. (1973). The maxillo-mandibular apparatus in the light of experimental investigations. In Morphology of the maxillo-mandibular apparatus (G. H. Schumacher, ed.) G. Thieme, Leipzig, pp. 13–25.

Simpson, G. G. (1953). The major features of evolution. Columbia Univ. Press, New York, xx + 434pp.

Spemann, H. (1936). Experimentelle Beiträge zu einer Theorie der Entwicklung. Julius Springer, Berlin, viii + 296pp.

Washburn, S. L. (1947). The relationship of the temporal muscle to the form of the skull. Anat. Rec., 99:239–248.

Weibel, E. R., and H. Elias (1967). Quantitative methods in morphology. Springer-Verlag, Berlin, Heidelberg, New York, viii + 278pp.

*Williams, G. (1966). Adaptation and natural selection. Princeton Univ. Press, Princeton, x + 307pp.

Zamachowski, W. (1970). Changes in the weight and length of the alimentary canal of the edible frog (*Rana esculenta*) in the annual cycle. Acta Biol. Cracoviana, Ser. Zool., 13(1):65–73.

2 ANALYSIS BY DISSECTION AND OBSERVATION: EGG EATING IN SNAKES

Facultative Egg Eaters

The Japanese rat snake, *Elaphe climacophora*, seems occasionally to vary its diet of adult mammals and birds by eating bird eggs; so do a number of other eastern Asiatic and indeed many North American species of the genus. This variation in the diet, together with the snake's special egg-eating mechanism, was discovered several times, independently and for different species, first by Oshima (1930) and then by Chernov (1945, 1957).

The egg-eating mechanism is associated with some modification of the vertebral column. Most species of *Elaphe* bear ventral projections or hypapophyses on the anteriormost forty or so vertebrae and none on more posterior ones (Figure 2-1). Such hypapophyses apparently serve as supplementary attachment sites for the vertebral muscles of the neck region, and in most of these species point ventrally and slightly posteriorly, ending in a rounded or flattened tip to which attach the tendons and aponeuroses of the axial musculature.

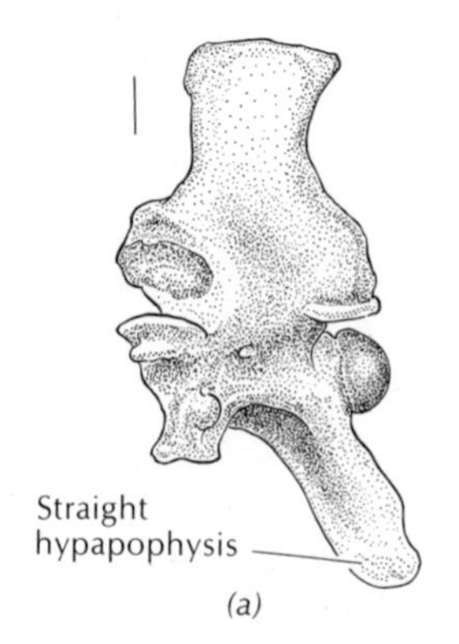

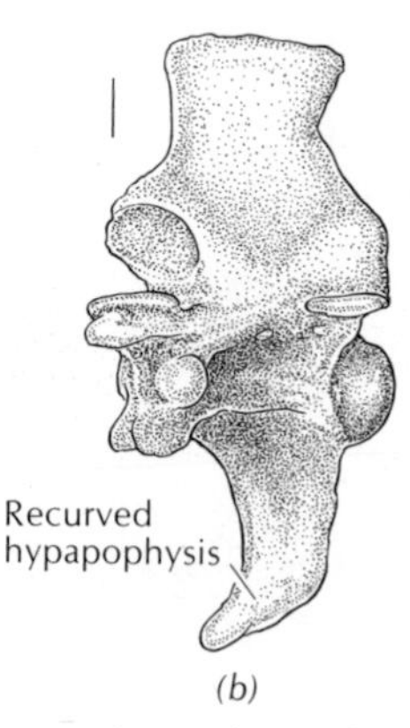

Figure 2-1 The neck vertebrae of rat snakes show ventral processes or hypapophyses, the tips of which point straight posteroventral in unmodified species such as *Elaphe taeniura* (a) but swing around to point anteriorly in the modified species *E. climacophora* (b). The bars are 1 mm long to scale.

In 1946 Dr. Masamitsu Oshima pointed out to me the curious egg-eating behavior of *E. climacophora* (Gans and Oshima, 1952), and upon dissection I confirmed that its hypapophyseal tips were modified (Figure 2-1). Rather than pointing straight back, as do those of North American species of *Elaphe*, the distal tips of the hypapophyses curved and terminated in an anteriorly directed point. The bone of these tips was irregular and of slightly different texture than that of the vertebral body and hypapophyseal base. The esophagus appeared closely adherent to the muscles inserting on the hypapophyses, and in some specimens (perhaps those somewhat emaciated) one could most obviously see the hypapophyseal tips pushing against the esophageal lining. No other noteworthy specializations were discovered in the head and body of *Elaphe climacophora*. One specimen of the Chinese species *Elaphe carinata* appeared to have the vertebral spines penetrating the esophagus, but this seems to have been an artifact of preservation (Dowling, 1959).

These rather limited modifications fitted the observed behavior of *E. climacophora*. The snake's pointed teeth were unable to grip the egg. Therefore, swallowing of eggs, even when these were little larger than the snake's head, had been observed to pose considerable problems. There ensued a juggling process until the jaws were sufficiently dis-

Figure 2-2 The long body of snakes is supported by more than a hundred vertebrae, each bearing a pair of ribs. Food objects, like the frog being ingested by the garter snake (*Thamnophis sirtalis* shown in this x-ray), are swallowed whole, and the ribs remain slightly spread until the food is digested.

tended to cause the teeth to reach the egg's widest point or indeed beyond it to the point where the egg narrowed again, thus producing an inward vector (Box 2-1). When the initial approach occurred in a cage, it was often ludicrous, involving successive bites at an egg that would roll away until it was accidentally maneuvered against a fixed object or coil of the body. This might be an artificial problem since eggs are unlikely to roll far in most bird nests.

Once the egg had been forced into the throat by action of the jaws, one could observe it being swallowed further. As the egg passed into the throat, the head would bend to one side and then reverse until a horizontal S-shaped curve was formed in the anterior part of the neck. The curve would elongate and progress posteriorly, shifting the egg backward. This rat snake was using the standard ophidian method for

Figure 2-3 Four views of egg eating in *Elaphe climacophora*, showing stages between the initial bite (*a*) and the crushing of the egg. The egg (diameter 39 mm) took 31 minutes to pass to the stage shown in the second photo (*b*). The snake's open mouth here shows the tongue tip and the glottis which lets the animal breathe while swallowing. Fifteen minutes later (*c*), the egg has just reached the anterior esophagus, but it was not heard to break for another 20 minutes. Within 5 minutes the snake became very active, and the shrinking egg was passing rapidly down the esophagus into the stomach (*d*). (From Fukada, 1959)

(*a*)

(*b*)

transporting large food objects, which is peristalsis involving the axial musculature as a whole rather than only the thin-walled alimentary canal.

As the egg entered the posterior esophageal region, the snake seemed to form a second S-shaped loop, starting this posterior to the egg and moving it anteriorly. The egg was then squeezed between the two loops and subjected to considerable pressure, judging from the obvious effort exerted by the snake. A final muscular effort was rewarded by a cracking sound followed rapidly by a deformation of the egg's outline in

(*c*)

(*d*)

the snake's body. Continued backward peristalsis then drove the crushed mass toward the stomach. If another egg was available it would be eaten next.

This method of swallowing and then breaking eggs appeared to be effective. The stiff shell of bird eggs poses an obvious problem to a predator. Slitting the egg's shell and drinking the contents, as does the scarlet snake, *Cemophora* (Palmer and Tregembo, 1970), seems possible only for soft-shelled eggs such as those of turtles. Hard-shelled eggs obviously have to be broken to facilitate optimal packing of the contents; by analogy a series of Ping-Pong balls in a sausage casing involves a significant waste of space. Unbroken eggs would also impede locomotion, since large inflexible objects would interfere with the curves formed by a snake's body. Clearly, the shell represents an impediment to digestion; until it is broken or digested away, the nutritious contents do not provide

Box 2-1 Vector Analysis

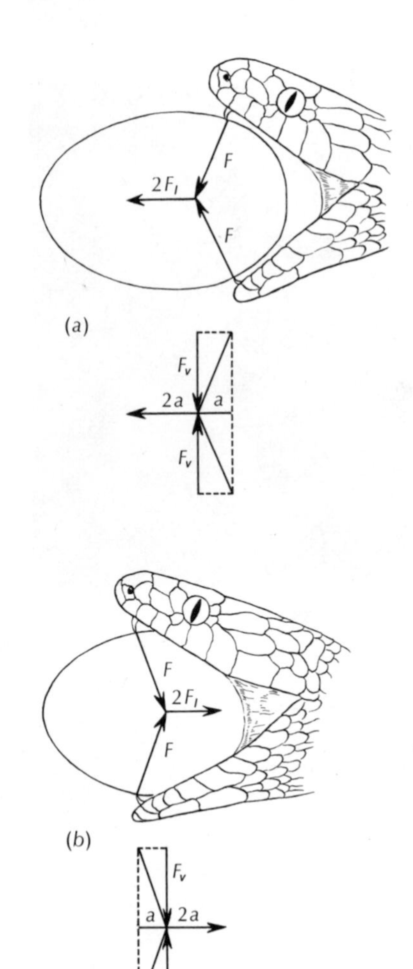

Many of the measures of interest in biomechanics have two descriptive attributes: magnitude and direction. Such parameters, examples of which are force, velocity, and acceleration, are called *vectors* in contrast to *scalars*, such as time, mass, and energy, which have only magnitude. By convention we show vectors graphically as straight arrows. The head of the arrow indicates the direction, and the length of its shaft the magnitude (on any desired scale). Vectors are assumed to be "free" (in other words, they can act anywhere along their line of action).

Addition and subtraction of vectors utilize a geometrical basis. They may be carried out graphically, or analytically by calculations utilizing simple trigonometric relations. To add two vectors graphically, one finds the intersection of their lines of action and here connects the vectors head to tail. The line joining the free ends of the two vectors is their sum or "resultant." It is similarly possible to "resolve" a vector into a pair of "components" lying on any pair of lines intersecting on the line of vector action. As the number of vectors that may be added to achieve a particular resultant is infinite, so is the number of component pairs into which a vector may be resolved.

In practice, some axes are obviously of more interest to the anatomist than are others. Take, for instance, sketch (*a*) showing a snake engaging in an initial bite against an egg. Assuming that the egg is smooth and the bite symmetrical, the teeth will exert two forces (F) at right angles to the curving surface. Each of these forces may now be resolved into a component F_v (subscript v for vertical) tending to exert a force at a right angle to the egg's long axis and a component F_l (subscript l for longitudinal) exerting a force along the egg's long axis. Since the two vertical vectors F_v, directed toward the axis, prove to be equal and opposite along the same line, they cancel each other. However, the two forces F_l act in the same direction along the same line and are hence additive. In sketch (*a*) the resultant of the imposed forces tends to drive the egg from the mouth. In sketch (*b*) the teeth have already passed beyond the egg's widest point and the horizontal forces propel the egg down the throat. (Gravitational forces are here omitted to keep the example simple.)

More complex cases, in which the resultant of one pair of vectors is added to a third vector, are seen in such figures as 3-7, which also show a secondary resolution of forces along a new system of axes. Further examples appear in other places in this book.

the snake with energy. The longer it takes to soften the shell before breaking it, the more protracted will be the digestive process.

The fundamental advantage of a breaking mechanism was hence clear. The mechanism of *E. climacophora* represents a significant improvement over the method used by American rat snakes (*Elaphe*) in which the shell of larger eggs is softened by digestive juices before being crushed. The modified hypapophyses of the Japanese rat snake are more pointed and, for a given muscular effort, would increase the stress concentration in a given limited area of the shell above the levels possible in those rat snakes with unmodified hypapophyses. The method also represents an advance over that of bull snakes (*Pituophis*) in which the egg's shell is crushed by forcing the body against hard objects (e.g., a tunnel wall).

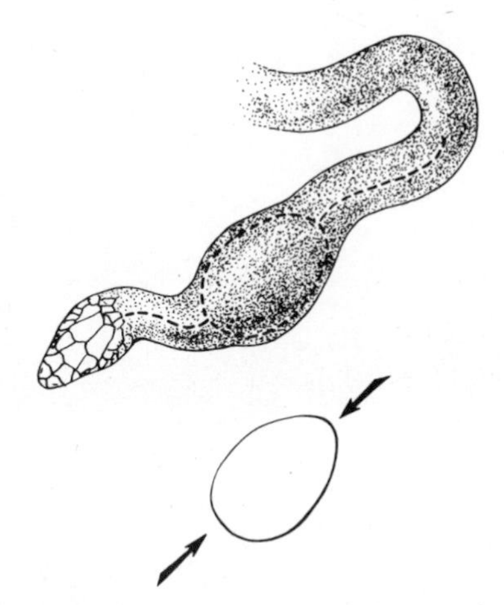

Figure 2-4 Tracing from a photograph shows how the egg is positioned and propelled by two S-shaped curves. Pressure exerted by the curved back of the snake's head forces the egg down the esophagus. The posterior curve controls the egg's movement and positions it for breaking.

The modification of the hypapophyseal tips showed some obvious ontogenetic differences between those of juveniles and adults. The latter often had more pointed tips that sometimes bore dense bony bumps. This raised the question of what might be the developmental pattern controlling the shape of the hypapophyses and the density of the tip. Did the bony excrescences of the modified tip develop as a response to the pressure imposed during repeated egg eating or were they inherited? Could this hardening process be a physiological adaptation, equivalent to that seen along the edges of mammalian incisor teeth after the enamel has been worn through? The problem proved to be a general one when it was discovered that other species, such as the more or less sympatric Japanese four-striped rat snake (*Elaphe quadrivirgata*), shared both modification and egg-eating habit (Goris, 1963). Subsequent observations now indicate, however, that the major shape and modification of the processes are mainly genetically determined. This is indicated by a colony of *Elaphe climacophora* raised in my laboratory; all were fed on soft-bodied prey only, all have anteriorly pointing hypapophyseal tips, and the larger specimens do show the characteristic bony excrescences.

The egg-eating mechanism in *Elaphe*, nevertheless, represents only a very limited level of specialization; it increases the effectiveness of utilizing one additional food item. It seemed appropriate to look next at the African egg-eating snakes (genus *Dasypeltis*) which had been reported in the literature as being obligate egg eaters, snakes that depend on eggs as their only source of nourishment.

Obligate Egg Eaters

Obligate egg-eating habits have been described for the genera *Dasypeltis* from Africa and *Elachistodon* from India. The African egg-eating snakes are apparently very common and widespread; some six species occupy all but the great northern deserts and the Mediterranean fringes of the African continent. The single species of Indian egg eater shares some of the structural characteristics of *Dasypeltis*, but hardly the evolutionary success. It is known only from fewer than a dozen specimens taken at a few places along the border of India and Bangladesh, close to Nepal, Sikkim, and Bhutan (Gans and Williams, 1954).

The six species of *Dasypeltis* differ in scalation, color pattern, and size, as well as ecology, but hardly in their treatment of eggs. All appear to be obligate egg eaters. Neither laboratory experiments nor field observations support any other interpretation. Hatchlings show the basic pattern of modification characterizing the genus; stomach contents of juveniles and adults consist entirely of egg fluid with occasional bits of shell.

Dasypeltis specializes by ingesting relatively large eggs, cracking them within the esophagus, squeezing out their liquid contents, and regurgitating the shell (rolled up into a cigar-shaped mass). The literature dealing with this process (cf. Gans, 1952) is extensive and combines careful anatomical descriptions of the snake's structures with behavioral observations of the egg-eating process; with a few exceptions these are

(*a*)

(*b*)

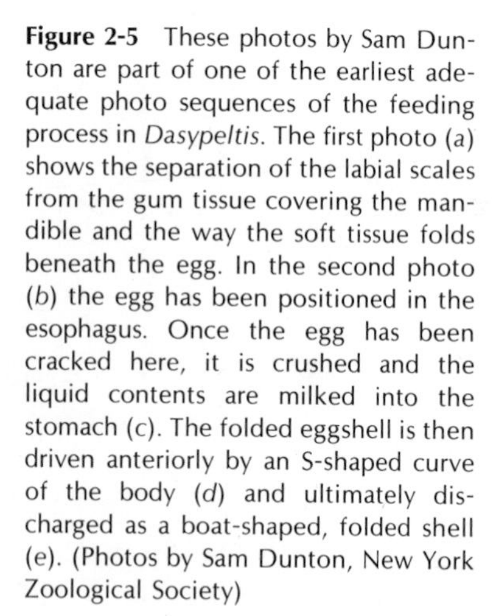

Figure 2-5 These photos by Sam Dunton are part of one of the earliest adequate photo sequences of the feeding process in *Dasypeltis*. The first photo (*a*) shows the separation of the labial scales from the gum tissue covering the mandible and the way the soft tissue folds beneath the egg. In the second photo (*b*) the egg has been positioned in the esophagus. Once the egg has been cracked here, it is crushed and the liquid contents are milked into the stomach (*c*). The folded eggshell is then driven anteriorly by an S-shaped curve of the body (*d*) and ultimately discharged as a boat-shaped, folded shell (*e*). (Photos by Sam Dunton, New York Zoological Society)

(c)

(d)

(e)

furnished by different investigators. Several photographic sequences of this process have been published, but many were obviously posed.

Analysis of miscellaneous reports and various series of photographs, particularly one taken by Sam Dunton (1945), suggested that egg ingestion proceeds by a relatively simple sequence (Gans, 1952). The snake bites the egg and the jaws distend over it, seemingly with less effort than had been described for *Elaphe*. The stretching of the skin appears to occur along a continuation of the angle of the mouth rather than by spreading at the middle of the chin. Indeed the chin scales seem to remain together, separating from the diverging mandibles as a unit and then sliding under the egg in direct opposition to the upper jaw. Once the egg has reached the throat and the mouth is almost closed around it, the snake's vertebral column bends sharply, apparently forcing some part of the hypapophyseal structure to break the shell. Subsequently, the egg becomes deformed and flattened while the snake undergoes various lateral and dorsoventral bends of the spine. Peristaltic waves of the body then draw out the egg's contents, apparently forcing fluid into the stomach. Finally, an S-curve of the spine forms behind the empty shell and moves anteriorly, forcing the empty shell out of the widely gaping mouth. Once the shell has been ejected, the snake proceeds to the next egg.

Previous authors provided a considerable series of speculations attempting to answer such questions as: How long does it take a snake to ingest an egg? (We have found that it takes 5 minutes to an hour, depending upon relative size of snake and egg.) Can such snakes be induced to take artificial eggs? (Yes, if the eggs have picked up the proper odor from resting in a nest.) Do they start to ingest the egg from the pointed or the blunt end? (Either will do.) Will snakes swallow eggs containing developing embryos? (Yes.) And what do they do in this case? (Squeeze out the liquid contents and reject embryos and shell.)

The literature on *Dasypeltis* also contains repeated references to four distinct sets of questions. The first resulted from the almost complete reduction in the number and size of the teeth, and asked how egg ingestion was managed. The second was the nature of the egg-breaking mechanism. Did this involve a shift (or functional replacement) of the enamel of the teeth to the esophageal egg-breaking structures? The third dealt with food sources. Were there alternates to eggs? What did the juveniles eat? Were eggs available to the adults throughout the year? The final question was: What might be the function of the peculiarly reduced, angled, and serrated scales along the sides of the body? (The original name of the egg eater was *Coluber scaber*, "scratchy snake.")

Method of Analysis

The egg-eating behavior of *Dasypeltis* posed some obvious questions for functional analysis. Of these, the simplest was: What method does *Dasypeltis* use to eat eggs, break the shell, squeeze out the content, and reject the shell? A second might be: Which structures of this animal are modified for egg eating?

Limited solution of the first question was possible by careful observation, in this case by careful analysis of the descriptions already available in the literature. The second question was more complex. After all, some of the specializations seen in these egg-eating snakes are related to the present egg-eating habit; others might be residues from earlier forms in the evolutionary line from which these snakes derived. Thus the Indian egg-eating snake *Elachistodon*, unlike *Dasypeltis*, retains a grooved tooth on the back of each maxilla. This suggests past or present occurrence of a venom-injecting mechanism. If *Elachistodon* alone were available, one would be tempted to relate grooved teeth to egg eating.

The obvious method of answering the second question was comparison, and it becomes important to consider which forms should be compared. Unfortunately, there were no poorly developed egg eaters furnishing intermediate conditions. Only the previously described egg-eating habit of the Asiatic rat snakes offered some clues, but it had quickly become apparent that their specialization was restricted to hypapophyseal modification. For this reason it seemed appropriate to defer questions about the closest relatives of *Dasypeltis*.

It seemed likely that *Dasypeltis* was derived from some part of the large assemblage classified as the family Colubridae, the largest and possibly the most diverse grouping of snakes. Such other snakes as boas and worm snakes, as well as vipers and cobras, were too modified to use as a basis for comparison. The behavior and structure of *Dasypeltis* were, therefore, compared directly to the behavior and structure of some more "typical" members of the Colubridae. Hence the pattern in *Elaphe* and its relatives was used as an example, as these forms were also egg eaters. A detailed functional analysis of feeding mechanics in rat snakes—describing the roles of individual subunits (quadrate, pterygoid, maxilla)—was not published until a decade later (Albright and Nelson, 1959; cf. Gans, 1961b). Consequently, the first analysis of the feeding mechanism had to proceed by reference to the actions of major mechanical units (such as left and right upper jaw arches).

In analysis, dissection of *Dasypeltis* proceeded in parallel with dissection of various colubrids, including *Elaphe*. Emphasis in description focused primarily on those features that seemed to differ between the two groups. In the same way, the feeding behavior of *Dasypeltis* was compared to the feeding behavior of other colubrid snakes, and differences rather than similarities were emphasized. After all, the differences might be assumed to be associated in some way with egg-eating specializations. The similarities were presumably due to the fact that all of these organisms were snakes.

The final step in this very simple kind of analysis consisted in correlating the observed structures with the behavior and noting the differences. If an unusual movement occurred, could one identify a bone or muscle that seemed to make it possible? If something bent, was there a region of unusual flexibility?

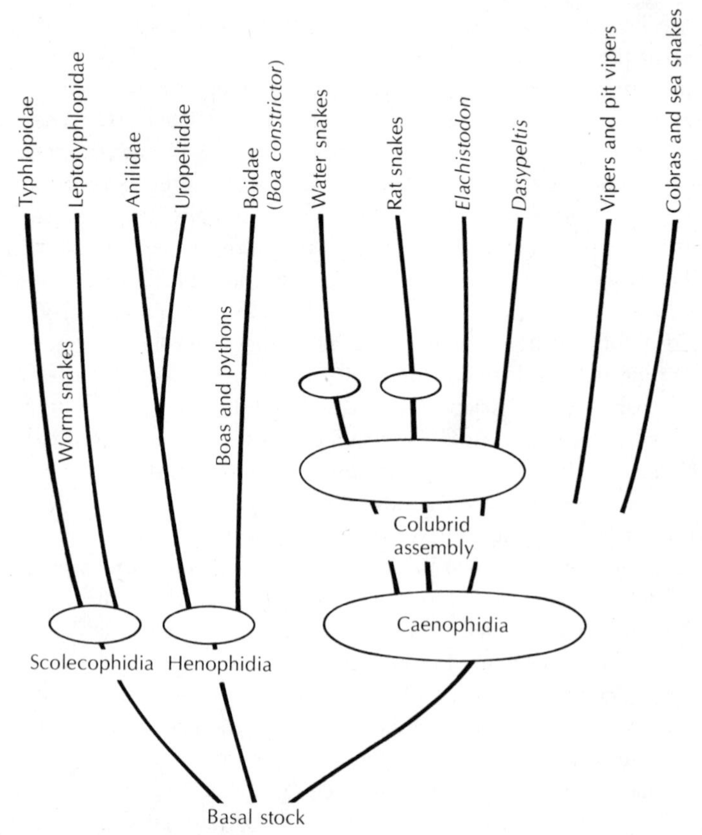

Figure 2-6 Sketch showing an arrangement of the major groups of snakes mentioned in this book. Some authors, notably Underwood (1967), further subdivide the family Colubridae into four families on the basis of details in the structure of their skulls, eyes, and copulatory organs. See Figure 3-1 for the place of snakes among the reptiles.

Structural Modifications of Dasypeltis

Modifications of the Skull

The skull of *Dasypeltis* is curiously light but at the same time much more tightly joined together than those in other snakes. ("Other snakes" here refers primarily to the basic colubrid pattern; see Figures 2-7, 2-8.) Particularly striking is a scalloping of the roofing bones of the skull: the nasals, frontals, and parietals. Their free edges appear to be scalloped in most specimens. The palatine, premaxilla, and maxilla are fairly broadly and tightly joined, rather than being loose and movable against the bones that protect the brain and sensory organs as in other snakes. The premaxilla and the two maxillae form a continuously joined horse-

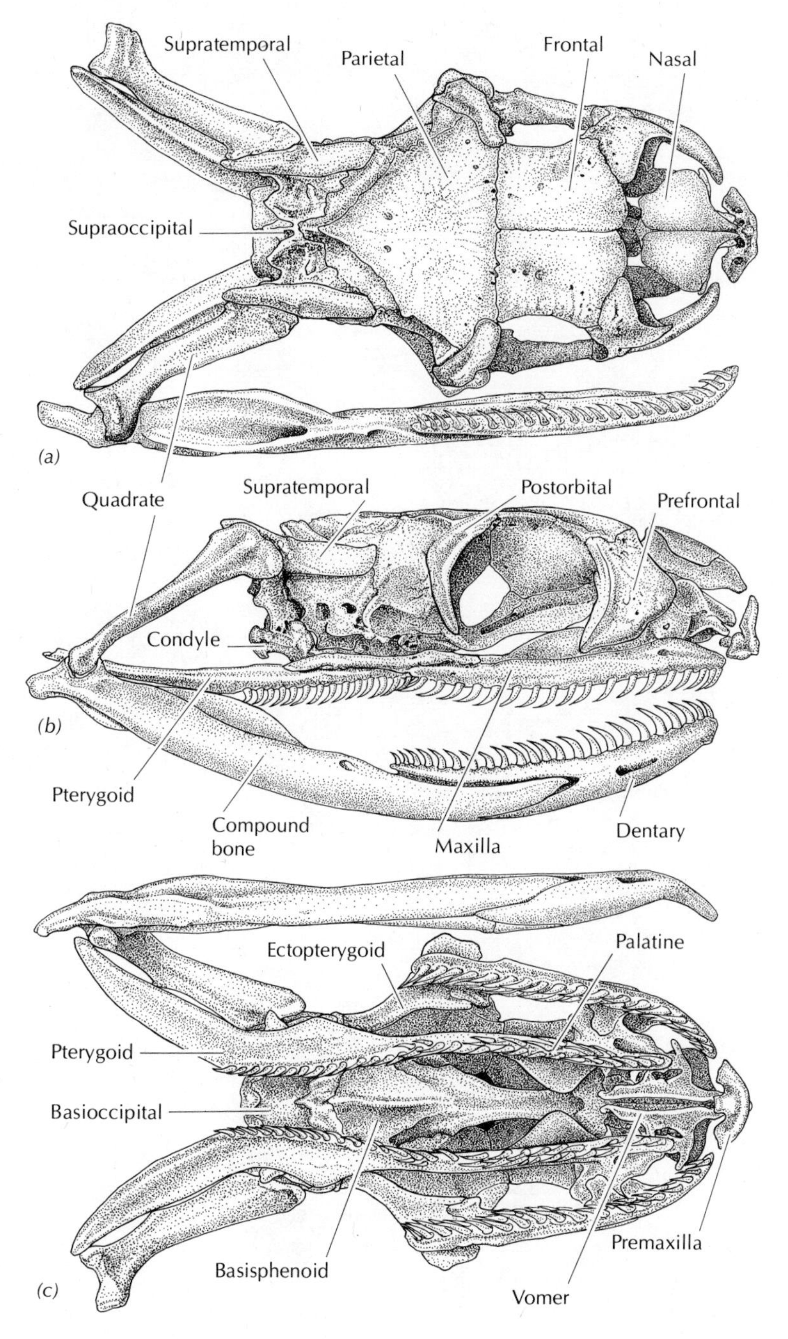

Figure 2-7 Skull of the indigo snake, *Spilotes pullatus*, in dorsal (*a*), lateral (*b*), and ventral (*c*) views to show the arrangement and names of the component bones in a typical snake. Although the braincase is solid, the eye is framed by relatively small bones. Much of the nasal capsule lacks bony reinforcement and in life is shielded by plates of cartilage and connective tissues. Four tooth rows occur in the upper and two in the lower jaw. All tooth-bearing elements can move relative to each other.

Figure 2-8 The skull of the African egg-eating snake, *Dasypeltis scabra*, in dorsal (*a*), lateral (*b*), and ventral (*c*) views. Note the much thinner framing of orbit and nasal capsule, the lack of a significant number of teeth, and the wide posterior projection of the jaw suspension. Obviously, this skull is not effective in restraining the actions of struggling prey.

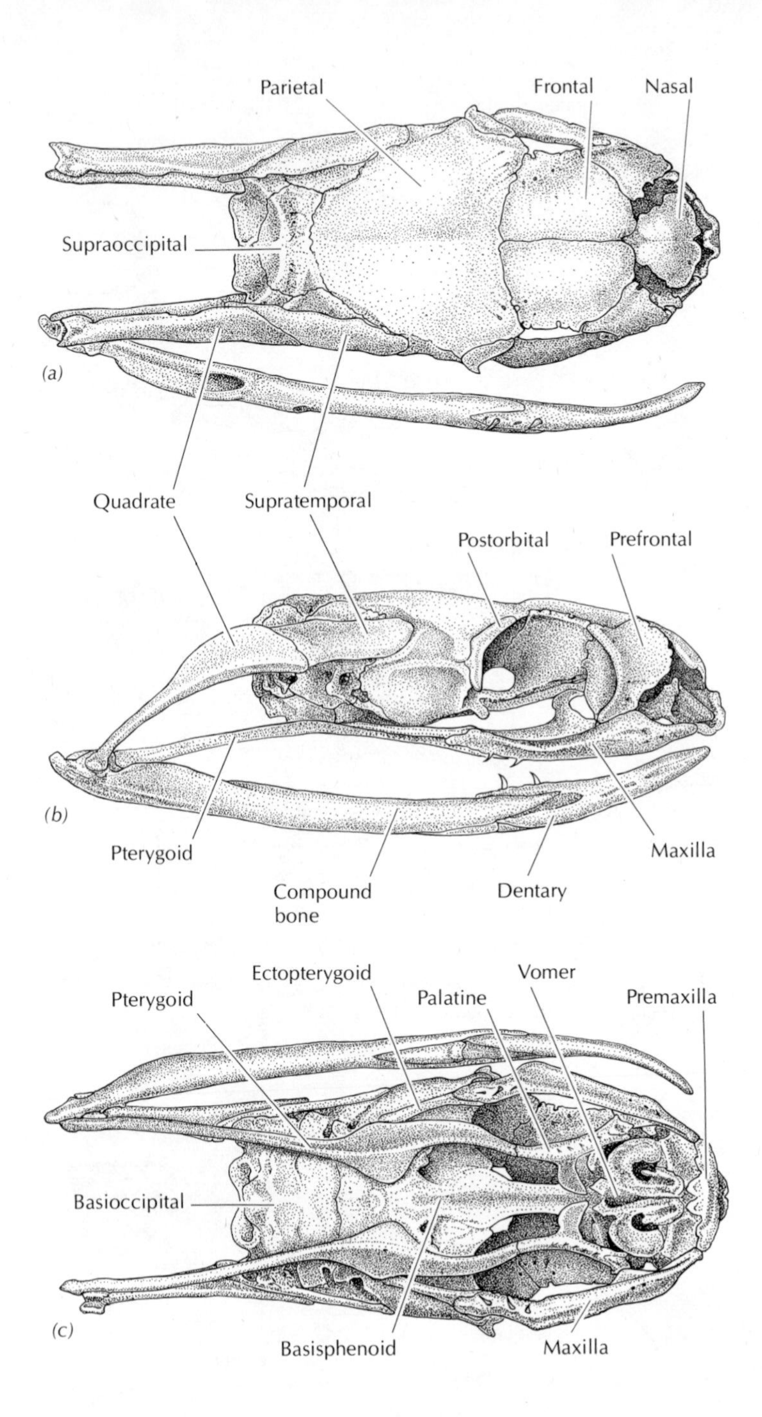

shoe around the ventral surface of the upper jaw. Even the anterior portion of the pterygoid is fairly tightly attached to the cranial base, but its posterior process is long, slender, and cylindrical rather than being stiffened by a lateral flange as in other snakes. Only the posterior tips of the maxillae and the entire ventral borders of the palatines bear the small, pointed teeth, which are very few in number and relatively larger in juveniles than in adults.

Most colubrid snakes have a platelike supratemporal attached firmly to the back of the braincase; its posterior end bears the quadrate articulation. The dorsal tip of the quadrate is thus articulated at the level of the first cervical vertebrae, extending the distance around which the jaw may swing and, consequently, enlarging the gape. In *Dasypeltis* the elongate supratemporal is very loosely articulated to the side of the braincase. Its posterior edge, furthermore, lacks a movable joint with the head of the quadrate; instead, the supratemporal and quadrate form a single mechanical unit that can slide on and rotate about the braincase. The ventral end of the quadrate provides articulation for the mandible and is in loose ligamentous connection with the posterior tip of the pterygoid. The mandibular articulation with the quadrate is saddle-shaped so that the quadrate can rotate slightly about its long axis. The compound bone of the mandible is relatively long and loosely connected with the short and extremely slender and flattened dentary. Its tooth-bearing surface is short and edged like that of the maxilla; also like the maxilla the dentary bears only a few small teeth on its posterior aspect.

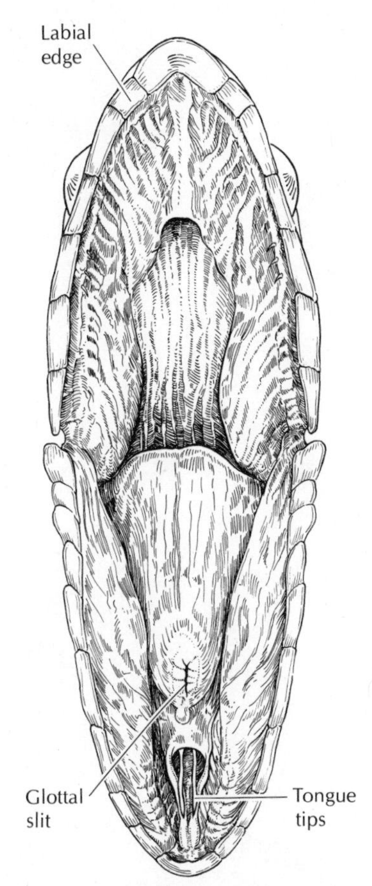

Figure 2-9 The gaping mouth of an egg eater shows the many folded mucous membranes, as well as the glottal slit of the air passage and the tongue tips in their sheaths at the bottom of the lower jaw. Note how the scales of the labial edges are closely connected and tied to the shelf of connective tissues. This restricts deformation of the upper jaw and forms it into a single unit.

Modifications of the Soft Tissues

The soft tissues of the head of the egg eater show several modifications. The lining of the mouth is most remarkable (Figure 2-9). Unlike the mouth of other snakes in which the teeth are the most prominent aspect, the mouth of *Dasypeltis* seems almost edentulous (toothless). The flexible mucous lining of the mouth is ridged rather than smooth. Numerous anteroposterior ridges pass between the weakly toothed bones. The supralabial scales form a pronounced and relatively stiff upper lip; a medial U-shaped shelf ties this lip closely to the maxilla. This shelf and its adjacent tissues exhibit virtually no folding, which suggests that the labial edges form a mechanical unit supported by the premaxilla and maxillae. Deep pockets of gum tissue lie between the mandibles and the edges of the chin shields (Figure 2-10). The ridged tips of the dentaries are crossed by multiple, fine, transverse folds of gum tissue. The entire anterior aspect of the dentary ridge is padded by a mass of soft, spongy tissue, and the investing mucous membrane is here thickened. The small dentary teeth on the posterior portion of the bone project slightly through the membrane.

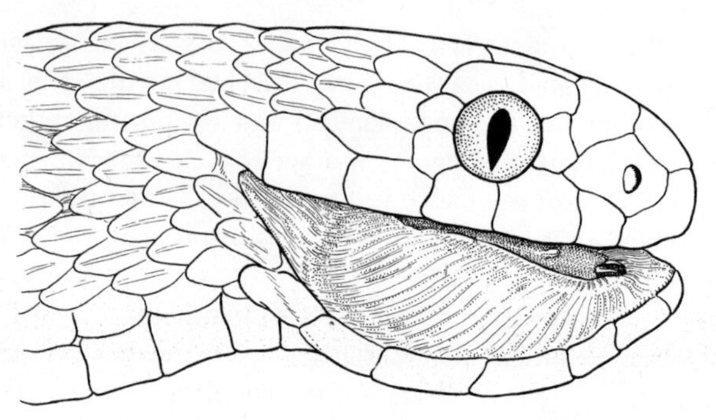

Figure 2-10 Lateral view of a specimen of *Dasypeltis* with the lower lip pulled down to expose the deep pocket of gum tissue between it and the mandible (recognizable by the small, projecting teeth). The capacity for stretch is restricted almost entirely to the edge of the gape.

The supralabial scales are tightly attached along the upper jaw, but the skin becomes looser toward the angle of the mouth. This is coincident with the beginning of a deep outpocketing of gum tissue which makes the connection of labials to the mandibles a very loose one. A fan-shaped muscle, inserting near the angle of the mouth, connects the skin to a wide region on the side of the neck. The last enlarged infralabial scale does not reach the angle of the mouth. Rather, this region is covered by a crowding together of several scales of the body's dorsal covering. The scales themselves appear to be mounted on central pedestals and their dorsal, ventral, and posterior edges are free, overhanging the intermediate skin which is arranged in a pattern of fine folds and striations. In fresh specimens the skin may be stretched so that the individual scales become isolated on the distended intermediate regions

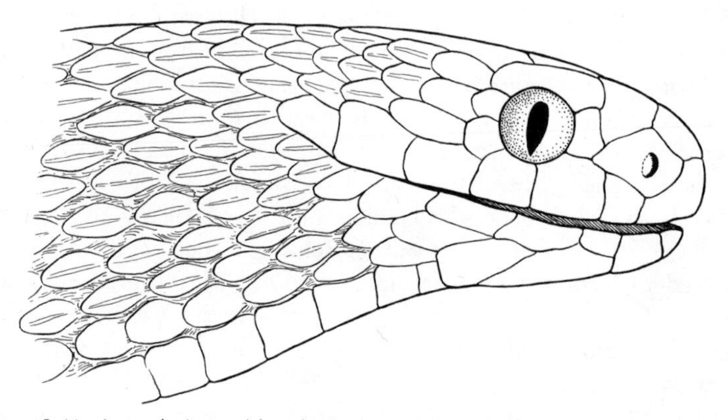

Figure 2-11 Lateral view of head of *Dasypeltis* with throat slightly distended to show the separation of segments along the region of the gape. Even some of the small-keeled scales of the dorsal region can slip and slide past the large scales along the back of the upper lip.

of naked skin. This pattern is in considerable contrast to the condition in other colubrid snakes in which the individual scales at the angle of the mouth attach fairly closely to one another so that the gape is well defined.

Another, even more obvious, modification of the soft tissues is in the very close association of the chin scales covering the lower jaw. All of these chin shields are very tightly joined, forming a continuous plate with the more posterior and widened shields that cover the body's ventral surface. Unlike most other colubrids, *Dasypeltis* has no mental groove. Such a groove normally forms an expansion line down the middle of the chin. The groove splits in the gular region and then passes to the two sides of the body. Its absence reflects a major change in the way the scales move when *Dasypeltis* gapes. In most other colubrids, the expansion line is down the middle of the chin, passing off to each side; in *Dasypeltis* the chin scales remain together and expansion occurs at the angle of the gape. When the scale-covered portion of the chin is

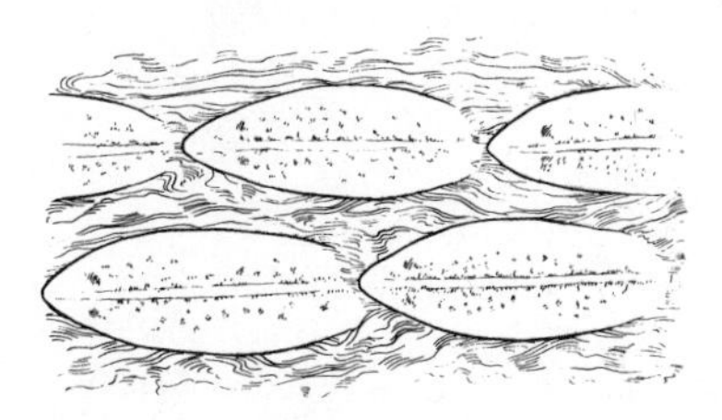

Figure 2-12 Detail of two scale rows to show the finely folded skin that allows the rows to separate when stretched by the passage of an egg.

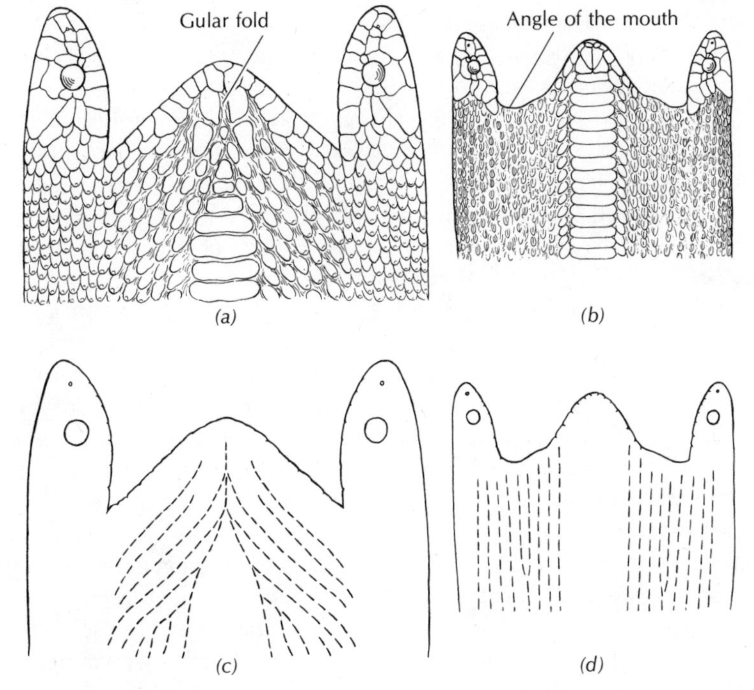

Figure 2-13 Details of two shed snakeskins split along the middle of the back to show the region where scales will separate when the snake distends itself around prey. In *Elaphe climacophora* (a, c) the scales separate along the so-called gular fold of the midline, and the stretched space then radiates outward to the sides. In *Dasypeltis* (b, d) the middle of the chin shows no stretching. All widening occurs in parallel lines, radiating from the angle of the mouth and separating the rows of scales along the back (see Figure 2-12).

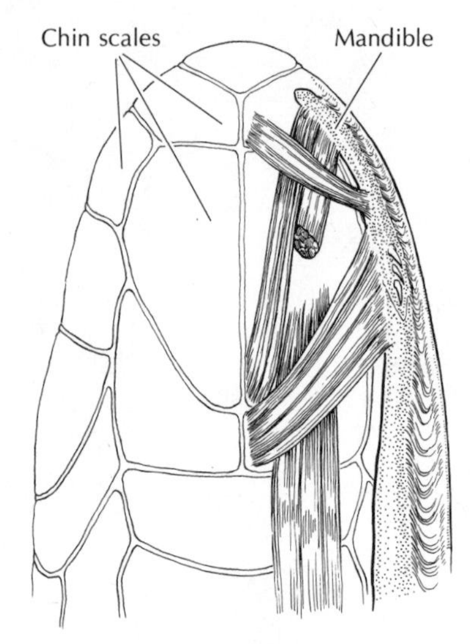

Figure 2-14 The muscles that anchor the tip of the mandible (lower jaw) to the inside of the chin scales cross each other and attach alternately to the dentary's dorsal and ventral edges. Their contraction could then cradle this bone, not only pulling it toward the midline but also changing its angle by twisting it along its length, shifting it in its loose connection to the rest of the mandible and rotating the mandible in the saddle-shaped articulation at the jaw joint.

moved away from the mandibles, the spongy tissue on the dentary ridge freely slips off the dentary tip; only posteriorly in the vicinity of the small dentary teeth is the mucous membrane attached.

When the skin covering the mandibles is loosened, one can see a thick boss of connective tissue stiffening the anterior tip of the chin and providing attachment sites for the base of the tongue. More posteriorly, muscle fibers run laterally from the central region of this boss in a V-shaped pattern to attach on the dorsal and ventral borders of the dentary. The gular shields show no other muscular insertions until well past the level of the quadrate, where the lateral edges of the eighth and ninth ventral scales receive the attachment of the elongate constrictor colli muscle, which originates from subcutaneous connective tissue in the vicinity of the supratemporal bone.

The Harderian (or tear) gland is enormous in *Dasypeltis*; its volume is equivalent to or larger than that of the eye. As in other snakes it connects with the subbrillar space (the chamber between fused eyelids and cornea). From here it drains via a duct running to Jacobson's organ (the accessory olfactory system) into the roof of the mouth. The labial glands are not particularly enlarged and open in a regular pattern along the lips lateral to the maxillae.

The mucosal folds of the buccal region grade continuously into the longitudinal folds of the esophagus. Some short distance back of the head, the esophagus becomes attached to the vertebral hypapophyses; the attachment zone becomes progressively more obvious posteriorly. On or about the 18th vertebra, the mucous lining of the esophagus becomes bunched up and hypertrophied over the head of each rounded intermediate hypapophysis; in adult specimens these areas have a central perforation leaving the bony surface open to the esophageal lumen. At about the 28th vertebra, the posterior hypapophyses become spinous and project far into the esophageal lumen. Their covering folds can readily be slipped back over the neck of the hypapophysis near the base of which they are attached fairly tightly in a ring-shaped constriction. The entire arrangement reminds one of the gingival seal around the bases of teeth. More posteriorly the esophagus swings away from its subvertebral attachment to lie loosely in the dorsal portion of the visceral cavity. In this region there is a marked thickening of the mucous membrane and a local hypertrophy of circumferential smooth muscles, the contraction of which should provide an extremely effective restriction of this portion of the alimentary canal.

As in most snakes the air passage or trachea of *Dasypeltis* ends anteriorly in a cartilaginous glottis. Located midway in the lower jaw this tube may be pushed beyond the tip of the jaw so that the snake can breathe even when its mouth is stretched around an egg. In the throat the trachea differs from that found in most snakes. The tube, normally circumscribed by U-shaped cartilaginous reinforcements, has opened up into a wide and flat sac of connective tissue incorporating a limited amount of smooth muscle and capable of extreme lateral distention. Posterior to the muscular band constricting the esophagus, this sac narrows again and the trachea then enters the lung.

Cartilage
Tracheal opening
(a)

Modifications of the Vertebral Column

The vertebrae of the anterior portion of the body are also modified. The most obvious modifications concern the hypapophyses, of which the anteriormost are as long as the vertebral centra and laterally compressed into flat plates; their ventral edges are slightly thickened so that they form a continuous ridge anterior to the more highly modified ones that begin at about the 17th or 18th vertebra. Here the intermediate hypapophyses seem to thicken; they are wider and swollen into enormous protuberances only faintly set off from the vertebral centra. The 26th to 28th vertebrae show transitional conditions and the 29th to 38th posterior vertebrae bear elongate, anteroventrally pointing, spinous hypapophyses, the tips of which penetrate the lumen of the esophagus and bear variously shaped excrescences. Their distal penetrating tips tend to be irregular. More posteriorly, the vertebrae of the main part of the trunk lack hypapophyses.

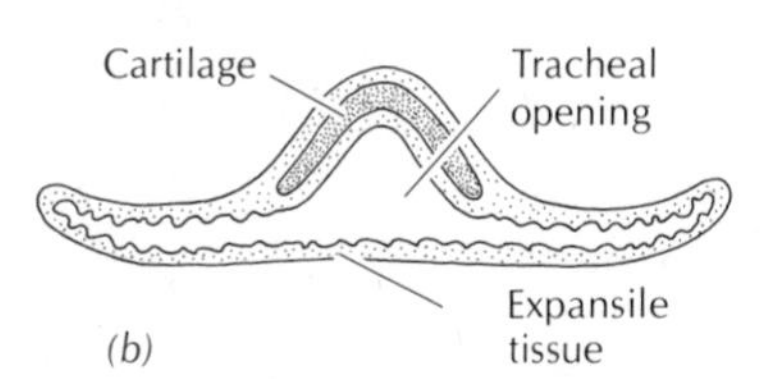

Figure 2-15 The trachea of a normal colubrid (*a*) is circular in cross section, whereas that of *Dasypeltis* (*b*) is wide and contains much loose connective tissue. The air then passes through a wide but low gap. Inflation of the tracheal sac, by forcing air forward from the lungs, distends the neck of *Dasypeltis*.

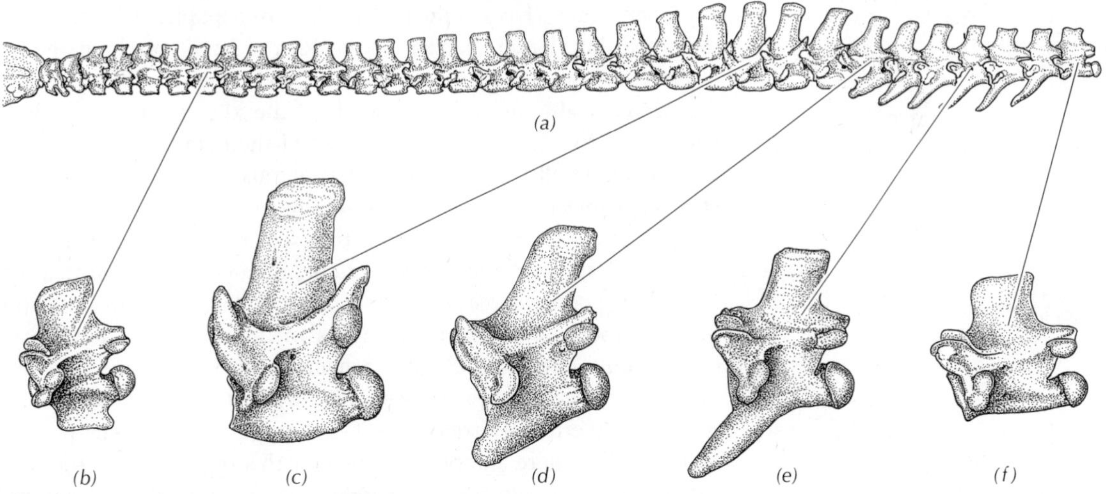

Figure 2-16 Diagrammatic view (*a*) of the vertebral column of a specimen of *Dasypeltis*. Note the unmodified anterior (*b*), the highly modified intermediate (*c*), the transitional (*d*), the spine-bearing posterior (*e*), and the normal (*f*) trunk vertebrae. (The number of anterior vertebrae has been reduced to simplify the illustration.) Three simultaneous changes occur in the sequence. These obviously affect the height of the neural spine, the degree of modification of the ventral hypapophyses, and the angulation of the lateral zygapophyses which have tilted dorsad in the intermediate series (see Figure 2-17).

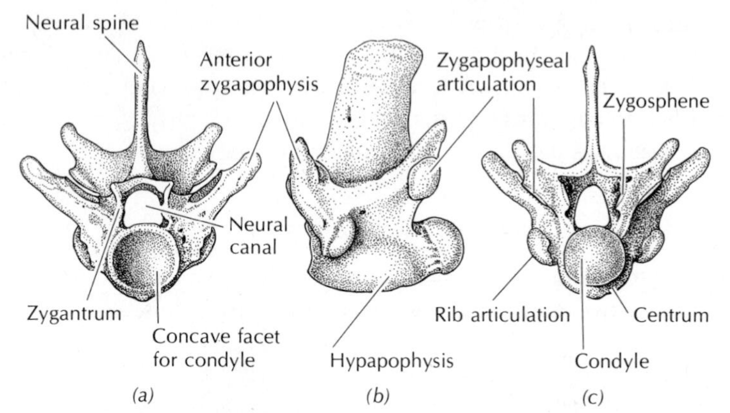

Figure 2-17 Anterior (*a*), lateral (*b*), and posterior (*c*) views of an intermediate vertebra of an adult *Dasypeltis*. Note the sharp inclinations of the zygapophyseal articulation as well as of those between zygosphene and zyganthrum, the rounded hypapophysis and the projecting neural spine.

A second series of changes may be seen in the massiveness of the vertebrae and the height of the neural spines, both reaching a peak within the intermediate hypapophyseal series. A final set of vertebral modifications concerns the plane of the intervertebral (zygapophyseal) articular facets. In other snakes those of both sides ordinarily lie in a single transverse plane that crosses the vertebrae slightly dorsal to the level of the intercentral articulations. In *Dasypeltis* the medial ends of the interzygapophyseal surfaces of the intermediate hypapophyseal region appear to tilt and shift dorsally, so that by the 20th to 22nd vertebrae, the facets of each side lie in a plane some 40 degrees from the horizontal. The displacement also affects the position of the zygapophyses themselves, which now rise sharply so that the roof of the neural arch rises on each side and the locally hypertrophied epaxial musculature rests in a concave channel bounded by bone ventrolaterally as well as medially.

The peak of axial modification is thus seen to be greatest between the 18th and 26th vertebrae. These vertebrae are enlarged, and their enormous hypapophyses may penetrate the esophagus, protruding into it between the loosely bunched mucous membranes. The dorsally concave neural arches and elongated neural spines accommodate the regionally enlarged epaxial musculature, whereas the altered inclination of the intervertebral articular surfaces suggests a different pattern of bending. The more anterior vertebrae of this region show little modification from the pattern seen in other species, but the more posterior vertebrae are mainly distinguished by their projecting hypapophyseal tips that seem to penetrate the esophageal lumen.

Dissection disclosed a considerable amount of individual variation, some of which reflects ontogeny (the growth of the specimen).

Kathariner (1898) and Haas (1931) noted seemingly larger teeth in juveniles. Juvenile skulls also appear more kinetic (i.e., their parts are more movably attached); in particular, the maxillopalatine arches are more loosely connected to each other, to the nasal capsules, and to the braincase in juveniles than in adults. Some individual variation was originally considered to reflect possible taxonomic differences, but a later review (Gans, 1959) still suggested that much of the variation is individual.

After dissections revealed the modifications of *Dasypeltis*, it remained to be decided which of them might be associated with egg eating.

Ingestion Sequence

General Pattern of Ingestion in Colubrids

The fundamental prey-ingestion pattern of typical colubrids (indeed of most snakes) is characterized by flexibility of the mechanical bonds that position the tooth-bearing maxillae, palatopterygoids, and dentaries of one side of the skull relative to those of the other (cf. Figure 2-7). The tips of the mandibles do not join in a junction or symphysis; each of the mandibles can consequently move independently of its fellow. The rotation of the quadrate around the supratemporal articulation may depress and protrude the quadratomandibular joint so that either mandible may slide anteriorly (relative to the braincase) as part or independent of the opening movement. The mental groove, or expansion line in the skin of the lower jaw, permits this motility, and each mandibular ramus carries with it the separate labial scales of its own side.

The maxillary and palatopterygoid arches are the tooth-bearing elements of the upper jaw. They are loosely suspended from the nasal capsules and anterior part of the braincase but are cross-connected by bony flanges intrinsic to them as well as by the ectopterygoid bone. The anterior portion of this system serves as a series of rather tightly constrained links; mainly the pterygoid bears muscular attachments that can lift it slightly and shift it forward and backward. The posterior tip of the pterygoid is also linked to the medial aspect of the quadratomandibular junction so that protrusion of the latter will tend to protrude and widen maxillary and palatine arches, depending on the contraction pattern of the pterygoid muscles. The complex application and transmission of compressive and bending forces via the pterygoid is reflected in the curious and extensive stiffening flanges along this bone (cf. Dullemeijer, 1956, for a structural analysis of the pterygoid of a viper). The general pattern, then, is one of four quasi-independent tooth-bearing elements (left and right maxillopalatine arch and left and right dentary). Each of the four units has some capacity for independent motion, but the major

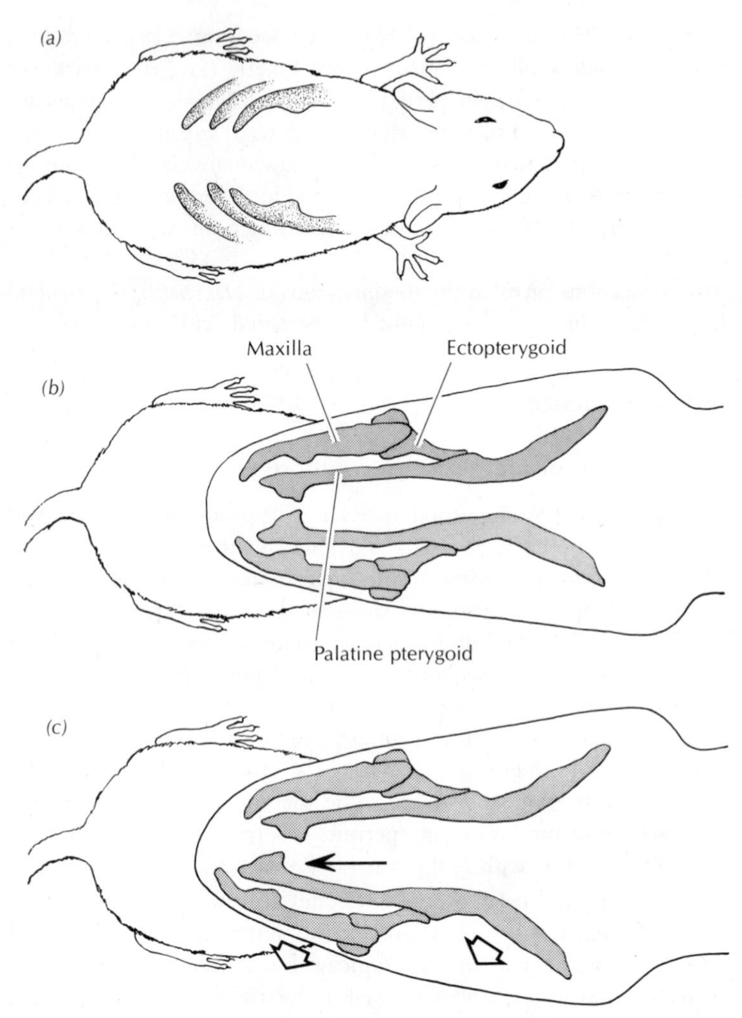

Figure 2-18 Outline of the major elements of the upper jaw of a colubrid snake sliding its head over the body of a mouse (*b-f*). Only the major tooth-bearing (maxilla, palatine, pterygoid) and linkage (ectopterygoid) elements of the upper jaw are shown. Note that the maxillary arches advance one side at a time. In retracting, they either pull the prey into the mouth or the head over the prey. The sketch of the mouse (a) shows the successive positions at which

coordinated movements are unilateral and associate the entire upper and lower jaws of one side.

The initial bite of most colubrids involves (1) a depression of the mandibles rotating about the saddle-shaped quadratic articulation and thus opening the mouth, and (2) some bilateral protrusion of the upper jaws extending the maxillae and widening the maxillary arch (if the aim has been accurate). The closure of the mouth clasps the prey and the

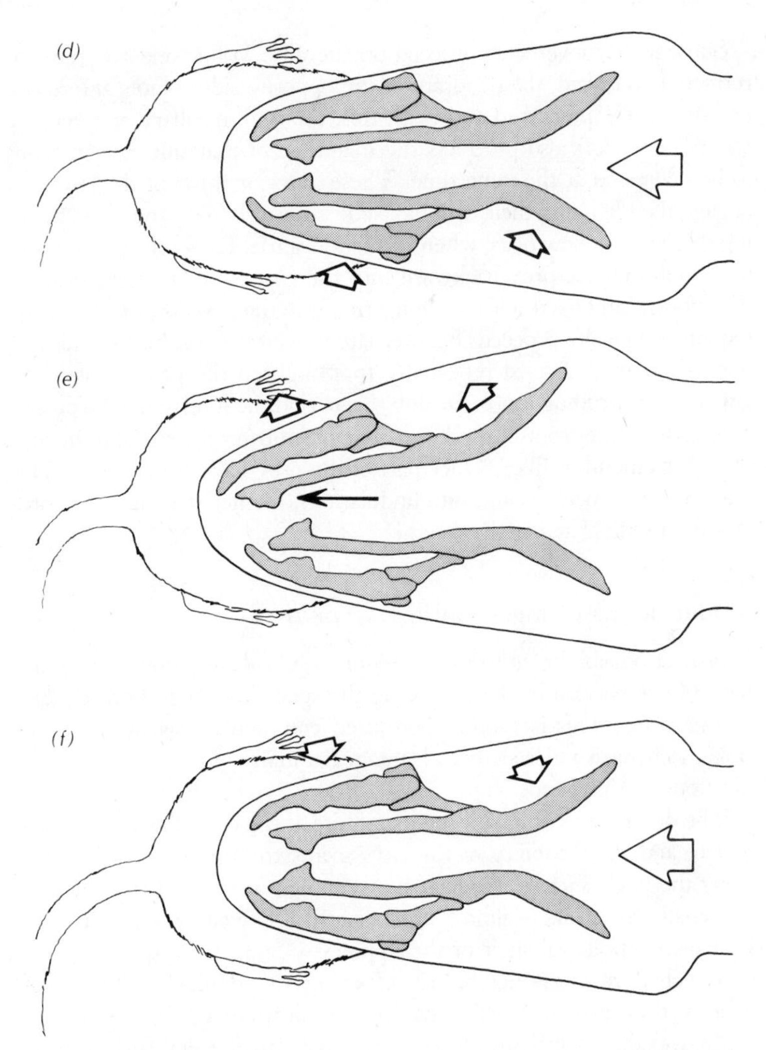

the teeth hooked onto the prey. Once the anterior end of the prey has proceeded into the neck of the snake, it will be pulled inward by undulation of the body without further action of the teeth.

retraction of the dentate elements of one side embeds the posteriorly curved teeth. As the prey is pulled backward by one side, the tooth tips of the opposite side tend to disengage partially. A detailed study of tooth-tip angles in pythons (Frazzetta, 1966) suggests that those of succeeding teeth show significant differences, an obvious adaptation for hooking into irregularly shaped prey. It is important to note that the tips of the dentary bones are insignificantly separated at this stage; there are no muscles that can spread them laterally when the mouth is empty.

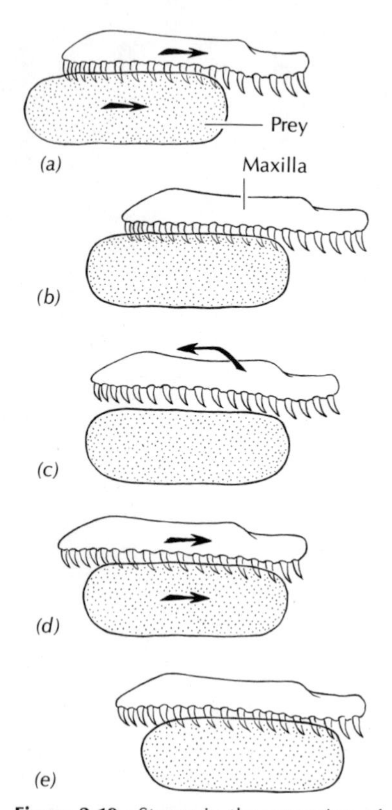

Figure 2-19 Stages in the retraction of a food object by the maxillary movements of a colubrid snake. Note how the toothed maxilla acts as a ratchet mechanism. Anteriad movements disengage the teeth, then a backward pull hooks them into the prey and draws the prey into the throat.

Subsequent movements proceed unilaterally. While one side remains retracted and fixed, the quadrate of the opposite side rotates anteriorly, protruding the pterygoid and, with this, the two maxillary and dentary arches. The shift also protracts the mandible of that side, which tends to be depressed at the same time. These anterior shifts of the jaws disengage the teeth and their tips may slide anteriorly over the prey, hooking them in at a new place when retraction starts. Each retraction movement will pull the prey backward into the esophagus, rotating it about the opposite and fixed side and elongating it in the process. The ingestion sequence typically proceeds by alternate movements of the two sides, or one side may be cycled repeatedly to straighten the prey or slide the jaws over protruding limbs. During this period the glottis may be pushed anteriorly to the front of the lower jaw, permitting the snake to breathe though its mouth is filled. Once part of the prey has entered the esophagus, the entire neck swings into undulations, further stretching the prey and then pulling it past the teeth.

Modifications of Ingestion in *Dasypeltis*

It proved possible to utilize the general colubrid ingestion pattern as a basis of comparison in characterizing the specialized ingestion sequence in *Dasypeltis*. This possibility permitted comments regarding the possible mechanical and selective advantages of many of the structural modifications of this genus.

The difference starts with the initial bite at the egg. The mouth gapes widely and the flexibility of the jaws seems to permit some deflection along the egg's surface. Contact between egg and jaws may start with the small teeth, but is almost immediately followed by contact of the U-shaped supralabial shelf of the upper jaw and of the spongy masses of bunched mucous membranes covering the dentaries. This initial contact thus promotes extensive communication and some adhesion of the mucous membranes of the oral cavity to the dry surface of the egg's shell. Experiments suggest that this soft-tissue contact with the egg increases friction and reduces the forward reaction forces that result when the initial bite involves contact of pointed teeth against smooth shell.

Although the forward, "out of the mouth" reaction forces—well known to those humans who have bobbed for apples—are thus reduced, they are not eliminated. Numerous observations (subsequent to my initial report in 1952) have shown that *Dasypeltis* often starts the attack by biting downward and then maneuvers the egg against irregularities in the nest, or ground, or a coil of its body. As soon as the reaction forces are balanced by some external forces, the snake will apply more force against the shell by curving its neck closer to the egg.

Box 2-2 Forces and Deformations

Forces imposed on solid bodies tend to deform them, and the direction of deformation reflects the direction of the force application. Five kinds of loadings occur. In *compression* the forces act toward the material which tends to bulge to the sides (bulge not shown). In *tension* the forces act away from the material which tends to stretch and narrow (stretch not shown). In *shear* two sets of forces act along parallel, closely adjacent but separate lines. Thus the blocks of material tend to slip past each other (along the dashed plane). Under certain circumstances couples (Box 2-4) will induce shear. In *simple bending* the forces are again parallel and opposed. As they act some distance apart, the transition is more gradual and the material deforms rather than tending to slip. In *torsion* the forces tend to twist the object, causing intermediate portions to deform or slip past each other.

Physical analysis of shear, bending, and torsion normally proceeds by considering the behavior of infinitely small cubes lying at different places within the loaded material. Forces can then be reduced to local tension and compression. For instance, the top of the beam here shown loaded in bending will then be seen to be stressed in compression and the bottom surface in tension.

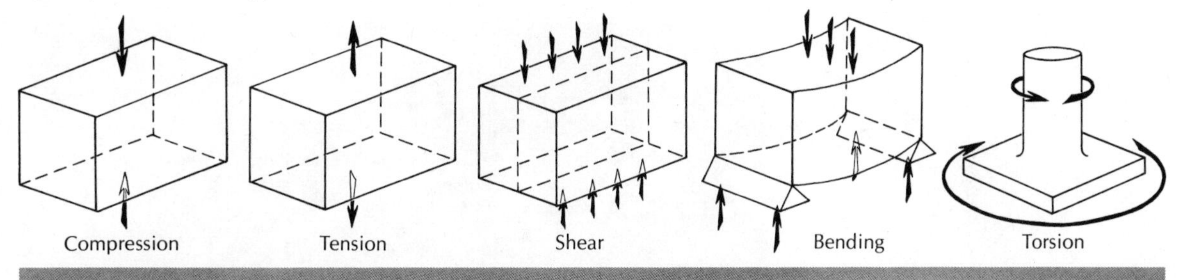

The initial contact is quickly followed by further ingestion movements. These are fundamentally different from those used by other colubrid snakes, as the elements of the two sides of the head move in concert rather than in left-right-left alternation. Indeed these continuing movements continue the pattern of the first bite in that the single unit of the upper jaw (i.e., both maxillae, palatines, and pterygoids) opposes both independently depressed mandibles. Each advance around the egg proceeds by an upward and forward slide of the cup-shaped upper jaw (with a slight return during which the labials can be seen to pull against the egg's surface) and a simultaneous lateral shift of the anterior mandibular tips. A further and even more striking difference from the "typical" snake ingestion pattern is the shift of sites of cutaneous expansion of the

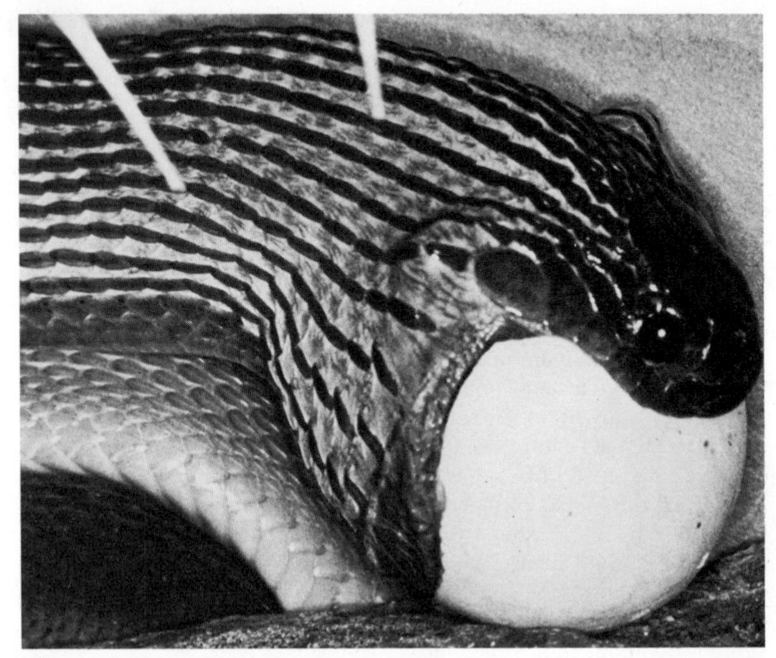

(*a*)

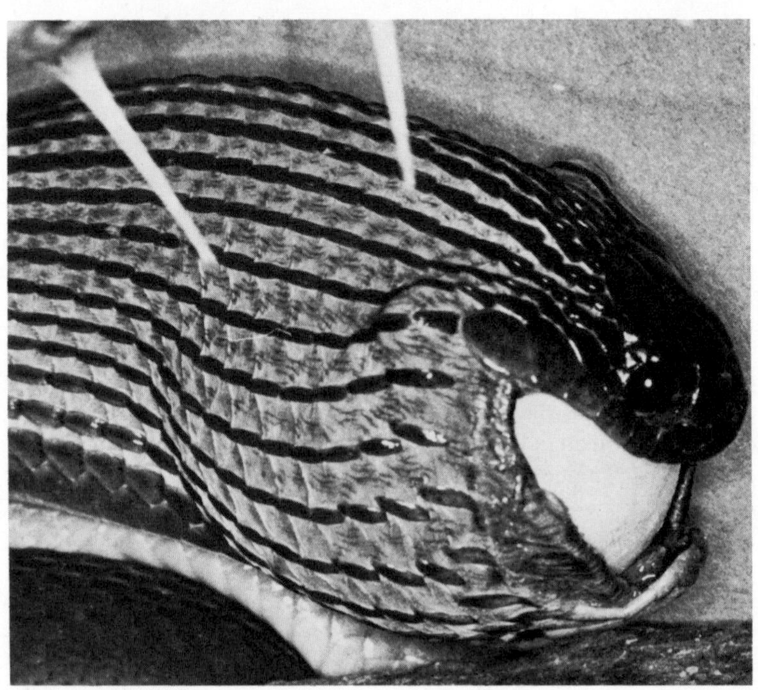

(*b*)

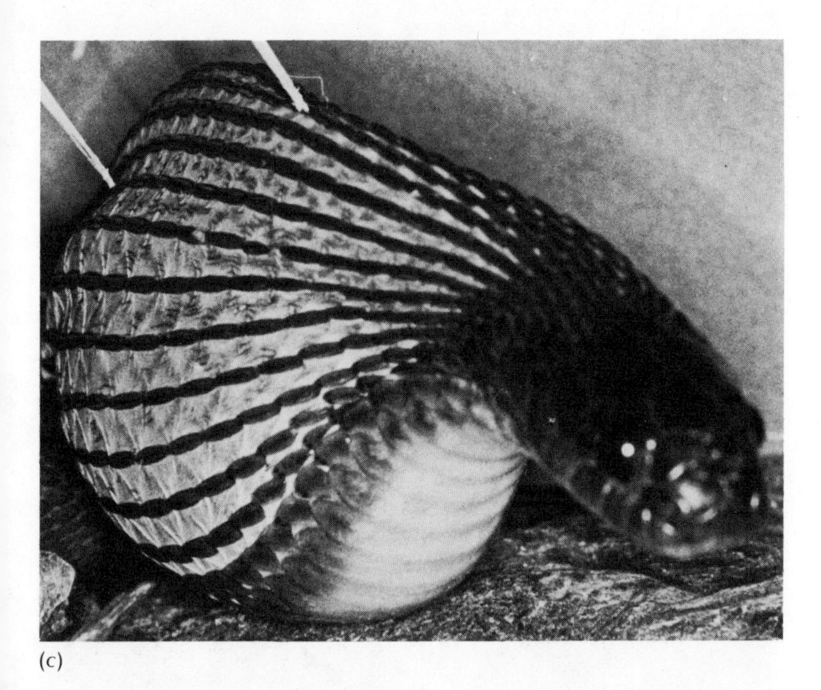

(c)

Figure 2-20 This specimen of *Dasypeltis atra*, feeding on a pigeon egg, nicely shows the separation of mandible from the scaly skin by exposure of the gum. The quadratomandibular junction is easily discernible through the stretched skin (*a*, *b*). The contraction of the constrictor colli pulls the ventral mass of the skin anteriorly and downward (*b*), so that the tip of the ventral blade of scales projects beyond the snout (*b*), carrying with it the tongue tip and the glottis. When the egg is large, relative to the diameter of the snake, one can see its ventral aspect stretch far below the level of the rib (compare with Figure 2-27). The anterior view (*c*) suggests that the ribs only extend down the side for the top six rows of lateral scales; the bottom five rows are stretched beyond the egg's sides and bottom. As the egg is being crushed, the lateral skin starts to contract at the animal's top; the ventralmost scale rows remain separated until the last.

lower jaw from the mental groove to the deep pockets of tissue lateral to the mandibles (Figure 2-10).

The firm connections of the chin shields and the marked expansive capacity at the angle of the mouth (as well as the underlying attachments to the hyoid) allow the chin scales to move as a fourth mechanical unit of the head. While the upper jaw slides over the top of the egg and the mandibles simultaneously move along its sides, this separate blade of scales is forced beneath the egg and lifts it off the ground. In doing so this blade can be seen to bend sharply near the angle of the jaws and then to move through an angle of more than 180 degrees to follow the egg's contour. The snake actually slides the four elements around the egg; it does not pull the egg into the mouth. This reduces need for the

Figure 2-21 This *Dasypeltis scabra*, gaping before ejection of the crushed egg shell, demonstrates the capacity of *Dasypeltis* to extend the buccal cavity by depressing the mandible and pulling posteriorly on the ventral musculature, thus flattening the neck and distending the throat as well. This posture provides minimal opportunity for damage of the mucous membranes when the egg shell is propelled from the mouth by anteriorly moving, S-shaped curves. The folds of the mucous membrane (and their attendant blood vessels) cross the mandibles at right angles. The tongue tip remains near the tip of the ventral blade of scales; the glottis is dilated and partially erected, showing how the snake can breathe by protruding this organ during ingestion.

longitudinal force vectors provided by jaw retraction in typical colubrids and much of the effort involved in shifting the jaws over the egg. Once the widest part of the egg has been passed, the tip of the snout bends downward, the mandibles bend inward, and the chin shields slide upward. The closing gape forces the egg into the esophagus.

The disappearance of the egg into the throat is often followed by one or two yawning movements. Similar movements occur in poisonous snakes that have struck at prey, and some authors have interpreted them as "displacement activities," implying something about the specimen's behavioral state. It is more appropriate to consider these yawns as repositioning stretches, as they are observed after all kinds of gaping movements and not just when a snake misses its strike. The head muscles of snakes differ from those of lizards in being much longer and in stretching in long arcs to attach to chains of bones lying at acute angles (Haas, 1973). Such muscles will not automatically return to their resting position when active contraction ceases. One or several repeated yawns are needed to reposition all of the complex elements.

Adaptive Features of the Skull

Which of the modifications in the head of *Dasypeltis* may be explained as devices involved in the implementation of such a radically distinct ingestion pattern? Certain obvious ones are the reduced intracranial kinesis, tight junctions between bones within the maxillary (and palatine) arch, the bracing of the supralabial scales to the marginal bones, and the development of semicircular gum ridges (Figures 2-8 and 2-9). All of these provide an (upper jaw) unit allowing maximum contact with the convex surface of the egg and more frictional resistance to anterior than to posterior slippage of the egg. The hypertrophied Harderian gland found in *Dasypeltis* (and in the Asian egg eater) may provide a specialized lubricant facilitating the forward slide of the upper jaw. The lower jaw is subdivided into three other elements of the ingestion mechanism, the two mandibles and the ventral blade. The mechanical bases of the ventral blade are formed by the stiff boss of connective tissue, the tight junctions between the chin scales, their bracing by the hyoid elements, their loose connections to the mandibles, and the contracted folding of the mucous lining of the mouth. Anteriorly, gum folds substitute for the teeth, which are consequently reduced in number and size. This change appears to reflect a shift from point contact by the teeth to surface contact by the gums; it reduces the outward reaction forces.

The anterior shift in the articulation point of the jaw suspension, from a quadrate-supratemporal to a supratemporal-braincase articulation, increases the radius along which the quadratomandibular joint swings and consequently increases the distance and position that the mandible may be moved along the egg's side. The need for a lateral displacement of the quadratomandibular joint and of the connected posterior tip of the pterygoid (in order to accommodate the width as well as the height of the egg) apparently conflicted with the need for tight junctions between the elements of the upper jaw which fixed the anterior part of

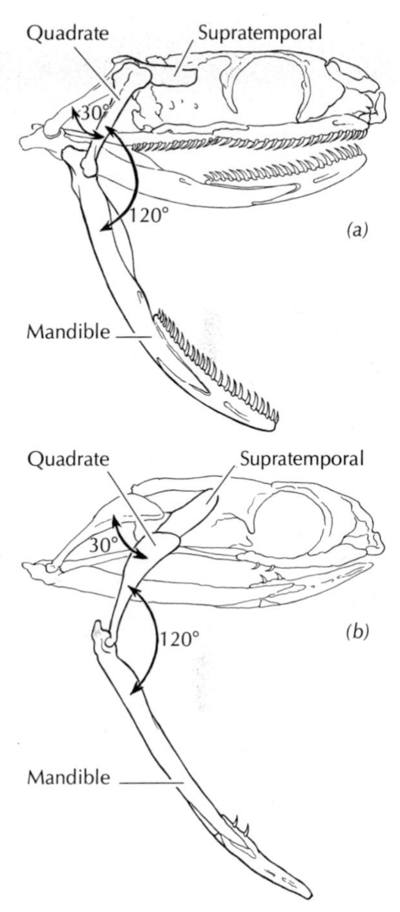

Figure 2-22 Sketches to document the effect of moving the jaw from the articulation between supratemporal and braincase rather than that between quadrate and supratemporal. For two snake skulls of equal length (from snout to occipital condyle), this will result in an almost 20 percent larger gape. When the skulls are not to proportion, the gape may be further increased by elongating the quadrate.

the pterygoid. Unlike typical colubrids, *Dasypeltis* does not have to transmit anterolateral forces to the edentulous anterior parts of its upper jaw. Indeed, the anterior blade of its pterygoid is fixed to the base of its braincase; its posterior wing is slender and rod-shaped so that its bends can follow the quadratomandibular displacements (Figure 2-8).

Certain other features of the head are less obvious. The slender dentary is movably articulated on the compound bone and can hence be bent along the surface of the egg, thus providing shaped support for the soft-tissue contact along the sides of the egg. The V-shaped muscular bands inserting on the central raphe of chin shields will be stretched around the arc of the egg's curve. Their origins on the dorsal and ventral borders of the dentary form a muscular cradle that may align this bone parallel to the surface of the shell. The bone then occupies the mechanically most advantageous position in resisting irregularities of the shell and facilitating the pull of the chin shields; this should become particularly critical when the widest part of the egg has been passed and muscular contractions narrow the gape to propel the egg through the esophagus.

Another complex specialization is in the mechanism that forces the strip of reinforced chin shields beneath the egg. The muscles can depress the mandibles but cannot induce a lateral widening of the gape. Rather, these elements are deflected laterally from the median axis of the body by forcing them outward along the three-dimensional wedge represented by the proximal half of the entering egg (Figure 2-23). As the mandibles shift laterally they stretch their loose connections with the strip of median chin shields. These shields show close interconnections

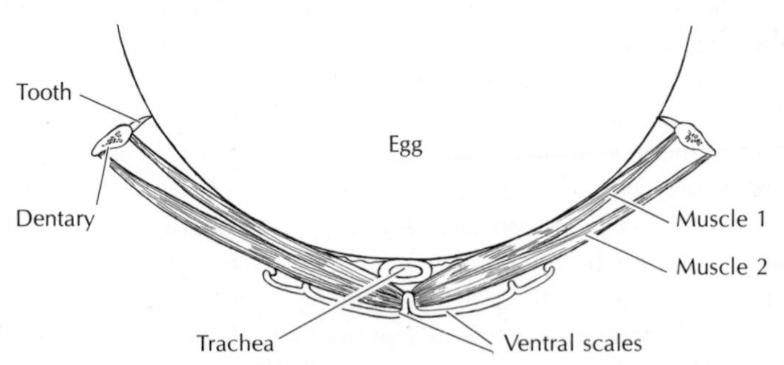

Figure 2-23 The loose attachment of mandible to the mass of chin scales further increases a snake's capacity to swallow large eggs. This cross section through the bottom of an egg during swallowing shows how far the mandibles can be separated from each other and from the chin shields. The position of the muscles is somewhat exaggerated to show the cradling effect which keeps them in position along the surface of the egg. Since the mandible is narrow but high, it has much greater resistance to bending when placed on edge rather than with its inside against the egg.

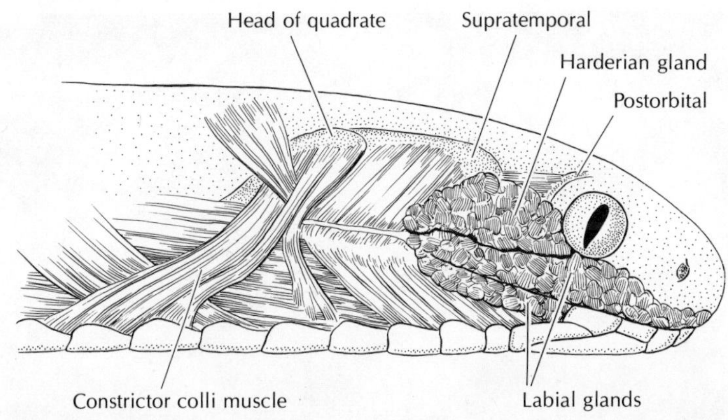

Figure 2-24 Head of *Dasypeltis* with skin removed to show the constrictor colli muscle which connects the sides of the ventral scales to the head of the quadrate. Contraction of this muscle will push the strip of reinforced chin shields away from their association with the mandibles and beneath the egg. The enormous Harderian gland and the large labial glands provide secretions that lubricate the egg's dry surface during swallowing.

between successive scales and a general stiffening of the anterior chin; both adaptations promote resistance to compression along the length of the transversely flexible strip, thus letting it bend. The fibers of the constrictor colli muscle, which insert at a very acute angle along the sides of the eighth and ninth ventral scales, push the band of scales forward. As the mouth opens, this band of throat scales is bent sharply just back of the jaws, but its anterior portion is applied against the bottom of the egg. Although the band of scales is curved through an acute angle, the resistance to longitudinal deformation causes the band to continue slipping through the curve when the constrictor colli contracts. It thus moves much as does a steel tape when pushed around a constrained angular bend.

One aspect that does not yield an immediate functional explanation is the general lightening of the bony skull of *Dasypeltis*. Not only are most elements remarkably slender, but the edges of the roofing bones show a characteristic scalloping that in other species would seem almost pathological. (Perhaps for this reason the edges have been smoothed out in some published views of such skulls—FitzSimons, 1962.) Cartilage and connective tissues bind the bony frame of the skull together even more tightly, further increasing its rigidity. There is no obvious aspect of egg eating that demands thinner bones in the head. Perhaps all snakes encounter an intrinsic disadvantage to a heavy skull, though relatively large and rigid bones may be selectively advantageous for snakes that habitually encompass struggling prey. *Dasypeltis* avoids such problems by feeding on immobile prey. According to this reasoning, the modifica-

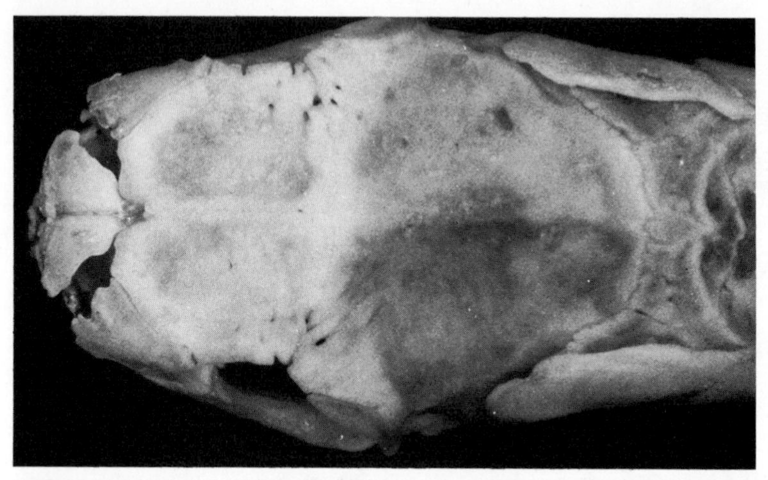

Figure 2-25 The edges of frontals and other elements of the skull are curiously scalloped in *Dasypeltis*. Has the skull been lightened, or does the scalloping serve some other function?

tion is permitted by, rather than facilitating, egg eating. Clues to such questions may well be found when examining other specialized colubrids.

All of these specializations, with the possible exception of the lightening of the bony skull, have facilitated egg ingestion; they allow *Dasypeltis* to swallow a larger egg than can a colubrid of equivalent size. Most critical is the unusual gape. The distance the mandibular tips may be depressed defines the height of the gape in typical colubrid snakes. In *Dasypeltis* the mouth may be opened wider than the jaws can distend, because of the introduction of a fourth element represented by the chin shields (which carry the ventral edge of the gape far below the level of any bony support). The relatively small species of the genus *Dasypeltis* may then deal with a spectrum of eggs ranging to sizes far larger than those suitable for colubrids of equivalent size. Not only does the increased gape reduce the need to struggle, but the volume of the egg and its contents increases by the third power (of the absolute gape). Thus adaptation widens the range of available food, and its selective advantage is obvious.

Separating Out the Contents

The egg gradually slides inward until the end posteriorly directed in the snake reaches the level of the spinous hypapophyses perforating the esophagus. If the egg is small, the mouth will be almost entirely closed at this time. If the egg is large relative to the snake, the sliding may be stopped with the mouth still partially distended. Only the closing action of the jaws and perhaps contraction of the muscles connecting the lower jaws to the chin shields appear to assist in pushing the egg into the esophagus. When the egg is small, the snake will be able to form a posteriorly traveling S-shaped lateral curve in the vertebral column anterior to it. Since a rigid object cannot bend to conform with S-shaped curves, this bend effectively pushes the egg farther down the esophagus to the spinous hypapophyses. Such traveling curves enable other snakes to move and stretch various kinds of large food objects within the body; the curves serve as a substitute for true peristalsis (alternate narrowing and widening) which is insignificant in the snake's thin-walled anterior gut.

The extreme distensibility of the gape continues to be reflected in the neck region as well. Particularly when the egg is large, one may see the scale rows separate sharply. Indeed, most species of the genus show some half-dozen accessory scale rows in the neck region that interdigitate between the ones found farther down the body. Only in the mid-dorsal region does the skin of *Dasypeltis* appear to be fixed to the muscles of the trunk; elsewhere the skin slips loosely over the underlying tissues. Thus the entire skin, and not just that of the sides, expands as the swallowed egg proceeds down the throat. This is apparent when a small snake swallows a large egg; the contours of the vertebral column are

Figure 2-26 This lateral view of *Dasypeltis atra*, in the process of ingestion, shows that several rows of scales stop in the neck. The more posterior skin apparently can stretch less widely than that in the front. The photograph also shows that the skin is only anchored along the mid-dorsal line so that it can distend all the way down the side.

easily distinguished from those of the attached sheet of ribs and muscles. The skin and underlying muscle slips may become enormously stretched so that the sides of the egg can be discerned through a translucent sheet that connects the sides of the midvertebral region to the ventral scales.

The distension induced by a large egg inevitably arches the vertebral column over it, forming a ventrally concave arch; a reverse arch (dorsally concave) forms behind the egg's posteriorly facing end. A similar but less concave curve anterior to the egg reaches maximum curvature just before a tensing of the dorsal muscles and a cracking sound signal the egg's rupture. The snake next goes through a series of wriggling motions. It gradually straightens the induced vertebral curvature and then engages in vigorous low-amplitude lateral undulations, first of the zone just posterior to, and then the zone directly over, the egg. Simultaneously there may be further cracking sounds, and the egg's outline becomes visibly slimmer. Toward the final stages of this squeezing the snake completely reverses the curvatures over the egg and the neck seems to push back into the egg's mass.

This squeezing process appears to be the most variable aspect of the feeding sequence. The types and sequences of movement are influenced by such factors as relative egg size, shell thickness, and consistency of the contents. Behavioral factors such as previous experience, most recent feeding (or the number of the egg in the series eaten), and extraneous disturbances will affect the length of the process. Snakes being disturbed or watched are particularly likely to cease squeezing, move to another location, or remain immobile for as long as an hour. These observations serve to explain some of the divergent reports in the literature.

In most cases an undisturbed snake will completely empty an egg's liquid contents in five to ten minutes. At this time, shell and shell membranes will have been compacted into a dorsoventrally flattened, boat-shaped mass with the membranes providing an extension pointing to the snake's tail. The bottom surface of this "boat" is normally only slightly deformed; the cracked pieces of shell remain glued to their underlying membrane in a reasonable semblance of their original position. In contrast, the top of the boat will often be covered with much more irregularly splintered shell segments.

The beginning of shell ejection is indicated by several major movements. The head and anterior body are generally lifted off the substratum and, as first reported by Rabb and Snedigar (1961), extended over an edge, whether it be off a branch, the side of a cage, or presumably the bird's nest. (There seem to be almost no observations of the behavior of *Dasypeltis* while actually "robbing" a bird's nest in the wild.) One next sees the muscles tense and an S-curve form at the level of the 40th to 60th vertebrae, posterior to the site at which the egg is located. The snake simultaneously gapes in a curious fashion, combining protraction (anterior rotation) of the quadrates with simultaneous, or sometimes alternating, depression of the mandibles. The strip of the chin scales is never depressed below the lowest of the two mandibles. On the other hand, there appears to be a posteroventral pull originating from the region of the ventral scales. Not only the buccal cavity but the entire throat becomes distended in this gape; this behavior differs from the opening of the mouth seen during intention movements before a bite or repositioning stretch (yawning) movements after swallowing or shell ejection.

The S-shaped curve travels anteriorly, driving the crushed egg before it. Simultaneously, the neck and gaping head swing from side to side; one next sees the crushed shell and its contents appear in the open mouth and then drop from it. Shell and membranes are always wet, suggesting copious secretions from the labial and enlarged Harderian glands. The egg membranes trailing the shell package may occasionally catch in the throat; they are released by acceleration of the lateral movements into a violent shaking of the head. The snake then proceeds to make two or three yawning movements and to move away, either to deal with the next egg or to rest.

Certain of the structural modifications seen in *Dasypeltis* may reasonably be explained as obvious specializations for egg eating. Other structures facilitate the emptying sequence although they have not become morphologically modified. Unspecialized structures are exemplified by the short, flat anterior hypapophyses (Figure 2-16). These bones are in a position to provide a guide that keeps the egg from pressing too closely against the vertebral bases and disturbing their muscular attachments. This function is not unique to egg eaters; as already noted, such anterior hypapophyses generally serve as attachment sites for the subvertebral muscles of most snakes.

The nature of the egg's shell explains the breaking mechanism. A structurally optimal container should involve a minimum amount of material encompassing a maximum volume; it will be limited by its ability to withstand imposed forces (cf. Frost, 1967). The optimum shape of such a container is a sphere of constant wall thickness, and eggs represent an approximation to this. The hard shell of the bird's egg is very brittle, which means that the elastic limit at which deformation under load changes from an elastic to a plastic (permanent) state is very close to the ultimate strength at which the minimum continuously applied force would cause the shell to rupture (see Box 2-3). Since the ultimate strength is a function of the stress generated at the contact site, the force required to induce initial rupture will vary inversely with the diameter of the punch used, or rather with that portion of the punch actually in contact with the shell. Hence the sharper the contacting hypapophysis (the punch in this case), the smaller the area in contact

Box 2-3 Strength of Materials

We know that a thick line will support a heavier fish than a thin one made of the same material. This indicates that the strength of an object is determined not only by the nature of the material and the force applied to it but by its cross-sectional area. One often combines measures of force and cross-sectional area by talking of *stress*, or load-per-unit area. Similarly, we know that under a given load a long rubber band will stretch farther than will a short one. Deformation must again be characterized as *strain*, or a percentage change from previous size.

Different materials show different deformation rates when loaded. In general, one talks of *elastic* or reversible changes and of *plastic* or irreversible ones. A rubber band is indeed elastic; the term, however, does not refer to its enormous stretchability but to its tendency to return to the original shape. This is also a property that one expects from a well-designed rail. It bends when a train passes, but then returns to its original shape. Normally, elastic regions of materials provide a straight line when stress is graphed against strain. In short, the material obeys Hooke's Law, which states that up to a loading, dependent on the particular material, the stress will be proportional to the strain.

The actual relations will differ with the material being stressed. Obviously, it takes more stress to bend a bone than a cartilage of equal size. The behavior of materials differs even more when the load is sufficient to take it over the elastic limit—that is, out of the elastic and into the plastic zone. Here the deformation is no longer reversible. In some materials a slight overload or tempo-

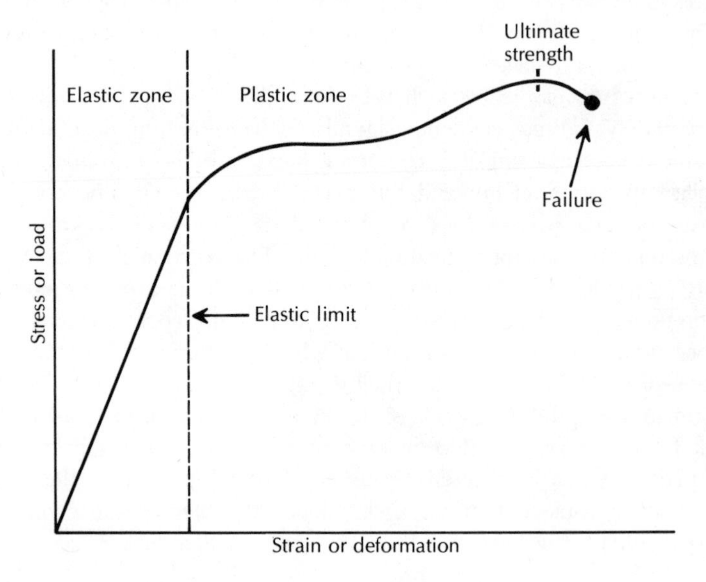

rary exceeding of the elastic limit will induce a slight permanent strain. If the load is then reduced, the material again seems to act elastically, though it has been permanently deformed. A material is called *ductile* if one can stretch it very far by pulling or rolling, causing deformation but not immediate failure. *Brittle* materials will fail very soon after their elastic limit has been exceeded. The material, then, determines whether failure happens suddenly or after considerable elongation.

Different materials show different strengths when loaded in compression, rather than in tension. Such properties become important in understanding the relative distribution of cartilage, bone, and collagen bundles in animal skeletons. This theory also explains the importance of locally applied forces. Obviously the stress will be greatest at the zone of smallest cross-sectional area. Irregularities such as sharp notches in the side of a bar cause further concentration of the internal stresses and, with this, increased strain and increased risk of failure. This relationship gives us physical reasons for the disadvantage of pointed objects projecting from the dashboard of a car and the selective advantage of pointed hypapophyses to an egg-eating snake.

and the higher the concentration of stress on the shell for a particular applied force.

The site at which this stress concentration is induced was once explained to my satisfaction, but the mechanism has recently been questioned. In my first study (Gans, 1952), I assumed that the initial piercing of the egg was produced by the most obviously perforant spinous hypapophyses of the 29th to 38th vertebrae. It seemed reasonable that the cracking involved a localized stress concentration on the egg's shell. Once the shell had failed at one site, the crack would be expected to travel in response to loadings much lower than those that could be borne by an intact shell. The more irregular the hole, the greater the local stress concentration and the more rapid the travel of the crack. These loadings during the sequence following initial rupture were presumably imposed by the very much modified intermediate hypapophyses (of vertebrae 18 to 25), so that the shell would be folded in along the line pressed by the hypapophyses.

This initial analysis, as summarized above, had been based primarily upon dissection, literature records, one good series of photographs, and a single, personally observed feeding sequence. Only much later did it become possible to observe, photograph, and analyze movie records of many more egg eaters feeding on eggs of various sizes (Gans, 1970). These studies suggested that the posterior spines not only induced rupture of the shell, but also of the shell membranes of the eggs; some of the records appeared to show that the spines were forced into the egg's posteriorly pointing end as well as its dorsal aspect. Lateral undulations of this portion of the vertebral column would often be noted at this time. These provided the basis for persistent comments in the literature that the egg was "sawed open," but were interpreted by me as a secondary rupturing of the shell membranes, which appear to be quite tough in some species of birds.

The risk of conclusions from the sort of indirect evidence just covered was brought home to me by a letter from Dr. A. Mittelholzer of Unterkulm, Switzerland, noting that he and a colleague had succeeded in taking a series of dorsal and lateral x-ray views of a specimen of *Dasypeltis scabra* ingesting, cracking, and crushing a very large egg (snake head diameter about 10 mm, egg diameter 43 mm as read off the x-ray). In a popular article based on this (Mittelholzer, 1970), it was pointed out that the egg's initial perforation occurred by means of the anterior (10th to 18th) hypapophyses and that the spinous posterior ones serve primarily to keep the mass of egg shell from proceeding to the stomach.

Through the courtesy of Drs. W. Güntert and A. Mittelholzer, I have been privileged to study an excellent series of some twenty prints made by Mr. R. Burger from the original rapid-sequence x-rays. They used a cassette changer that did not disturb the animal when they took one picture every few seconds, rather than cinefluoroscopy which provides a continuous record of 24 or more frames a second. Exposures eight and nine would seem to show the initial perforation by the hypapophyses *anterior to* the knob-shaped intermediate series; they also show that the crack progresses from back to front as earlier supposed. These pictures even suggest that it is the last of the anterior series (vertebrae numbers 14 to 19) rather than the heavy middle series that continue to crack the shell. It might be possible to argue that the actual moment of perforation had been missed. Certainly the second picture shows that the posteriormost portion of the indentation (initially opposite vertebrae 14 through 19) had become repositioned when the hypapophyses of vertebrae 10 through 14 forced the shell inward. The print, however, does show a thin vertical crack in exposure nine. Later prints of the series seem to show that the posteriorly directed end of the shell broke off as crushing proceeded; it is most unlikely that such a capping of the egg shell would occur if the posterior part of the shell had been weakened initially by penetration of the spinous hypapophyses in the center of the terminal area rather than the shell being broken by the anterior ridges.

The series, however, still left some questions for further analysis. Whether broken anteriorly or posteriorly, the egg seems definitely to be cracked by a local stress concentration due to contact of the shell with a zone of small cross section. The egg used by Dr. Mittelholzer and shown in Figure 2-27 was four times the diameter of the head, and thus

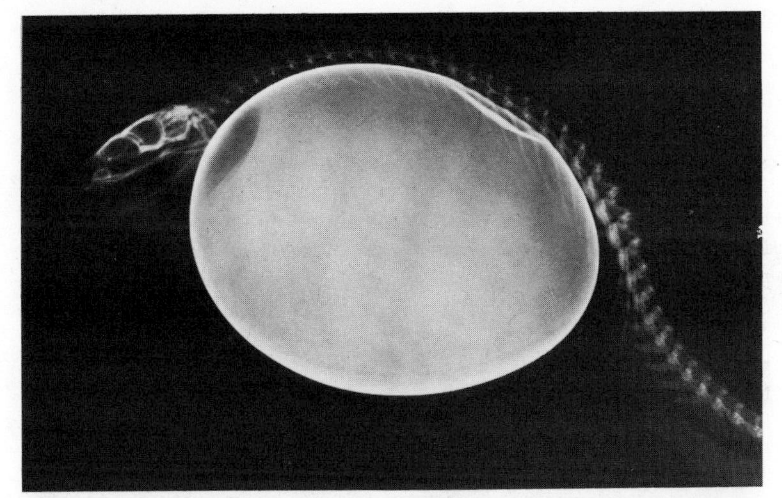
(a)

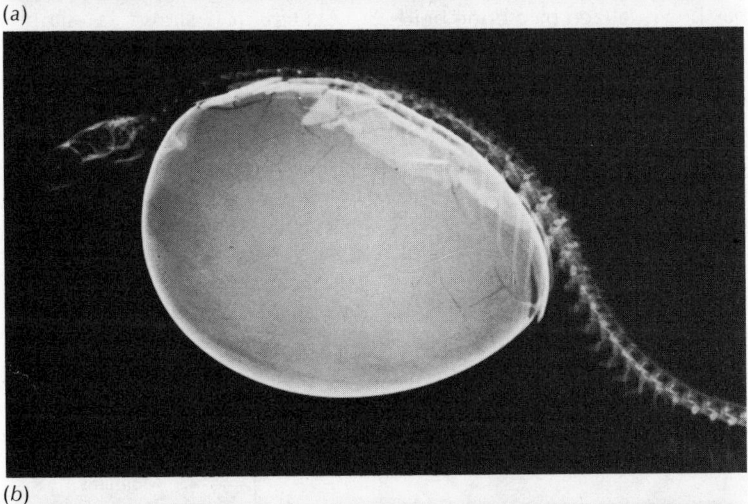
(b)

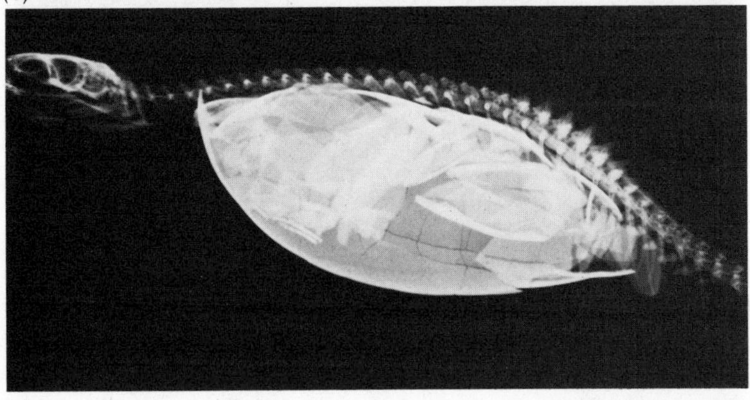
(c)

Figure 2-27 X-rays 8, 11, and 15 of Dr. Mittelholzer's sequence show a stage in the first rupture of the egg (*a*), the cracking of the dorsal shell and initial infolding (*b*), and a stage during which most of the liquid contents have passed down the esophagus (*c*). Note particularly the position of the spinous hypapophyses and the bulbous ones on the intermediate vertebrae. The eggshell is clearly breaking in a nonrandom pattern.

Box 2-4 Couples and Moments

Vectors need not act upon a point but rather upon a multidimensional object. When such an object is met by two equal, parallel, but noncoincident forces, we call the imposed load a *couple* and note that it imposes a *moment* tending to rotate the object. (Moments are defined as forces inducing rotation; their magnitude is a product of the force and the perpendicular distance of its line of action to the center of rotation.) The effect of a couple may be visualized by a bone held in equilibrium between two teeth (*a*). If we assume the bone to be weightless, we see that the downward force is exactly counteracted by the upward force, both directed along the same line. If the lower jaw is protruded slightly (*b*), its force may remain parallel to that imposed by the upper but will be noncoincident. The moment produced by this force couple then rotates the bone out of equilibrium until it contacts other teeth.

The easiest way of analyzing a complex of forces acting upon a three-dimensional object is to resolve them into components along two or three axes and carry out a summation separately for each. The greater the number of simultaneous equations one may write about such a system, the greater the number of unknowns for which the system may be solved.

Sketch (*c*) shows a simple analysis of an element loaded by a couple. The element is a human digit pressed downward against an edge which it overlaps only at its very tip. In theory, a force, R, equal and opposite to that exerted by the edge would balance it. However, the only place where such a force may be exerted is the joint between the phalanges. The forces F_1 and R would then produce a couple tending to lift

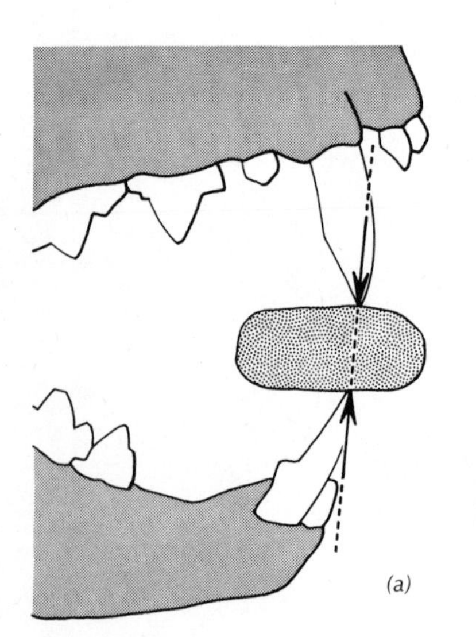

(*a*)

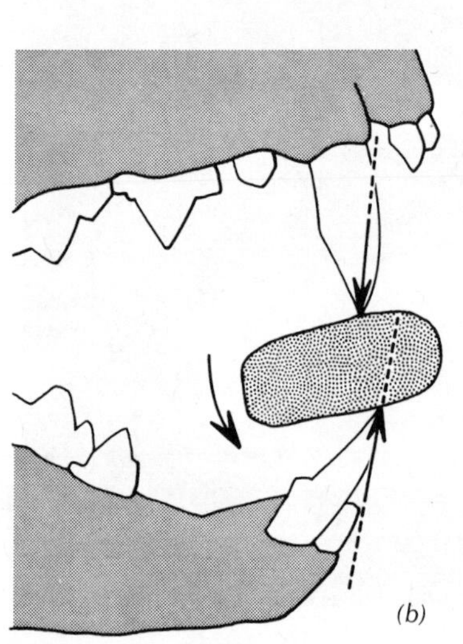

(*b*)

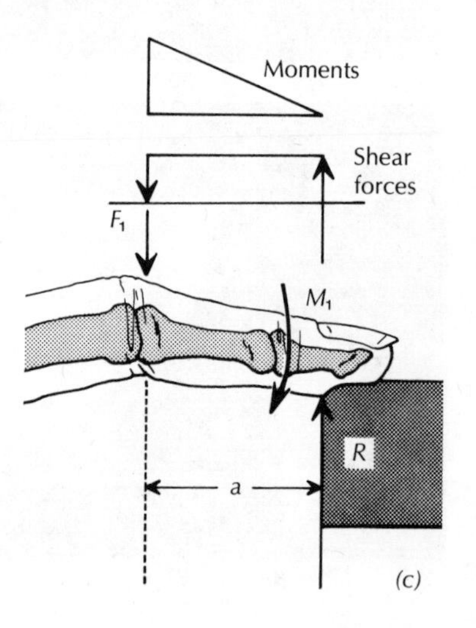

(*c*)

the finger away from the support. If the system is to be balanced, a moment, M_1, must be transmitted by locking the finger joint with tendons and ligaments. The graphed moment at the top of the figure indicates that the magnitude of this moment will increase with the length of the digit or rather with the distance (a) between the point of force application and the first functioning joint. The longer the digit, the higher the moment that has to be counteracted at the joint. This very simple example is, of course, a cantilevered beam turned upside down, and thus one of the simplest situations in structural design. For more examples, see any engineering text on "strength of materials."

far larger than those normally taken by *Dasypeltis*. Did this affect the results? It would certainly be easier to bring the pointed spines of the posterior hypapophyses into position for piercing an egg only twice or three times the head's diameter. Yet analyses of other films available to me tend to confirm the conclusion that the initial perforation occurs at or anterior to the modified vertebrae of the middle series, and the egg is then shifted farther back for breaking. Also unclear is the mechanism by which the fluid contents could be directed posteriorly in the snake if the initial opening were induced along the egg's front or even top. Do the spinous posterior tips only rupture the membranes *after* the shell has been cracked?

Although the instant of cracking and indeed the vertebrae inducing the initial perforation are still in doubt, the functional explanation of egg ingestion and of the emptying sequence are supported by these new data. The egg shell is initially weakened by the perforation caused by the imposition of maximal forces upon a restricted zone. A further crack moves anteriorly along the hypapophyseal line. Failure of the shell cannot be restricted to this zone but will be general, and the x-rays show numerous cracks down along the shell's sides as well as longitudinal breaks of the upper quadrants.

The modification of the intermediate vertebrae appears to be the key adaptation for the emptying cycle. The widened, rounded hypapophyses serve as a device for crushing inward the dorsal aspect of the shell, thus inducing an increase in the hydrostatic pressure of the fluid contents that forces the posterior membranes of the egg to bulge into contact with the projecting spinous hypapophyses. The series of rounded processes can also be seen to fold the shell inward as the contents escape through the posterior rupture. The cracking patterns in partially emptied shells (seen, for instance, when *Dasypeltis* had fed on eggs containing partially developed embryos) suggest that the ventral half of the shell is cushioned in the adherent ventral portion of the esophagus.

The dorsal shift of egg and esophagus is apparently achieved by the

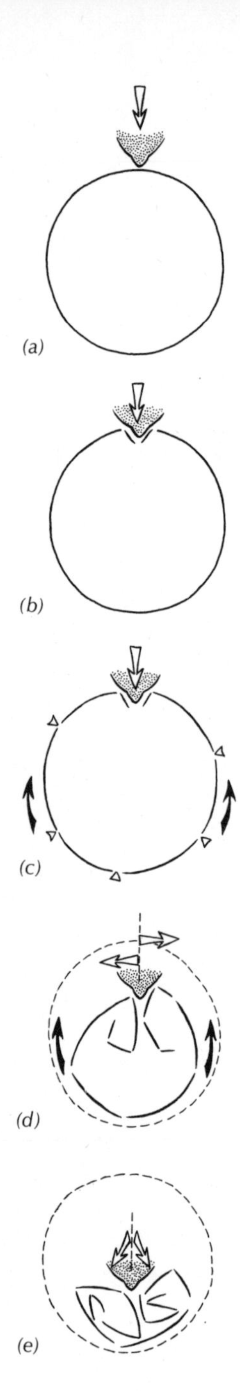

Figure 2-28 The initial break of the dorsally pointing aspect of the shell is followed by further, more-or-less parallel, fractures that allow the shell to fold as the contents are squeezed out. The lateral movements of the bulbous intermediate hypapophyses not only break the shell but roll up its edges, held by the egg membrane, into a boat-shaped mass containing minimum liquid. The arrows show major forces exerted on the egg; the triangles show points of rupture.

sheets of muscle reaching from the sides of the ribs to the ventral portions of the skin. In small specimens this action may be seen from the outside. Muscle contractions are indicated by downward shifts of the rib ends between egg and loose skin; the interscalar skin first becomes folded near the dorsal zone so that the portion adjacent to the ventral scales contracts last. The effect of this upward shift of egg—or rather the downward shift of vertebral hypapophyses—is to restrict the failure zone to the egg's upward surface. There is no opportunity for ventrad or anteriad escape of egg fluid; the nutrient contents exit through the rupture torn by the spines in the shell membranes. They continue into the stomach past the spines and the muscular esophageal valve.

The inward movements of the intermediate hypapophyses soon become mixed with simultaneous, low-amplitude, lateral undulations that clearly must break the lateral edges of the shell and compact the fragments as emptying proceeds. Finally the snake may be seen to bend its back to a dorsal concavity, forcing the bulbous hypapophyses into the shell where they displace the remnants of the fluid, simultaneously squeezing the shell into the boat-shaped mass which is ready for regurgitation (Figure 2-5).

These observations are most suggestive regarding the probable functional bases for the structural differences seen between the dasypeltine vertebral column and that of the generalized colubrid. The slight enlargement of the anterior hypapophyses may reflect the demand for a smooth sliding surface for the egg and a stiff protection for muscle insertions on the cervical vertebrae; yet the difference between *Dasypeltis* and such colubrids as *Elaphe* is probably too subtle to permit decisions not based on morphometric analysis, which involves statistical analysis of measurements taken of many specimens.

The function of the spines of vertebrae 29 to 38 is clearer. Whether or not they serve as stress concentration processes that perforate the shell, they certainly must rupture the membranes. Thus they ensure and enhance the posteriad flow of liquid nutrients. The x-rays now available indicate that these spines also serve to arrest progress of the shell toward the esophageal valve. This function is clearly advantageous as it reduces the risk that unusually energetic movements might cause the shell to shift into the stomach.[1]

[1]It is possible for an egg's shell to slip past the esophageal valve, as witnessed by a snake in a zoo that was force-fed lizards (*Anolis*) before the snake was identified as an egg eater.

The modification of the intermediate group of anterior vertebrae is by far the greatest; their modification is also most clearly associated with observed functions. The enlargement and rounding of the hypapophyses, which in the adult occupy the entire ventral aspect of each vertebra, obviously permit the animal to do more than crack the shell. As the vertebral column swings through lateral movements while the egg is being crushed, the shell's hard parts are packed to the two sides. Large shell fragments break further into many small pieces, and while stuck to the egg's membranes may roll up along the sides or remain grouped along the midline (Figure 2-28). The lateral stresses imposed at that time presumably require the ontogenetic thickening of these intermediate hypapophyses from a pattern in which only the ventral edge is enlarged to one in which the enlargement encompasses the whole vertebral spine, which is then widened and shortened (Figure 2-29).

The extreme lateral and dorsoventral bending, and particularly the reverse bends during the final movements of emptying, presumably explain the modification of the epaxial region of the intermediate vertebrae. This modification also reflects the significant forces that have to be applied to empty and fold the shell. The hypertrophied dorsal musculature permits crushing and emptying sequences involving as many as six eggs, one after another. The enlargement of the neural spines into high and thin plates deepens the trough (Figure 2-17), and the flaring of the zygapophyseal roof extends the trough laterally. The contracting muscles remain stretched along the arc of the ventrally concave spine rather than slipping downward to occupy its chord. Certain other modifications are seen in this region, such as almost plane (flattened) rib articulations. This characteristic would presumably give the ribs in this region a different kind of motility and perhaps prevent the ends of the ribs from being stressed by the egg during the crushing movement.

The perforation of the hypapophyses through the esophageal wall has attracted attention ever since their French discoverers speculated in 1820 that these spines might represent nature's compensation for the teeth lost from the mouth. Various studies, the most recent that of Schmidt (1958), have documented that "nature's compensation" to the contrary, these spines lack an enamel covering. Bare bone is of advantage as the stress concentration increases when the tip is not cushioned by soft tissues; hence the animal need impose less total force to crack the shell. These bare spines, however, do become longer with age; how then do they grow?

Bone grows by accretion, and this requires that the surface must be covered by a membranous periosteum; growth is limited to periosteum-covered surfaces. Internal (interstitial) growth cannot occur in osseous tissues; internal growth only occurs in a bone when cartilagenous zones separate bony regions, as happens, for instance, in the long bones of the

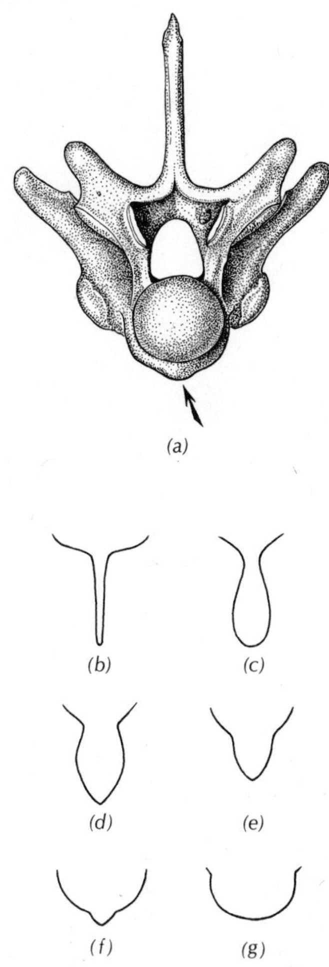

Figure 2-29 The hypapophyses of the intermediate vertebrae change during development from plates similar to those on the anterior vertebrae into rounded bosses in the sequence shown here.

human arm. No such pattern is seen in hypapophyses, which are in bony continuity with the vertebral centra. Any periosteum underlying the thin mucosal lining of the esophagus and loose connective tissue would, furthermore, suffer damage when intense localized pressure was exerted thereon. Differential growth, in this case growth occurring only at the dorsal surface of the vertebrae, would require very major resorptive restructuring. One plausible growth mechanism would be a seasonal one, if these snakes estivate. This explanation assumes that the hypapophyses would become gradually uncovered during each feeding season. The periosteum would be displaced from the hypapophyseal tips during the repeated feedings. As the animals started to estivate, during seasons when bird eggs were uncommon or drought inhibited activity, the periosteum would grow back over the tips and their osteogenesis could proceed. It would be possible to continue such speculation for some pages, referring for instance to other cases of seasonally exposed bones such as the thumb spikes of the male frog, *Babina*, and the perforating ribs of the newt, *Pleurodeles*. I desist because this question could better be resolved by a combination of observations and some simple experiments to check the actual sites of bone deposition.

Ecological Consequences

The pattern of egg eating here described may be termed an adaptive syndrome. *Dasypeltis* uses a specialized food source of high energy content. The feeding process not only permits the snakes to ingest eggs of varied, sometimes extreme, size rapidly and effectively, but also to separate the nutritive material from the packing. Not only may the liquid contents then be held in the stomach in the least obtrusive format, but the nutritionally disadvantageous portions of the food "package" are separated and discharged. The snake neither expends digestive effort on them nor carries them about. The snake can also obtain the remaining liquid nutrients from partially incubated eggs. It need not choose among eggs encountered in a given nest.

All of this is related to the total dependence of *Dasypeltis* on bird eggs. Occasional and apocryphal remarks about the eating of plant material or lizard and snail eggs by these snakes remain speculative and unsupported. Juvenile *Dasypeltis* can apparently subsist throughout the year on eggs obtained from birds' nests; indeed, the feeding mechanism is ideally suited to assure that the egg eaters may start to feed on eggs just as soon as the first ones are deposited. The snakes need not cease feeding until just before the last clutches of the season hatch, a time at which no nutrient liquid remains in the eggs.

The greater the amount of nutritious liquid packed into the stomach at the end of the last meal of a season, the less problem the snake has in fasting until the start of the next laying season. This may not be too important in some of the wet-zone tropical areas where birds breed virtually all year long. Here there might be selection for behavioral shifts by *Dasypeltis* relating to the kinds of eggs available during each month. In other zones there are one or two wet seasons, each with an associated bird breeding season, while some species of birds breed during the dry season. All we know is that *Dasypeltis* eats nothing but bird eggs; no study has yet shown which bird eggs it prefers, whether and how long it estivates between breeding seasons, and whether it shifts its microniche or behavior with the seasons.

Unfortunately we are also ignorant about the interaction of *Dasypeltis* and the population of bird eggs. Only a few peripheral remarks in the literature refer to such interactions in nature; in no case are there data on the relation of bird behavior to snake attacks, or whether the snake can deter the parent birds from attacking. Nor do we know why *Dasypeltis* drops the empty shells out of the nest; does this keep the birds, at least those of some species, from abandoning the nesting site? If the widespread distribution of African egg-eating snakes has long persisted, there would have been time for the development of anti-*Dasypeltis* strategies in some birds. Are the complicated nests of the African weaver bird with their blind "entrances" responses to egg predation? Is there natural selection for thicker-shelled eggs (the way there has been among domestic chickens)? Could distastefulness of eggs have been an advantageous strategy? We know almost nothing about such topics.

We do have some firsthand data on a different question. The high degree of egg-eating specialization is correlated with a reduction in mandibles, teeth, and apparently fangs (Duvernoy's gland, the "venom" gland normally associated with the short fangs of colubrid snakes, is retained in *Dasypeltis*—Kathariner, 1898). Once these structures were no longer needed for food procurement, there was no longer a selective advantage for the developmental and metabolic energy needed to produce and maintain them. Yet jaws and fangs serve in defense and deter predators. How then does an intrinsically defenseless egg predator deal with a potential predator on itself?

In developing answers one rapidly moves into the areas of behavior and population dynamics, though biomechanics continues to provide explanatory schemes for the movements observed. The defensive strategy is perhaps best understood by considering the nature of predators. The various snake consumers reflect a spectrum of specializations. *Dasypeltis* is a highly specialized predator on eggs; similarly there are highly specialized predators on snakes. The facultative Asiatic egg eaters (*Elaphe*) include eggs in their diet, but eat other prey as well; similarly some snake eaters are facultative and may only include some snakes in their diet. Facultative snake eaters may hence be expected to show less specialization in dealing with problems of ophiophagy, such as the defenses of snakes; other snake eaters may be more specialized at counteracting these defenses. The existence of two classes of predators, facultative and highly specialized, suggests that *Dasypeltis* uses two classes of defense, and such a mixed strategy is actually observed (Gans, 1961a).

Various wild pigs and birds are probably the most frequent predators of snakes. Some will attack and subdue any snake with impunity; cobra, viper, or egg eater are of equal interest to a secretary bird. Cryptic coloration (camouflage or imitation of the background) appears to be the best defensive strategy against such predation, and this is used by *Dasypeltis* and many other snakes. All species of *Dasypeltis* inhabiting open regions such as subdesert or savanna show very effective cryptic coloration. The overall color of their dorsal surface resembles the color tone of each local region they inhabit. Beyond this there is some disruptive coloration due to the vertebral blotches that alternate or link with lateral bars occupying the sides of the animals (see specimen in Figure 2-5). All of these markings break up the snake's outline and may preclude recognition by potential predators.

Lest it be assumed that this protective coloration is clearly understood, it should be noted that populations of the species *D. scabra* and *D. atra* also show pattern dimorphism, indeed polymorphism in various localities. This means that several subgroups of each local population differ in their color patterns. Some individuals show the disruptive pattern whereas others are unicolored in various shades of brown (depending upon the general color of the substratum). *D. atra* also has a melanistic variant that occurs at frequencies of 20 percent in the montane forests of central Africa, with another 20 percent of the population being unicolored brown. Could it be that selection for background matching is less critical in these disrupted and shaded habitats, or does the disrupted background provide a selective advantage to a species with polymorphic coloration?

The strategy directed at generalized (facultative) predators consists of two aspects—aposematic (warning) behavior and mimicry. The warning behavior may in this sense be a mimicry of a whole class of predators; it is effective because many animals have found it advantageous to withdraw from any sudden and unusual event or behavior as this may indicate the presence of a predator. The aposematic behavior of *Dasypeltis* (Gans and Richmond, 1957) combines a movement sequence with a striking visual color pattern and sound production. When an egg eater is disturbed by the close approach of a predatorlike organism, it pulls its head back to the side of its trunk and forms the second and third fifths of the body into a C-shaped curve with the open side facing the disturbance (Figure 2-30). The anterior portion of the neck lies closely adjacent to the side of the C. Simultaneously the snake inhales

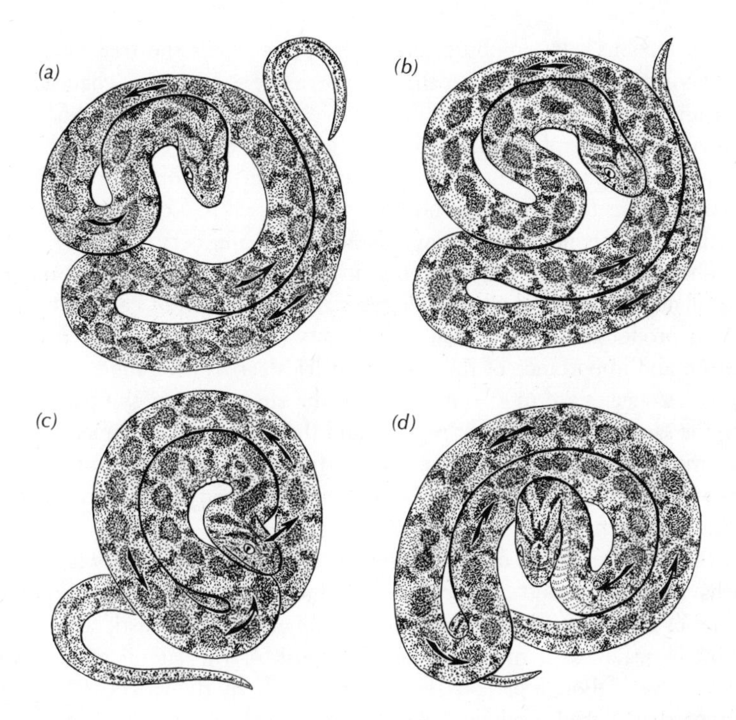

Figure 2-30 Formation of the C curve by an egg eater in the act of stridulation. Note how adjacent portions of the body move past each other in opposite directions. Thus the inclined and serrated scales along the snake's sides will rub each other, producing the hissing sound.

deeply, inflating the posterior air sac and lung and distending the neck and trunk. (The connective-tissue sac in the throat then permits the snake to puff up its neck.) The C-shaped position described is adopted only briefly as a new loop immediately forms in the neck and the body forms a multiple C-shaped path. This movement of the snake's body along a series of closely adjacent C-shaped tracks causes the sides to rub against each other, bringing modified lateral scale rows into contact. With the air sac distended, the inclined, serrated, and keeled scales (Figure 2-31) are situated on a stretched membrane so that the clicking contact of their tips is summated into a sustained hissing sound. Such a sound can also be produced when the snake's neck is distended by an egg. As the adjacent loops travel rapidly in opposed directions, the snake's outline becomes confused. Slight shifts in the movement pattern cause the entire coil to move forward to back, and a series of darting movements of the widely gaping mouth provides a most startling effect that may well frighten an unsuspecting predator.

The defensive strategy known as mimicry relies on the fact that certain organisms are truly noxious, so that potential predators will either die as a result of attack or will find the experience otherwise distasteful. Any interaction between the noxious "prey" and the potential predator is likely to lower the fitness of both. Consequently, there is selective advantage both to the development and the recognition of the signal

Figure 2-31 View of the side of *Dasypeltis* showing the region of scale inclination and keel serration. (After Gans, 1959)

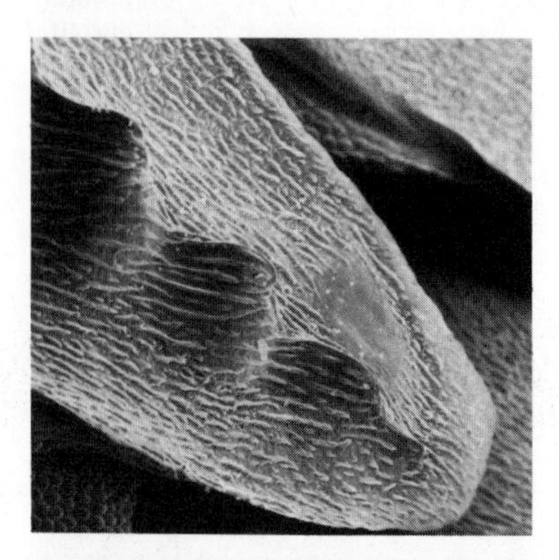

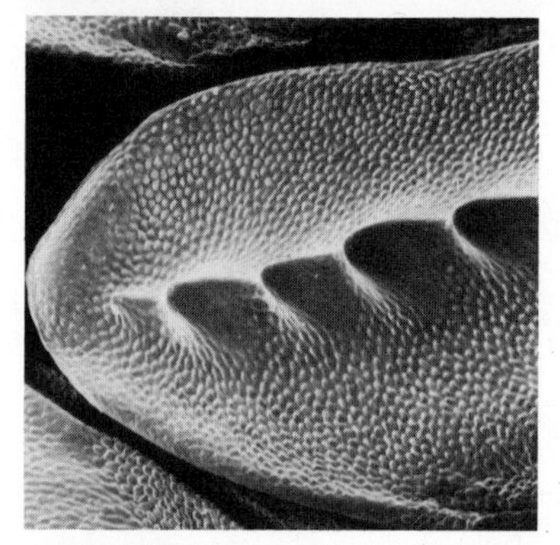

Figure 2-32 Scanning electron micrographs of a serrated scale of the egg-eater, *Dasypeltis scabra* (top), and the saw-scaled viper, *Echis carinata*, respectively mimic and model. Although the surface ornamentation of the scales differs between these two genera, the functional similarity is clear.

"I am noxious." In simplistic terms, the prey avoids the trial bite, the predator the distasteful or painful experience. Since the chances of learning (or reinforcement of instinctive recognition) of a signal by individuals or members of a group (Wickler, 1968) increases with the frequency of that signal, there may be further advantage for several noxious forms to share a common signal addressed to a common class of predators (a phenomenon known as Müllerian mimicry).

The small vipers of Africa represent such a group of Müllerian mimics; all are noxious and all share a general combination of color pattern, sound production, and sudden movements (Gans, 1965). The overall nature and importance of these signals varies across the range; those of any two species are most similar where the species occur sympatrically (in the same environment or region) and their differing ecological zones are most closely adjacent. The vipers differ in size, and comparisons show greatest similarity of color pattern between specimens of the same diameter.

On the other hand, major differences occur in the way in which the behavioral and acoustic effects are produced by vipers. For instance, some of the species hiss and the others generate the acoustic signal by rubbing scales with modifications like those of the lateral scale rows of *Dasypeltis*, though perhaps developed initially by the vipers as digging rather than sound-producing devices. The egg eaters, which are not noxious, derive benefit by mimicking this system. Potential predators may then be misled, either desisting from attack or being distracted long enough to permit the harmless *Dasypeltis* to escape. Such a set of resemblances designed to mislead a predator is referred to as Batesian mimicry. The specific color pattern of *Dasypeltis* is always matched to that of one of the local vipers; matching is closest to a specimen with an equivalent body diameter, but not necessarily of the same absolute length. There is also the curious situation that the aposematic behavior of *Dasypeltis* is very similar to the "Müllerian mimicry" signals of the vipers. Does the behavior in *Dasypeltis* act as an independent deterrent of independent origin or derive from its existence in various small vipers?

It is interesting to note the presentation of the various aspects of the signal in these different snakes. The auditory signals are induced, in some cases, by the forcible expulsion of air, in others by the variant methods of scale rubbing; apparently the noises are perceived equally by the potential predators. The various movements share the characteristics of suddenness, more properly high values of acceleration and jerk,[2] not only between snake and ground but also between parts of the animal. The adjacent loops travel in opposed directions when the snake is in a stridulating (rasping) coil; this inhibits the formation of a visual

[2]Acceleration is rate of change of velocity with time; jerk is rate of change of acceleration.

image and the total effect (to man) is one of a swirling coil. Only short sections of the animal are perceived at a time; this would appear to explain why the color-pattern similarities occur for snakes of equal diameter rather than equal length. The egg eater uses the widened trachea (Figure 2-15) to blow up its throat, further increasing its similarity to a viper.

Both vipers and egg eaters have the color patterns, presumed to serve as pattern signals, expressed more clearly on the interscalar skin than on the scales themselves. The inflation during the warning cycle separates the scales and stretches the skin, thus emphasizing this pattern. This again suggests the duality of strategy—all of these snakes hide (and remain inconspicuous) to avoid the attention of the snake-eating specialists. In contrast, they become more conspicuous as they inflate to warn (or bluff and run) in a deterrent to general predators such as baboons.[3] Such dual strategies, affecting diverse aspects of the animal's morphology and behavior, emphasize the complexity of solutions that facilitate survival.

The preceding discussion has attempted to document an analytic approach to the various morphological specializations in the African egg-eating snakes of the genus *Dasypeltis*. Analysis suggests that the numerous modifications apparently resulted from monophagy. The method of analysis has been one of deducing likely function from simple comparison, both of structure and of behavior, between the specialized animal and a seemingly more generalized one. This approach yielded a first approximation, the accuracy of which decreased rapidly as internal structures, beyond the reach of simple observation during life, were dealt with. Another area of uncertainty stemmed from the laboratory rather than the field approach to the several levels of predator/prey interaction. Supplementation of direct observation by cinematography (p. 57) and then by x-ray analysis (p. 58) in each case permitted refined understanding of the scheme and simultaneously generated new questions. The mimicry situation indicates how a seemingly unrelated aspect of the snake's biology may have to be modified as a secondary response to a major functional specialization, necessitating the reduction of other defenses. The warning behavior explains the selective advantage of the tracheal air sac.

Yet even this restrictive approach by simple techniques into the specializations seen in a single genus of snakes did circumscribe some of the areas of uncertainty in the earlier literature. The approach also suggested some possible selective advantages of the adaptations observed

[3]It has also been suggested that the defensive strategies of vipers of the African savanna may have evolved to discourage trampling by ungulates in the same way that the rattlesnake's rattle possibly evolved on the American prairies (Gans and Maderson, 1973). Whatever the driving effect, the mechanisms remain the same.

and with this brought out various areas in which additional questions could best be asked. Finally, it documented the necessity of a general knowledge of the snake as an organism when attempting to interpret the complex effects of a particular specialization.

REFERENCES

*Albright, R. G., and E. M. Nelson (1959). Cranial kinetics of the generalized colubrid snake *Elaphe obsoleta quadrivittata.* II: Functional morphology. J. Morphol., 105:241–292.

Chernov, S. A. (1945). On the adaptations of snakes to feeding on bird eggs [Russian]. Pap. Biol. Inst. Acad. Sci., USSR (1941–43):67.

———— (1957). On the adaptations of certain of our snake species for eating bird eggs [Russian]. Zool. Zhour. Acad. Sci., 36(2):260–264.

Dowling, H. G. (1959). Egg-eating adaptations in the Chinese ratsnake, *Elaphe carinata* Günther. Copeia, 1959(1):68–69.

*Dullemeijer, P. (1956). The functional morphology of the head of the common viper, *Vipera berus* (L.). Acta Neerland. Zool., 12:1–111.

Dunton, S. (1945). Egg-eating snake. Animal Kingdom, 48(6):188–189.

FitzSimons, V. F. M. (1962). Snakes of Southern Africa. Macdonald and Co., London, 423pp.

*Frazzetta, T. H. (1966). Studies on the morphology and function of the skull in the Boidae (Serpentes). Part II: Morphology and function of the jaw apparatus in *Python sebae* and *Python molurus*. J. Morph., 118(2):217–295.

*Frost, H. M. (1967). An introduction to biomechanics. C. C. Thomas, Springfield, Ill., 167pp.

Fukada, H. (1959). About the egg eating habits in *Elaphe climacophora* (Boie). Bull. Kyoto Gakugei Univ., B14:29–34.

*Gans, C. (1952). The functional morphology of the egg-eating adaptations in the snake genus *Dasypeltis*. Zoologica, 37(4):209–244.

———— (1959). A taxonomic revision of the African snake genus *Dasypeltis* (Reptilia: Serpentes). Ann. Mus. Congo Belge, ser. in 8°, Zool. 74:1–327.

———— (1961a). Mimicry in procryptically colored snakes of the genus *Dasypeltis*. Evolution, 15(1):72–91.

*———— (1961b). The feeding mechanism of snakes and its possible evolution. Amer. Zool., 1(2):217–227.

———— (1965). Empathic learning and the mimicry of African snakes. Evolution, 18(4):705.

———— (1970). Beobachtungen an africanischen Eierschlangen. Natur und Museum, 100(10):460–471.

————, and P. F. A. Maderson (1973). Sound producing mechanisms in Recent reptiles: Review and comments. Amer. Zool., 13(4):1195–1203.

————, and M. Oshima (1952). Adaptations for egg eating in the snake *Elaphe climacophora* (Boie). Amer. Mus. Novitates (1751):1–16.

————, and N. D. Richmond (1957). Some notes on warning reactions in snakes of the genus *Dasypeltis*. Copeia, 1957(4):269–274.

————, and E. E. Williams (1954). Present knowledge of the snake *Elachistodon westermanni* Reinhardt. Breviora (36):1–17.

Goris, R. C. (1963). Observations on the egg-crushing habits of the Japanese four-lined rat snake, *Elaphe quadrivirgata* (Boie). Copeia, 1963(3):573–575.

Haas, G. (1931). Über die Morphologie der Kiefermuskulatur und die Schädelmechanik einiger Schlangen. Zool. Jahrb., Abt. Anat. Ontog. Thiere, 54(3):333–416.

———— (1973). Muscles of the jaws and associated structures in the Rhynchocephalia and Squamata. In Biology of the reptilia (Carl Gans and T. S. Parsons, eds.). Academic Press, London and New York, 4:285–490.

Kathariner, L. (1898). Ueber den Verdauungscanal und die "Wirbelzähne" von *Dasypeltis scabra* Wagler. Zool. Jahrb., Abt. Anat. Ontog. Thiere, 11:501–518.

Mittelholzer, A. (1970). Wenn der Bissen grösser ist als der Mund . . . Das Röntgenbild enthüllt den "Trick" der Eierschlange. Aquar. Mag. (Stuttgart), 1970(7):313–315.

Oshima, M. (1930). Reptiles. In Ogô Dôbutsu Zukan (Illustrated Zoology) (S. Uchida et al., eds.). Hokuryukan, Tokyo, viii + 786 + xvipp.

Palmer, W. M., and G. Tregembo (1970). Notes on the natural history of the scarlet snake *Cemophora coccinea copei* Jan in North Carolina. Herpetologica, 26(3):300–302.

Rabb, G. B., and R. Snedigar (1961). Notes on feeding behavior of an African egg-eating snake. Copeia, 1961(1):59–60.

Schmidt, W. J. (1958). Natürliche Färbung von Reptilien-und Fischzähnen durch Eisenoxyd. Zool. Anz., 161(7/8):168–178.

Underwood, G. (1967). A contribution to the classification of snakes. British Museum (Natural History), London, x + 179pp.

Wickler, W. (1968). Mimicry in plants and animals. McGraw-Hill Book Co., New York and Toronto, 255pp.

3 ANALYSIS BY LIMITING THE QUESTION: LOCOMOTION WITHOUT LIMBS

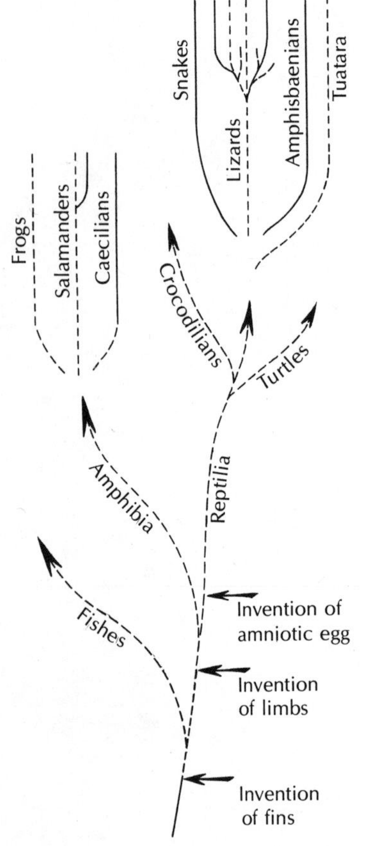

Figure 3-1 Limblessness has arisen many times. Those groups of vertebrates shown by solid rather than dashed lines contain limbless species.

Problem

The development of paired fins seems to have been a key to the widespread radiation of fishes in the Devonian. Later tetrapody (the shift to two pairs of limbs) permitted the invasion of land, which led to a similar successful radiation. Yet a surprising number of "tetrapods" lack functional limbs. In this category are some 160 caecilians, 2230 species of snakes, about 130 amphisbaenians, and diverse lizards. How do such animals move, and what may have led or permitted them to become limbless?

Historically, the first question has been a source of confusion; thus Proverbs (Chapter 30, verses 18/19) lists the way of a snake on a rock as one of four phenomena truly "too wonderful." The dilemma was primarily one of description. How does one characterize the curious curves assumed by a snake? Why do they keep shifting in irregular patterns, and how are these related to the surface across which the animal moves? Which of the interactions between ground and snake actually causes the animal to progress?

It quickly becomes obvious that the observation-dissection sequence so useful to the analysis of feeding mechanisms (Chapter 2) is inadequate to resolve limbless locomotion. Limbless forms are elongate and have a relatively high number of trunk vertebrae ranging from near 100 to 300 or more. Each vertebra supports a pair of ribs, and the dissections of Mosauer (1935) and Gasc (1967) suggest that there are as many as 20 discrete muscles on each side of a single snake vertebra. These muscles connect vertebra to vertebra, vertebra to rib, rib to rib, and both rib and vertebra to the skin. Beyond this, there are muscles attaching segmentally to longitudinal tendons, and this tendon-associated system is remarkably complex in those species that can form and control the shape of long curves. In such snakes there is a further displacement of the segmental muscles which may span fourteen to seventeen vertebrae. Although all elements are discrete, numerous connective-tissue links occur between the various muscles.

It is possible to resolve such a system by comparing the contraction of two sides or of epaxial (embryologically dorsal) versus hypaxial (embryologically ventral) groupings. Later, I shall derive certain general conditions on this basis. Yet such an approach is basically unsatisfactory

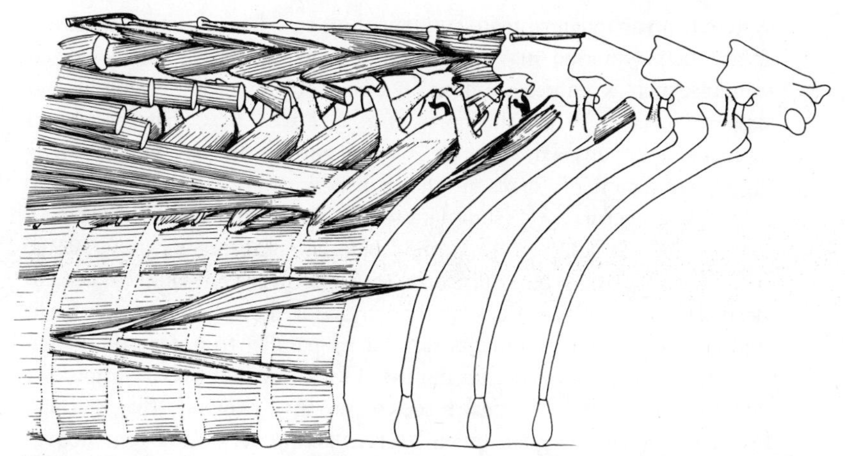

Figure 3-2 Sketch to show the complexity of the musculature on the outside of an amphisbaenian. Note the extended linkages bridging series of vertebrae. (Courtesy of J. P. Gasc)

because the complex muscles obviously have the potential for complex contractions. Unlike the regular and parallel packing of muscle fibers seen in the axial musculature of some fishes, that of most limbless "tetrapods" lacks parallel arrangement and seems to have a correspondingly more complex innervation. It may be possible to ascribe some of this complexity to remnants of a tetrapodal pattern, a pattern that remained after the limbs had become reduced. This does not explain the obvious increase in complexity that one observes, for instance, in colubrid snakes. Clearly, there are functional bases to this diversity and, as clearly, the number of elements is too large to let us deduce these bases from inspection by synthetic analysis.

Free-Body Analysis

How, then, should we approach the problem? The obvious way is by explicit application of the "free body" technique. This approach ultimately derives from Newton's first law of motion, generally stated as "An object at rest remains at rest and an object in motion remains in motion (at a constant velocity in a straight line) unless acted upon by an external force." In this case one can define the object as an animal (or one of its parts) floating in space (hence "free body") and ask what forces must be acting upon it to maintain the observed or momentary static or dynamic state (cf. Tricker and Tricker, 1967).

This method increases in utility when the magnitude or the direction, at least of some of the forces, may be defined; a selected number of measurements reduces the number of unknown factors and may move the system from an indeterminate (undefinable) to a determinate (definable) state. Such definition may be considered on the basis of an idealized example, such as a horse standing on a perfectly smooth, horizontal, and slippery surface. When the horse is at rest we know that forces and moments along any axis must be equal to zero. This may be written as ΣF_x (summation of forces in the x-direction) $= 0$, $\Sigma F_y = 0$, and $\Sigma F_z = 0$, as well as ΣM_x (summation of moments about the x-axis) $= 0$, $\Sigma M_y = 0$, and $\Sigma M_z = 0$. As the horse is in the earth's gravitational field, its weight may be assumed to act (downward) at right angles (or normal) to the surface. Because the surface is slippery we know that it will transmit neither moments nor forces parallel to it (assuming that the edges of the hooves do not dig in). We know that gravity imposes a downwardly directed vertical force, so we may conclude that this will be counteracted by equal upward forces acting across the four hooves (Figure 3-3). As the horse is bilaterally symmetrical we may then assume that the load carried by each forefoot is equal to that carried by its

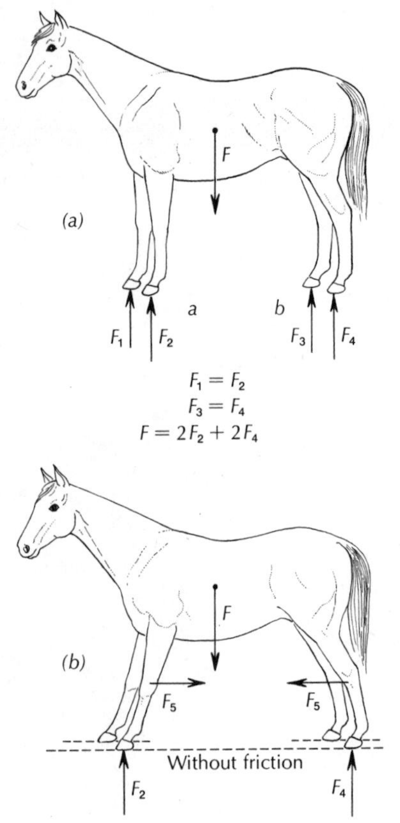

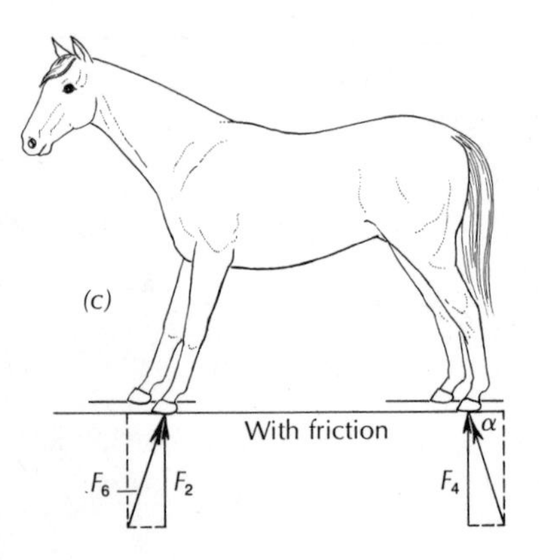

Figure 3-3 A horse standing immobile on a smooth surface may be considered to represent a free body (*a*). Analyzing the external forces imposed upon it, we see that it is in equilibrium between the force F of gravity acting downward through its center of gravity and the four opposed forces (F_1, F_2, F_3, F_4) acting on its hoofs. If we know that the animal is resting and assume that the left and right feet are placed symmetrically, we can calculate the weight at each foot from a knowledge of the horse's weight by the relations $F = 2(F_2 + F_4)$ and $aF_2 = bF_4$. Alternatively, we can calculate the weight of the horse by measuring the forces exerted by forefeet and hindfeet.

If the horse stands with its legs at an angle to a smooth surface (*b*), the forces at the feet will be equal to those in (*a*) as long as their relation to the center of gravity remains constant. However, we can now predict that to keep itself in balance the animal must exert an internal force, F_5, inward on each leg, the way a skater must when standing with one leg before the other; skilled skaters keep the center of gravity above one foot and do not have to exert this force when resting.

The position of the horse has not changed in (*c*), except that the horizontal force is now supplied by friction, which is the interaction between hoof and ground. The difference between conditions in (*b*) and (*c*) can be determined by specifying the kind of "joint" between hoof and ground. The horse in part (*c*) will remain stationary as long as the horizontal component of F_6 is equal to or less than the maximum frictional force between hoof and ground (see Box 3-2).

In this case the free body was a stationary object, but one could run a similar analysis on a sailboat, or a Portuguese man-of-war, moving at constant speed over a smooth pond. One can predict that the force of the wind against sail and boat would be just equal to the drag of the water against the boat's bottom. When the force of the wind exceeds the drag of the water, the boat will accelerate; when it is less, the boat will slow down. Since much of the force exerted by the wind acts on the sails, which are held by ropes, one could subdivide the original free body and carry out an independent analysis for each sail. By measuring the force at each rope one may also determine the effect of wind-induced forces.

mate, and that the same applies to the hind feet. A pair of smooth-surfaced scales (or two measurements in sequence) under one forefoot and one hindfoot will then indicate the load on each foot and the center of mass of the entire animal.

A similar analysis is possible for moving objects, even when the motion is curvilinear and involves acceleration components. If one knows the angle at which a frog's center of mass is moving just as the toes leave the ground, one can calculate its velocity from either apogee or distance traveled; if one knows velocity and the frog's mass, one can estimate the minimum force that was needed to induce the appropriate acceleration (Figure 3-4).

The nature of the surfaces across which the forces are exerted is obviously critical. If surfaces are plane, smooth, and do not interact, we may assume that only normal (or perpendicular) forces can be transmitted. If the surfaces are rough or show interaction, they may transmit parallel "frictional" forces. Various kinds of cylindrical or spherical joints may transmit forces in one or more directions but will not transmit moments around one, two, or three axes (Box 3-1). Selection of the right place may greatly simplify analysis of forces, especially when only a part of the animal is considered at a time—for instance, when the free body is a limb or portion thereof.

The horse, in our example, may extend its legs so that they are not directed normal to the ground (Figure 3-3*b*). They will then tend to slip forward and backward; only if the animal's ventral muscles contract will it avoid ventroflexion (dorsal concavity) of the vertebral column. Any amount of friction (see Box 3-2) between hooves and ground will reduce the internal tension needed to stabilize the system (Badoux,

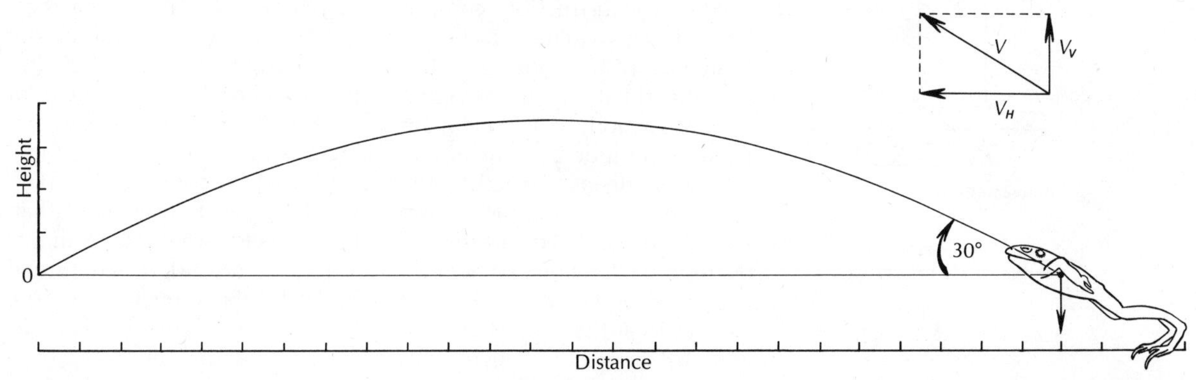

Figure 3-4 Approximate path of a jumping frog. Since we know the gravitational acceleration of the frog's center of gravity, we can calculate the trajectory from the vertical component of the initial velocity by determining how long the gravitational acceleration will take to reduce this upward velocity (V_V) to zero. The frog will obviously take twice as long to reach the ground again, and the horizontal component of velocity times the elapsed time ($V_H t$) shows how far the frog will have traveled relative to the ground.

Box 3-1 Kinds of Movement and Degrees of Freedom

(a)

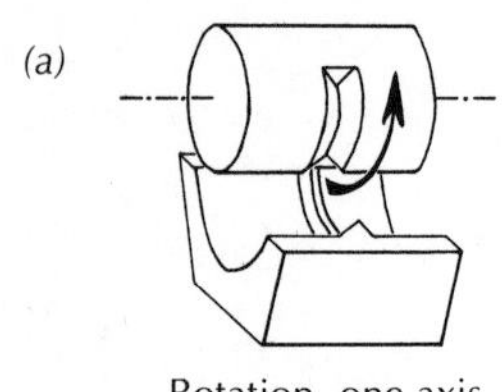

Rotation, one axis

(b)

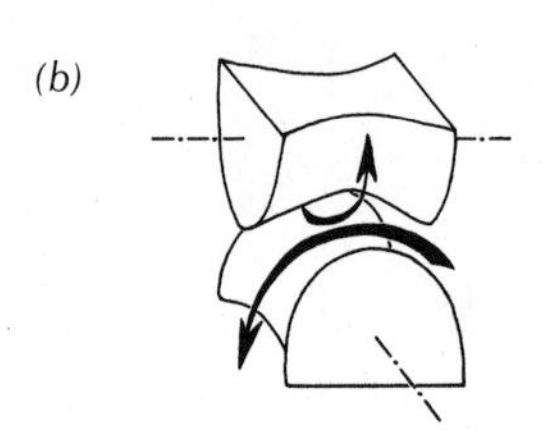

Rotation, two axes

(c)

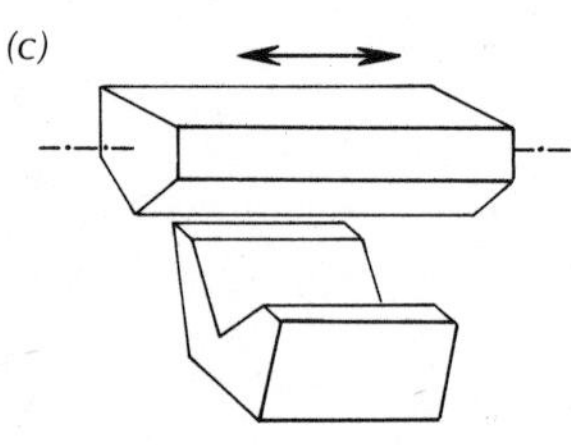

Translation, one axis

(d)

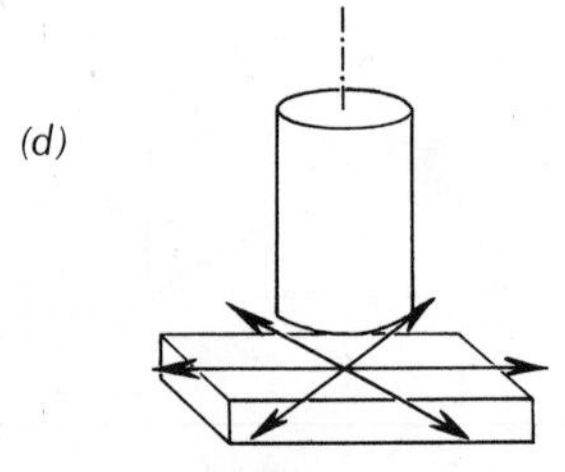

Translation, two axes

(e)

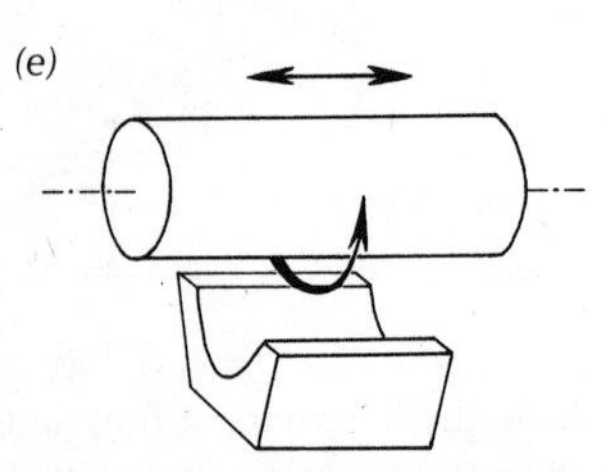

Uniaxial rotation and translation

The linear flight path of a bird may be described in terms of three axes (for instance, up-down, north-south, east-west). The bird can also twist or rotate while flying, performing barrel rolls to the east or south or a spin in a vertical path. The bird can also change its configuration by folding its wings, flapping, and moving its feet. Even if we omit such changes, we must describe the bird's motion in terms of three kinds of *translation* (travel along an axis) and three kinds of *rotation* (motion around an axis); we can state that its movement has six degrees of freedom. Such "liberty" allows the animal numerous options but complicates both the control and the description of movements. In general, the fewer the movements possible at a joint, the simpler will be the control or description. For example, the muscle bending the healthy human knee does not need special instructions controlling twisting, as joint surfaces and ligaments restrict torsion; contrast this situation with the complex instructions that an elephant needs to position its trunk or that an octopus needs to place its arm.

Description of motion is also easier when defined in terms of degrees of freedom. Whenever two solid bodies are in contact we may treat the contact zone as a joint. The nature of such joints determines the allowable motion, either by restricting translation or rotation, sometimes by restricting both. Thus the ankle of a goat permits uniaxial rotation only (*a*). In contrast, the joint between quadrate and mandible of a snake has two axes, as the jaw may simultaneously rotate downward and (for some distance) about its long axis (*b*). A skater with sharp-edged skates travels by translation along one axis (*c*), whereas a Hollywood comedian walking across a floor covered with ball bearings encounters translation along two axes (*d*), which may lead to translation along a third axis if his legs go in different directions. Some jaw joints permit both uniaxial rotation and translation; the jaw not only rotates downward but can also slide from side to side (*e*). What kind of wear patterns could such animals impose on their teeth?

Motion between two bodies may be limited. When the shaft in (*a*) turns to the end of the groove, it locks into position and the two bodies move as one until rotation is reversed. Similarly the leg acts as a unit when the knee is locked. The hand of *Bipes* (Figure 4-17) can fold at more than four levels when the animal brings it forward; when turned backward the joints lock and the hand acts as a unit to scrape dirt from the tunnel's end.

1964). The sum of the horizontal forces will also be equal to zero, but their existence complicates analysis, since the resultants at each foot are no longer vertical. More measurements have to be taken to see if friction is acting. Information about the magnitude of the frictional coefficient will permit one to circumscribe the uncertainty by establishing the boundaries between which the answer will lie; in our example it will indicate how far a horse will be able to tilt its limbs before they slip.

The nature of the interaction between two surfaces in contact determines the horizontal force that may be applied without causing slippage between the two surfaces. In the range of biological materials this will be a function of (1) the force holding the two surfaces in contact and (2) a coefficient depending on the contacting materials (and the direction of slippage). This coefficient of "static" friction will be the tangent of the minimum angle at which the force can be applied without slippage. This angle then determines the amount that the loading may be off from vertical in any direction and the static condition maintained.

A different kind of frictional interaction occurs when the surfaces move relative to each other. As soon as slippage starts, the interaction of the surfaces decreases; perhaps the surfaces separate slightly, freezing microscopic interlocking projections, or reducing adhesive forces. Once contacting surfaces begin to slip they will continue to move until the angle of the imposed force is increased or the lateral force equivalently decreased below a value corresponding to a coefficient of sliding friction. In animals the horizontal driving force must obviously be produced by muscular effort if the object, namely the animal or its part, is to be kept moving at a constant velocity. If the force is exceeded, the object accelerates; if the force is reduced, motion stops.

Consideration of frictional interactions should note that the contact area does not appear in the equations; it is the absolute force rather than the pressure (force per unit area) that is critical. This is true for an ideal system, but animals move in a real world. If a sliding animal encounters a surface irregularity, the coefficient of sliding friction will be temporarily increased; if the horizontal driving force is temporarily less than the sliding friction, the system may come to a stop. If the animal decelerates or stops it will remain stationary unless its momentum again builds up a force sufficient to exceed static friction. The chance of encountering a surface irregularity is likely to be a function of the size of the surface area encountered. Consequently, a snake might discover more irregularities per unit traveled if it moves sideways than if it crawls straight forward.[1]

[1]The word traction is sometimes used to define the amount of pure static friction plus other slippage resisting effects such as those due to claws or edges that search for or catch crevices. One may again show that this rather imprecise term refers to a phenomenon of a magnitude proportional to the forces pressing the surfaces together.

The free-body analysis, then, starts by considering an animal as a kind of "black box" of mysterious and unknown contents. The box deforms and progresses. The box exerts forces onto the environment and receives reactions from it. These externally noted events can be measured. They will now be used to establish generalities about the adaptive locomotor strategies used by particular groupings of animals. From these patterns I will attempt to define some hypotheses concerning the meaning of the internal organization observed; ultimately these hypotheses could suggest further experiments subjecting them to test in subsequent analyses.

Box 3-2 Friction

Two closely adjacent objects tend to show some interactions. These interactions will resist movement along the contact surface. We speak of the (maximum) *static friction* as the force that has to be overcome in order to induce motion along the contacting surfaces. The maximum static friction, F_{max}, has been determined as $F_{max} = \mu \cdot N$, where μ is the coefficient of friction (which is determined by the nature of the surfaces) and N the (normally directed) force pushing the surfaces together. In Figure 3-3 this is the weight forcing the horse's hooves against the ground. The equation does not include a factor for area; although the zone in contact obviously increases with the stress level, the increased area reduces the stress.

If the maximum static friction is exceeded, the objects start to move relative to each other. The resistance to further travel is significantly less thereafter, and we may talk of *sliding friction* as the force that has to be overcome in order to keep the objects moving at a constant velocity (*a*).

The reason why sliding friction is ordinarily lower than static friction may be seen by comparison of (*b*) and (*c*). Surfaces are never absolutely smooth; consequently their minor irregularities contact one another. When two such surfaces are at rest, their irregularities interdigitate and interact. To induce motion one first has to promote an initial separation. Thereafter the moving elements tend to remain separated. The sketch also suggests

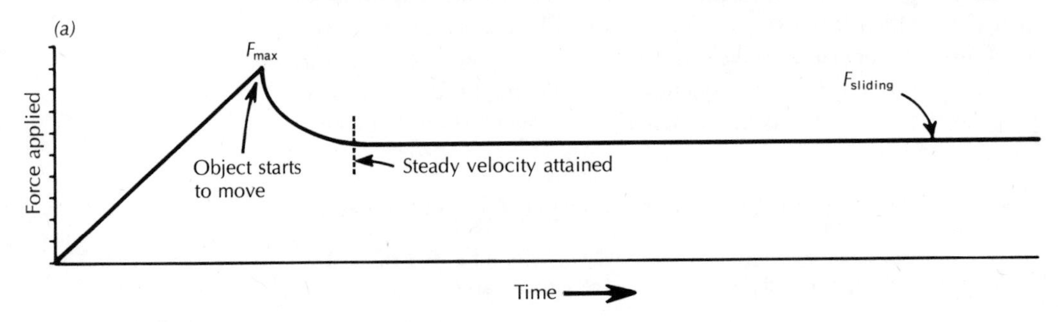

reasons why sliding promotes wear. A relatively few minor projections will support the entire normal load, and the local stress level on these projections may then exceed the elastic limit, leading to plastic deformation and local failures.

The introduction of multiple small particles or of a layer of fluid between two objects (*d*) tends to keep them from contact. If movement occurs, it is within such a "lubricating" layer. Lubrication consequently reduces both static and sliding friction. Their magnitude no longer depends on the nature of the surfaces but on that of the lubricant. Lubricated contact is a common biological phenomenon, as the cavities of joints and tendon sheaths are often liquid-filled. Another kind of lubricated contact is observed when snakes travel across dry, sandy surfaces. Slippage between sand grains requires special locomotor adaptations for such traverses (page 95).

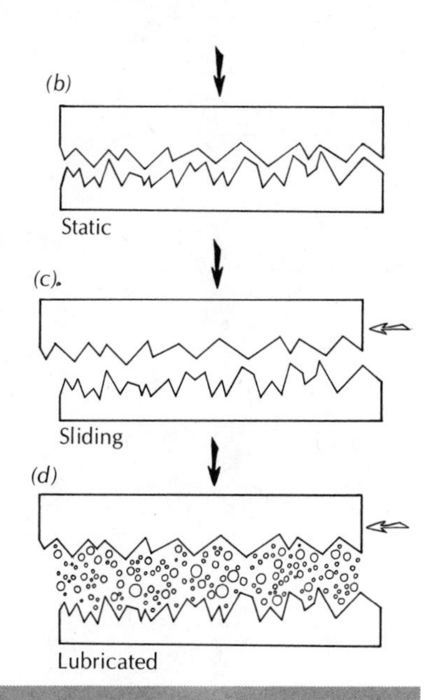

Lateral Undulation in Fishes

Most fishes propel themselves by some variant of lateral undulation in which the body is thrown into a series of roughly S-shaped curves that shift tailward. The concavities are produced by a shortening of the muscles (of the concave side); the stimuli pass alternately along the two sides of the spinal cord, producing alternating waves or contractions along the sides of the trunk. Fish muscles are arranged in serial myomeres, and some spreading of the wave of muscular contraction has been produced by alternate anteriad and posteriad deformations of the myocommata (intersegmental or intermyomere connective-tissue septa); these deformations lengthen the radius of the bend as contraction extends over a longer absolute zone. The muscles of each myomere are generally organized into numerous motor units, each unit consisting of a single neuron innervating some hundreds of muscle fibers. Generally the axons will be grouped into the single ventral root nerve pertaining to the segment. Gradation of contraction (further increasing the bend) is produced by some degree of intersegmental spreading of the innervation of each motor unit and by innervation of single muscle fibers by neurons from several spinal roots. Such multiply innervated fibers produce partial contraction in response to signals passing along nerves from one of several sequential spinal roots; complete contraction of the fiber occurs only when multiple stimulating signals reach a single muscle fiber within a specified time (Hudson, 1969). The signals may thus be summed to produce a necessary level of stimulation.

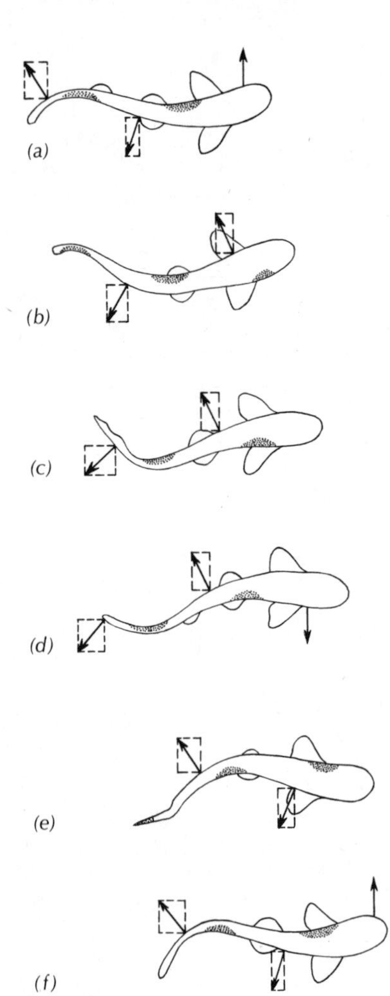

Figure 3-5 Series of sketches from a film of a swimming fish showing the direction of forces applied to the water. Note how the useful fraction of force increases as waves move posteriorly. (The vectors are all shown of equal size as we lack data for the instantaneous force generation in the sequence.) (After Gray, 1933)

The backward progression of undulations in the fish's body assures that alternate short sections will be tilted so that they face backward to the left and backward to the right. The portions of the body forming each of these sections change continuously. The net effect is that of a series of (curved) inclined planes, each displacing water posterolaterally and consequently incurring anterolateral reaction forces. As alternate planes face left and right we may assume that the lateral forces generally cancel each other; hence the anterior reaction forces are summed and induce forward propulsion of the body. The traveling waves generally increase in amplitude (lateral spread) as they pass posteriorly, and their angles to the direction of motion change from loop to loop. These changes relate to multiple factors such as the density and viscosity of the fluid, the velocity of the travel, and the surface characteristics of the fish. Ultimately the change involves the kinds of bends the animal is capable of forming and the contraction rates of the muscles, yet all these interesting topics are peripheral to the present analysis of limbless locomotion. Those interested are referred to Breder (1926), to Taylor (1952), and to the books of Aleev (1969), Alexander (1967), and Gray (1968).

The net reaction force, produced at any site and instant, results not only from the absolute motion of the surface (and from its dorsal or ventral inclination, which are here ignored) but also from the slippage of water in opposition to the direction of motion. This relative motion between the animal's surface and the boundary layer of water involves a local interaction referred to as *drag* and equivalent to friction. Inspection of Figure 3-7 shows that the effect of drag is to increase the lateral component of the posterolateral reaction force and, with this, to reduce the propulsive components of the anterolateral reaction forces. Consequently, two kinds (contour and profile) of drag effects must be overcome. Both generally oppose motion of the object through the water, but the drag also reduces the effectiveness by which propulsive forces are transmitted.

At least two additional aspects deserve attention. The first is to note that the backward travel of the undulations proceeds faster than the forward travel of the entire fish. Consequently, the waves do not stand (remain stationary) relative to the ground, and it is very difficult to observe and characterize the motion of the fish exactly, even if it is only swimming upstream as fast as the water flows down (Figure 3-8). Here and in most of the other analyses of animal movement, action proceeds too fast for unassisted observation. One is forced to analyze the movements on the basis of motion picture films or television tapes. Analysis is easiest when a naturally or artificially marked animal is recorded while moving against the background of some grid (Bernstein, 1967). A sig-

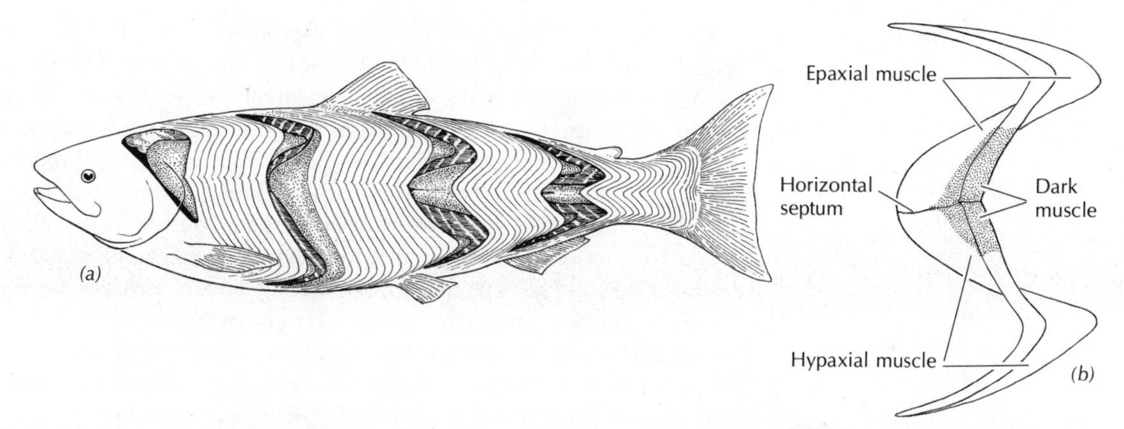

Figure 3-6 The axial musculature of a trout (*a*) consists of serial myomeres (*b*), each folded complexly so that their sequential contraction from head to tail causes a gradual, rather than a localized, shortening. In some species the region adjacent to the outside of the horizontal septum is occupied by dark muscles rich in mitochondria that permit them to contract more strongly. The peripheral position of this specialized tissue allows it to induce maximum moments in propelling the fish.

nificant amount of theoretical work combined with observations is now being marshaled to describe such animal movements (cf. Golani, 1969).

The final point is that the equal and opposite lateral forces keep changing in their location along the body of the fish. Rather than simply canceling each other, they form force couples inducing moments that also travel front to rear and reverse their direction every time a new bend forms. These changing moments induce some oscillations in the system; particularly important are the oscillations they cause at the anterior (and posterior) end of the animal (it is obviously disadvantageous to wave the sense organs back and forth or to wave the head when approaching prey). The cause of this oscillation can be visualized if it is remembered that the balance of lateral forces requires either (1) an equal number of posterolateral forces equal in magnitude directed to each side or (2) a temporary or continuous condition during which forces of greater magnitude on one side offset a greater number of forces directed to the other. One might assume that a new bend must form anteriorly just as an equivalently directed bend "runs off the tail." But even in this case the total imposed moment would be reversed since the direction of the first (and all subsequent) couple(s) has reversed as well.

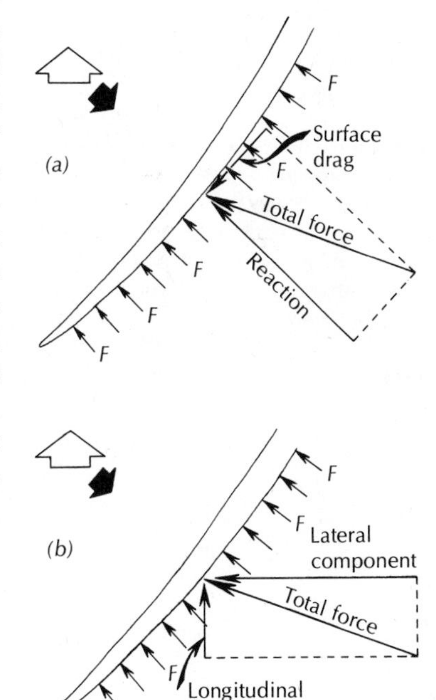

Figure 3-7 These sketches show forces on the tail of a fish that is propelling itself in the direction of the large, open arrow by sweeping its tail in the direction of the small one. Part (*a*) shows the resultant of the distributed reaction force induced by displacement of water and how the surface drag deflects it. The resultant force (*b*) may be resolved into lateral and longitudinal components. One can see that the longitudinal component representing useful propulsive force decreases as the drag increases.

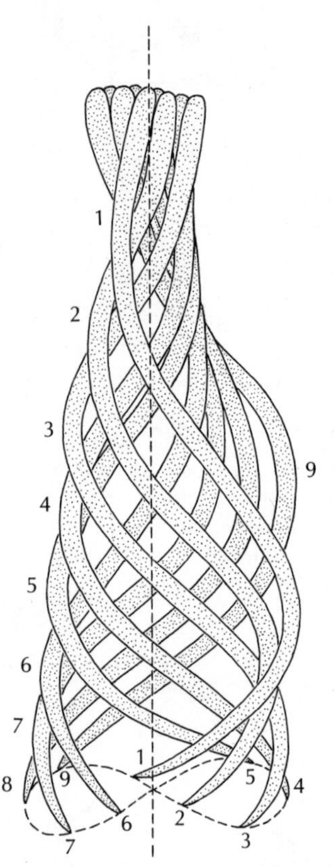

Figure 3-8 These superimposed sketches of a swimming eel show the lateral movement of different parts of the body. The numbers refer to successive stages in a movement cycle. Note that both head and tail oscillate and that the lateral excursions increase toward the animal's rear. (After Gray, 1933)

This oscillating effect will, for a given propulsion rate, be inversely related to the number of bends in the body at any instant. The fewer the bends, the greater is the absolute magnitude of each lateral component and hence of each couple. From this, one might predict a selective advantage for elongate body shape; an elongate fish could produce a greater number of bends at any one time and hence propel itself more smoothly. Most fishes have indeed evolved along this path, but other species have moved into niches in which extreme elongation would be disadvantageous. Thus one finds compensating devices such as a stiffening of the anterior trunk (the lateral waves in such cases only occupy the animal's posterior portion) and stabilizing devices such as fins or lateral flattening of the stiffer body. These devices dampen the lateral oscillations of the head and consequently permit a straight-line approach to prey or food. All of these alternatives are departures from the elongation representing a mechanical optimum for lateral undulation. Selection has thus effected compromises rather than perfecting any one aspect of the animal; we may talk of generally optimizing selection.

Lateral Undulation in Tetrapods

A number of limbless tetrapods also occupy aquatic habitats, and all of these use lateral undulations as the major locomotor pattern. Examples are a few caecilians (the limbless amphibians also called apodans), sea snakes, and certain water snakes. Such species tend to have flattened and elongated tails; often there is the analog of a caudal fin fold. In sea snakes and the elephant-trunk snakes (Acrochordidae), most of the body is laterally compressed, and a flat fold of skin extends ventrally along the entire length of the visceral cavity. Films show that the propulsion proceeds essentially as in the swimming of fishes. Certain sea snakes have apparently solved the prey-approach problems by reducing the size of the head and neck. These parts are small, relative to the rest of the trunk, and the trunk's inertia may be used to absorb the reaction forces when the neck is extended to bring head to prey. A slender neck has the further advantage that it may be maintained near the center line of progression. This is possible because its limited mass makes it easy to reposition. Not surprisingly, these snakes tend to form the first propulsive loop somewhat posterior to the neck. They also keep the first loop small initially, only letting its radius increase as it travels along the trunk (Figure 3-10).

Fish (and sea snake) swimming occurs with the body totally submerged and hence is much different from the progression along the surface of the water possible to most snakes and some limbless lizards. Most snakes hold the head slightly above the surface when swimming, keeping variable amounts of the back out of the water as well. The

Figure 3-9 Sketches from photographs of a swimming fish show that the waves of the body and, with them, the sites of force application to the water (or rather the sites of maximum lateral force) shift posteriorly faster than the animal progresses. In this case the fish simultaneously exerts either two or three such lateral forces. Since their lateral components are opposed, and parallel but displaced, they act as couples and produce moments. One can see, for instance, how the forces in (*a*) and (*b*) set up a clockwise moment acting to rotate the shark's head to the right. This reverses as the force on the left side passes from the region of the pectoral fin to the pelvic one and a new force is induced on the side of the head (*d*). After this reversal the moment acts counterclockwise to turn the head toward the left. Similar reversal occurs at the tip of the tail. The momentum of the fish and the effect of dorsal fins reduce the lateral movements. (After Gray, 1933)

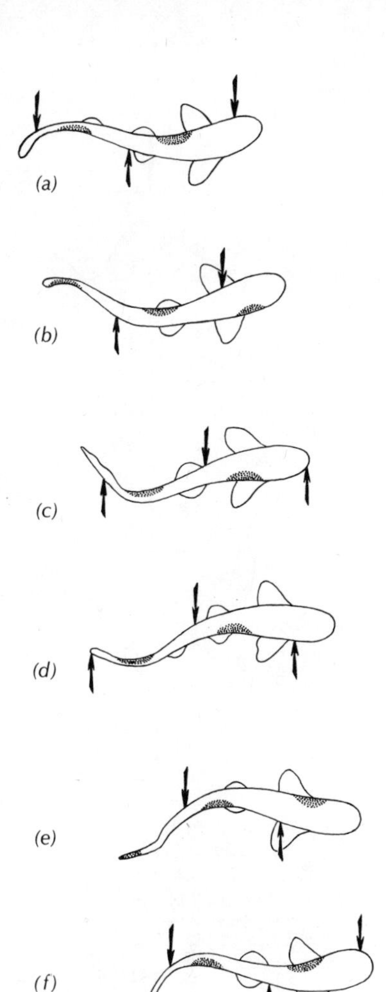

depth of immersion is a function of the degree to which the lungs are inflated (and of the extent to which they continue posteriorly into an elongate air sac). Although the profile and surface drag incurred by any submerged body may be less thereby, the resistance to progression by a partially immersed body is compounded by surface tension and wave effects (Van der Vaart, 1958). The propulsive forces are furthermore exerted only by the submerged portions of the body. Various kinds of torsional (twisting) movements of the snake's trunk apparently reflect an approach to counteracting the resulting imbalances in the postero-lateral forces directed to each side.

Actually we have only the most scanty and purely observational data on the surface-swimming behavior of most snakes. Theoretically there could be an energetic advantage to complete immersion, even though the drag is increased. Surface swimmers might be affected by factors such as exposure to predators, breathing problems when the water is rough, the need to navigate, and the effect of excitement. Even qualitative data for different species and factors might start one on more meaningful generalizations.

When limbless vertebrates move on land the force application pattern is quite different, a difference that more clearly reflects the nature of the substratum than difference in the sequences of movement. The animal now has weight; its mass is no longer floated in a medium of equivalent density, and unequal support induces torsion and shear forces between the vertebrae even at rest. In general it may be assumed that these weight-induced forces are transmitted via the ventral surface of the body, whereas the propulsive forces will continue to be transmitted by the sides in lateral undulation. In shifting its pattern of lateral undulation from the water to land, the animal no longer imposes uniform pressures onto the posteriorly sweeping loops of the body. Instead, the applied forces are confined to local regions of the body by contact with specific pressure sites (such as rocks, twigs, or grass stems). These sites are irregularly placed, reflecting particular *points d'appui* (points of force

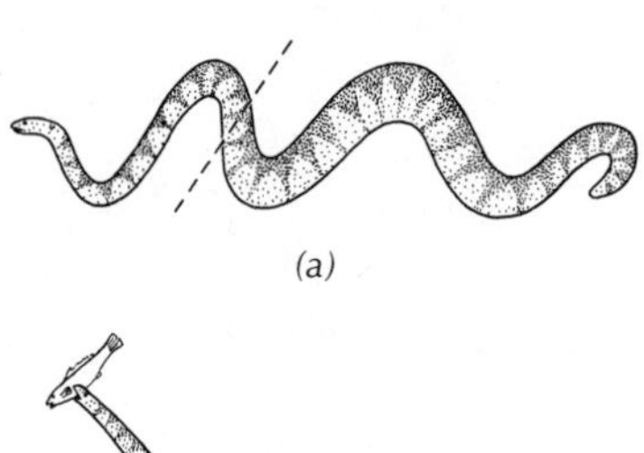

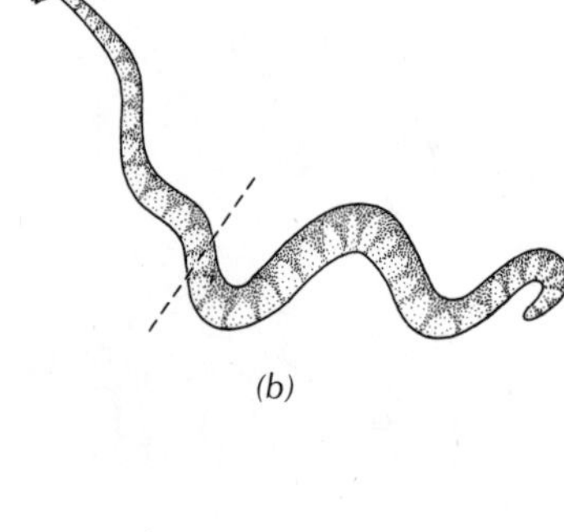

Figure 3-10 Sketch to show a strike at prey by a floating sea snake. The mass contained in the slender head and neck is less than a quarter of that of the trunk and tail. The reaction due to the sudden lateral shift of the anterior portion will be absorbed by the inertia of the rest of the body and by the resistance to lateral slippage produced by the snake's dorsoventral flattening.

application) that can withstand such imposed forces without failure (Figure 3-11). Consequently the waves move backward along the trunk just as fast as the animal progresses; the waves seem to "stand" relative to the ground.

It is very significant that a pilot study completed just after this chapter was written documents that lateral undulation on land requires less than one-half the metabolic cost of tetrapod locomotion, but about three times the propulsive cost of undulant swimming by a fish of equivalent mass (Chodrow and Taylor, 1973). Clearly there is a cost for supporting the body's weight, and even more cost when the body is lifted high rather than slid along the surface. Whatever the ultimate conclusions about motor energetics, the animals in lateral undulation propel themselves by the summed longitudinal resultants of the posterolateral forces.

The localized posterolateral forces may be applied to one or to a series of resisting objects. It is customary to refer to such sites as pegs with rounded surfaces so that the imposed forces are assumed to be normal to the peg no matter what the direction along which they are applied. In reality the forces of a single loop may be applied to varied objects and at varied angles. Thus, when a large snake undulates through grass, many stalks will be bent by each loop, as multiple points of contact can be utilized simultaneously in propulsion.

In the early forties Gray and Lissmann measured the direction and magnitude of the forces that moving water snakes (*Natrix natrix*) exerted on their environment. The experiments were ultimately directed to questions of proprioceptive (self-sensing) control and essentially involved very simple experimental apparatus. The snake was encouraged to move past an array of small, spring-loaded vertical and horizontal platforms that deflected as forces were applied to them. Perhaps the conceptually most elegant experiment involved an array of pegs suspended as pendulums (Gray and Lissmann, 1950). The array was filmed from above as a snake moved through it. The direction of peg deflection represented the direction of the applied force, and the extent to which

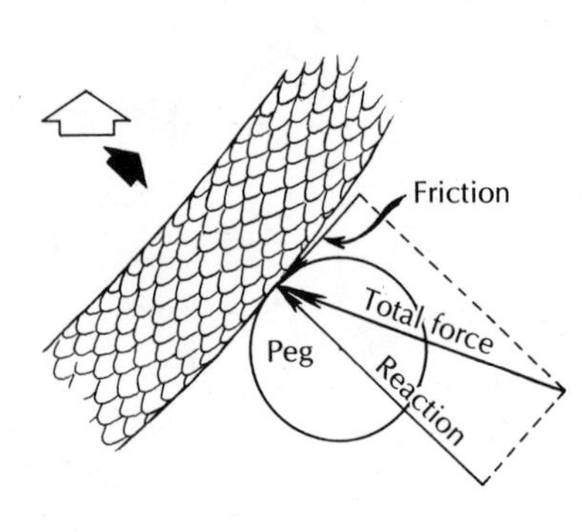

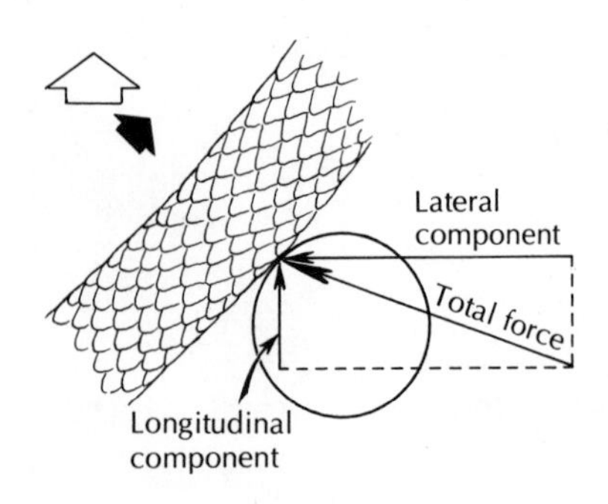

Figure 3-11 These sketches show the great similarity between the forces applied by an undulating snake to a solid peg and those applied by a fish to water (Figure 3-7). The large, open arrow shows the path of the snake's center of gravity, and the smaller one the path traveled by the body in moving along the peg. The fixed peg has been substituted for moving resistance of the water; sliding friction is substituted for surface drag. Both drag and friction decrease the useful (longitudinal) components of the reaction forces.

the peg swung when it was displaced indicated the force's magnitude (Figure 3-12).

Such studies have defined the parameters within which snakes can and do use lateral undulation. Lateral undulation involves continuous progression; that is, no part of the body is stationary while the snake moves. The animal's weight is carried via the ventral surface, which is generally in continuous sliding contact with the ground. Propulsive forces must be directed posterolaterally, so lateral undulation may be used only when there are anterolaterally facing surfaces on the substratum against which the forces may be applied. Hence this form of locomotion is useless in a straight-sided tunnel. It is equally inappropriate on plane surfaces, where the irregularities are too low to be reached by the loops. Certain snakes (such as the night adders, *Causus*) compensate for this by twisting the body, so that only the posterolateral part of the outward moving loop on each side contacts the ground. The sliding friction thus incurred provides reaction forces that propel the snake in this rather inefficient variant of swimming movement. Since the intermediate portions of the trunk are lifted slightly, the frictional component includes the entire weight of the body as a contact or friction-promoting force that drives the scraping edges into contact with the surface.

With this curious exception, friction appears to be disadvantageous to animals using lateral undulation for locomotion. As the trunk slides over the ground and past the lateral contact zones, both weight-induced forces on the ventral surface and propulsive forces at the contact zones incur the resistance of sliding friction. Since these frictional forces are opposed to the direction of motion, they have to be overcome by muscular effort. Consequently, lateral undulation might be expected to proceed most effectively in a frictionless system. I tested this concept by constructing a field of cylindrical, high-friction, rough-surfaced pegs, each of which rotated on adjustable pin bearings. A small indicator pin on each peg showed those in contact with the snake. The horizontal surface could be lubricated with wax and the bearings could be tightened to make the pins more difficult to turn. The effective friction of each peg opposing forward progress increased drastically when the propulsive contact between snake and pin shifted from rolling to sliding; snakes moved freely across the field when the pegs spun loosely, but were significantly slowed as the bearings were tightened or the floor roughened. Hence friction at the ventral surface and the lateral points definitely decreased the speed and increased the muscular effort needed for progression (Gans, 1962).

The disadvantage of sliding friction has been reflected in the texture of the skin. In caecilians, friction is reduced by keeping the skin moist and mucus-covered so that the contact is lubricated. Reptilian skin is covered with a layer of keratin. This horny material consists of dead

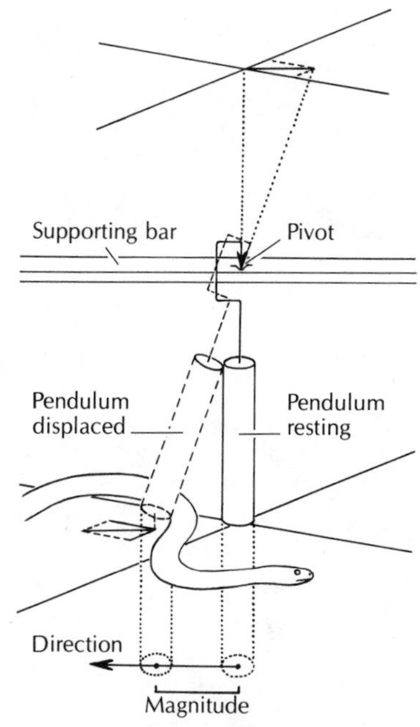

Figure 3-12 Sketch of a water snake moving past a pendulum peg of the kind used by Lissmann and Gray. One can see how its deflection signals to the observer both the direction and the magnitude of the imposed force.

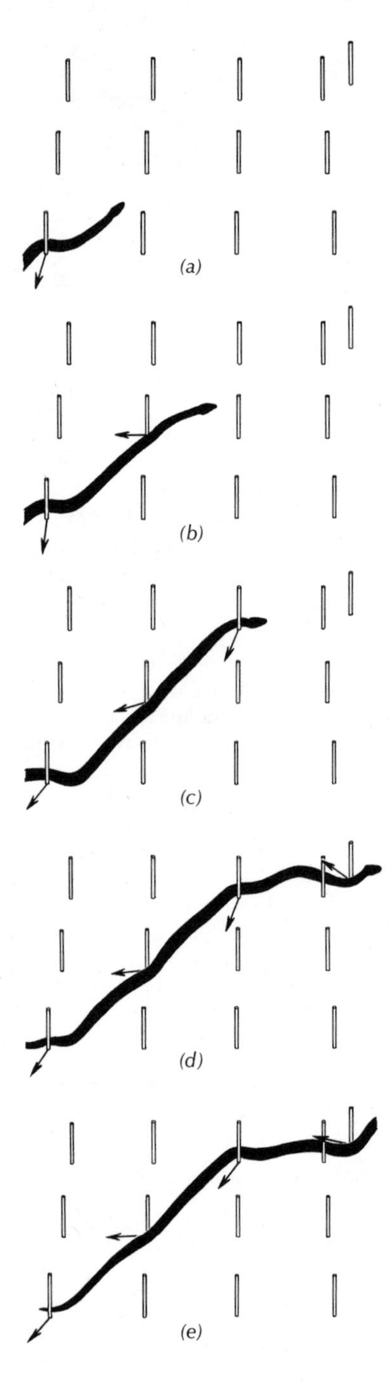

Figure 3-13 Tracings of an angled view of a field of rotating pegs being traversed by a rat snake in lateral undulation. The intervals are one-sixth of a second. The arrows indicate the direction in which forces are applied to those pegs with which the snake had contact. In general the waves move backward only as fast as the snake progresses; hence they "stand" relative to the ground. Note the slight but unmistakable deformations of the body at the pegs where the forces are applied and also the remarkably straight path of this slowly undulating snake.

cells, so that growth consists of proliferation of a deeper layer followed by shedding of the surface. The skin itself folds into a pavement of scales, scutes, knobs, and plates. The polished ventral scutes of snakes are well known; their frictional coefficient is .34 when a water snake is pulled forward on wood, and .65 when it is pulled over sandpaper. The contact zones along the sides of the body and those along the middle of the back of some tree snakes show similar low-friction devices. Equivalent patterns are seen in limbless lizards. The texture of the surface layer in reptiles is maintained smooth by regular replacement (shedding or ecdysis). The meaning of integumentary ornamentation could stand more study. Are the various smooth, faintly keeled, ridged, and knobbed scales seen in reptiles really functionless? Are the diverse types of scale keels related to terrestrial or to aquatic progression, to defense, or perhaps to mating?

To me the most interesting feature of lateral undulation, particularly of the highly sophisticated method used by tree snakes, is the decision-making process that must be involved in the placement of the loops. Gray and Lissmann (1950) showed that the total relative magnitude of the lateral-to-backward forces increased, seemingly exponentially, with the number of sites contacted (Figure 3-16). Since such lateral forces represent wasted energy, this suggests that the number of contact points be reduced toward three. Three vectors are the minimum that can balance each other in an undulatory pattern and still maintain stable progression. However, three points require either a relatively slow or a fast progression, since the transition from left-facing to right-facing tripod must not cause much spatial displacement. A slow transition allows time for a balanced shift; rapid progression may permit the stored momentum to carry the animal through the transition zone. All of this suggests size limitations; if the snake is too short and stout, lateral undulation may become less effective. African puff-adders (*Bitis arietans*) and gaboon vipers (*Bitis gabonica*) are obvious examples of snakes that are too short to form adequate loops.

The literature contains a single report that suggests that considerations of the energy expended are indeed appropriate in snake locomotion. Heckrote (1967) chased various-sized specimens of the garter snake, *Thamnophis s. sirtalis*, through an array of pegs spaced on one-inch centers. He found that the snakes' speed in terms of number of

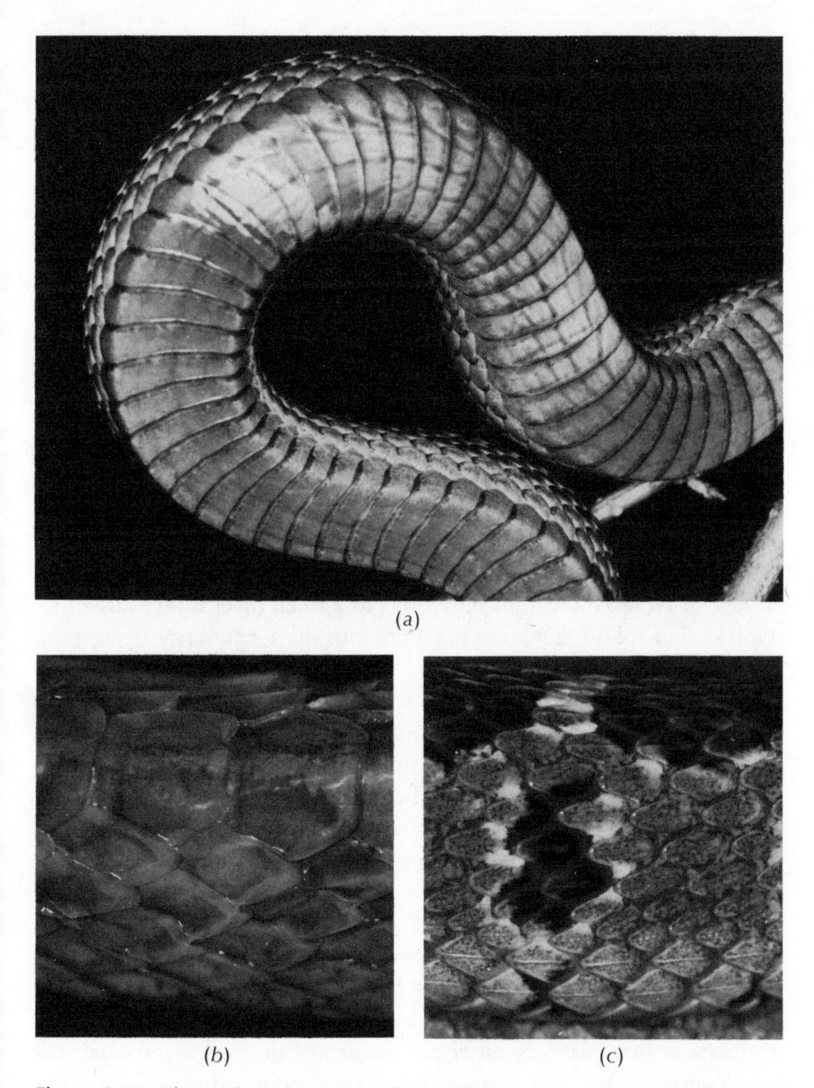

(a)

(b) (c)

Figure 3-14 The scales of snakes reflect their function. The ventral view of the Japanese rat snake (*Elaphe climacophora*) in (a) shows the widened ventral scales as well as the sharp lateral keels, often seen in arboreal snakes. The dorsolateral view of the South American snail-eating snake (*Dipsas indica*) shows that the scales covering the midvertebral line have become widened, perhaps in response to friction resistance, as these species may rub their backs against branches. Both of these snakes have smooth scales, whereas the side view of the African egg eater (*Dasypeltis scabra*) in (c) indicates that each scale is keeled (compare Figure 2-31).

body lengths per second increased as he used larger specimens, reached a maximum, and then decreased for the largest; this suggested that there was an optimum size for a snake. Unfortunately there are no observa-

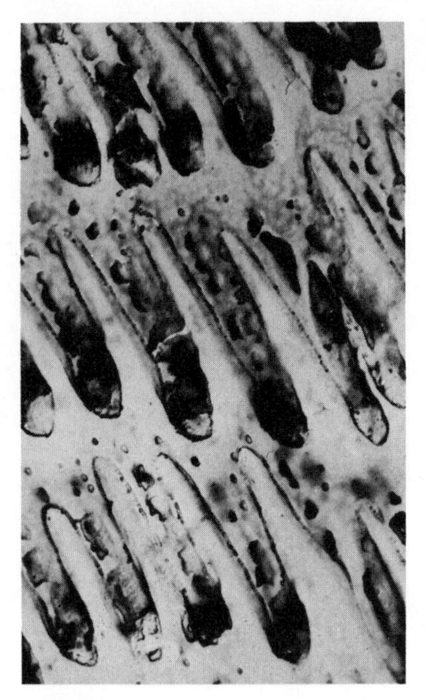

Figure 3-15 The surface sculpture of the ventral scales of *Dasypeltis scabra* shows a series of backward-pointing projections which account for the low-friction glide of scale surface over the ground. (Shadow electron micrograph courtesy of M. Silveira and P. de Souza Santos)

tions concerning the number of points a snake contacted at any moment (perhaps a function of the snake's size) nor the effect of a variable peg spacing. It seems possible that the one-inch peg spacing used provided the optimum number of contact points for precisely those animals that attained the maximum progression rate. Different spacings would then be expected to yield maximum progression rates for snakes of different absolute size. Such rates might, of course, be affected by the resistance encountered by the animal; thus they would reflect the frictional coefficients of the substratum. The overall effect parallels that in forced human walking. The energy cost per distance traveled reaches a minimum at a different velocity for each step length and for each bipedal pattern (heel and toe, walk, jog, sprint); even uninstructed athletes will tend to make efficient use of energy by shifting their locomotor pattern as they speed up. For instance, referees at a track meet must watch that the heel and toe walker does not shift into the less energy-consuming trot.

How does the snake "know" where to place the ever-changing curves? What is the source of the information that permits loop placement at optimum pressure sites? Is this information obtained by contact receptors in the snake's skin, by muscle spindles in perhaps the costocutaneous (p. 107) or intervertebral muscle system, or by some quite distinct proprioceptive mechanism? Does the snake spread its loops laterally to some generally preset position or does the resistance of the contact points determine the lateral displacement for each loop? As yet, we have only minimal information to answer such questions. Observations show that snakes undulating across coarse and sharp-edged gravel will lift a forming loop over a particular (often an unsteady) rock to place it in a more firm place. Similarly loops formed in grass or on sand will sweep outward, bending stems or piling up a sandy ridge, and, presumably, accumulate resistance up to some minimum at which lateral deflection stops or is reduced. The similarity to aquatic undulation becomes remarkable. Indeed, Gray (1968) notes that when a swimming snake

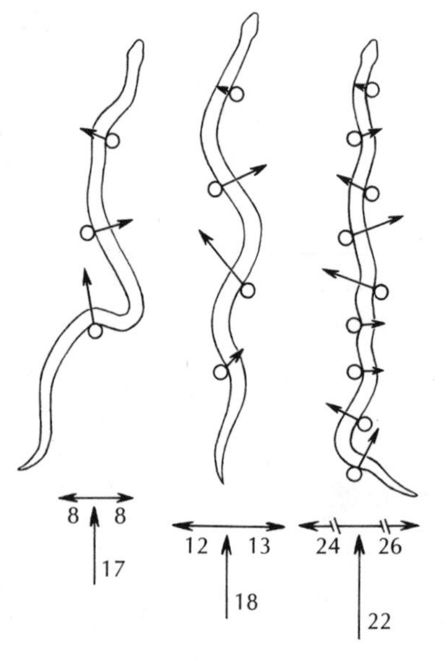

Figure 3-16 The sketches, redrawn from the data in the study of Gray and Lissmann (1950), show that the use by the snake of a closer spacing of pegs may provide a straighter path but will also increase the ratio of lateral to longitudinal forces. In lateral undulation all of the propulsive forces, induced by muscular contraction, are exerted against the pegs. The weight of the snake passes to the ground via the contact zone of the ventral scales and consequently produces sliding friction that must be overcome to maintain motion.

Figure 3-17 Some snakes have become short and stout; this reduces their effectiveness in lateral undulation. This specimen of *Bitis caudalis* from southwest Africa illustrates such a species. (Photo courtesy of H. Mendelssohn)

is gently in contact with two rigid objects it will immediately and locally use them as points against which to push, thus increasing the control at low speed and probably decreasing energetic efficiency of locomotion.

Obviously, lateral undulation is a method that may be utilized by almost all limbless vertebrates in traversing the ground. The highly cost-effective basic motion is that found in elongate fishes; only the control pattern has been modified during evolution to incorporate the "search and place" circuits. Elongation and fusion of muscles to span multiple vertebral sequences could have followed later. As the trunk is constantly moving, the method permits a relatively efficient, medium- to high-speed traverse of broken terrain. It produces a constant speed in most of the body mass so that only lateral accelerations need be reversed. Lateral undulation has limitations in that the animal must be relatively elongate, the method is generally unsuitable for traverse of flat surfaces lacking irregularities against which the animal may push, and the method cannot be used for traverse of a straight-sided tunnel such as a rodent burrow. Although friction helps sweep up ridges in traverse of sand flats, it remains uniformly disadvantageous in lateral undulation. The other methods of limbless progression may be presumed to have advantages in special situations of particular importance in the environment of certain animals. Examples are flat, open areas, tunnels, and tree trunks.

Lateral Undulation Off the Ground

Before examining the other methods of limbless locomotion, it should be noted that the almost-limbless African lizards of the genus *Chamaesaura* will undulate amid tufts of grass; certain snakes move similarly amid the branches of trees and bushes. Many such snakes also show a number of other specializations for arboreal life. Among these are vision with considerable anterior as well as ventral overlap of the visual field (often coupled with a double fovea [Mertens, 1970; Underwood, 1970]), an elongate tail, widened shields covering the dorsal as well as the ventral surface, and possible modification to the inner ear. Movement by undulation off the ground requires the capacity for forming variably sized bends, but more than that it requires control. The species capable of moving by this method are generally slender, elongate, and often relatively light-bodied. They form a spectrum of shapes and modes of effectiveness. Some normally move slowly, but most, for instance the mambas, will rapidly traverse swaying branches (cf. p. 100).

The most important difference between on- and off-ground undulation is in the placement of the weight-induced forces. On the ground these pass across the ventral surface while the propulsion forces move laterally. Off the ground both weight and propulsion forces often pass across a single site, placed some 45 degrees from the vertical. When such snakes come to a stop they presumably must shift the site of force application on each branch from that facing forward relative to the snake's direction of motion to one facing medially. The snake may also use the weight-induced friction to keep itself from slipping.

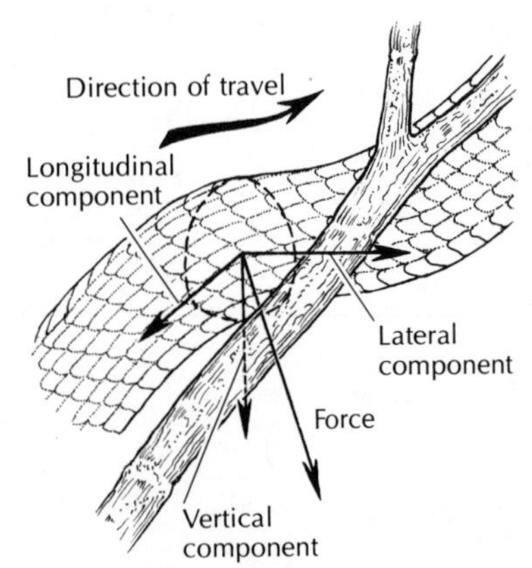

Figure 3-18 Some tree snakes manage to balance the edge of their trunk against branches, thus transmitting support and propulsive forces through the same site in a kind of lateral undulation off the ground.

It is unclear whether such snakes maintain some tonic level of muscular contraction when resting immobile or if the internal rigidity necessary to maintain the loops is produced by some innate resistance of the trunk to deformation; perhaps they lock their coils by using the resistance of the vertebral column to direct torsion. Many elongate arboreal forms will rest coiled into a terminal axil or the tip of a palm; some species show elaborate behavioral syndromes for forming a complex coil in such a suspended position. Since a coil would tend to be self-locking, much more stable, and less likely to be displaced by wind movements, this suggests that maintenance of the extended undulatory formation does have some energy cost. Direct recording of muscular activity (or metabolic rate) in both undulatory and stationary snakes may well be the simplest way of dealing with this question of energetics.

The control complexities should easily be visualized from the above. The snake must exert forces against twigs and branches to maintain and propel its trunk; yet the position, resilience, and relative response to external factors change according to the nature of each fixed point. Approach to arboreal prey may require steady progression. Field observations and those on captive snakes show that this requires multiple minor adjustments on the animal's curves and contact sites. These adjustments involve small localized movements and proceed very quickly, even when the snake is scarcely moving. Preliminary experiments suggest that these movements represent responses to local instability. Even when most of the trunk is fixed, shift in support for the remainder produces local adjustments in the fixed portion; are these mediated on the level of the spinal cord or in higher centers of the nervous system?

Some boas and vipers extend their body from branch to branch, stabilizing it by first anchoring the tail. A few elongate arboreal snakes, such as the tropical American *Imantodes*, extend their head and up to one-half their body in a straight line upward or forward without caudal prehension (grasping with the tail). In both cases the posterior portion of the body will be firmly anchored. These snakes then use a so-called concertina pattern to cross between branches.

Concertina Locomotion

A stationary animal may, without slipping, exert a horizontal force on the substratum just a trifle less than that necessary to overcome the static friction (which is equal to the animal's weight times coefficient of static friction). The animal may use this force to impart momentum to one portion of its body. Thus a rooster may stretch its neck and flap its wings and a man may wave his arms without, in either instance, the feet sliding along the ground. If man or rooster stands on ice, the static friction may be too low for vigorous stable activity.

The use of static friction may permit locomotion, however, where pure undulation is ineffective. Caecilians, snakes, and amphisbaenians commonly propel themselves by a movement based on this principle. A stationary site of the body in static frictional contact with the substratum serves as a base from which an adjacent part of the trunk is pushed or pulled. In its simplest form, when locomotion proceeds across a plane surface, the looped animal extends its head and neck toward a new resting site. At the new site, the trunk follows the head and neck to the ground in sinuous loops. This alternate folding and extension has led to the name "concertina" progression.

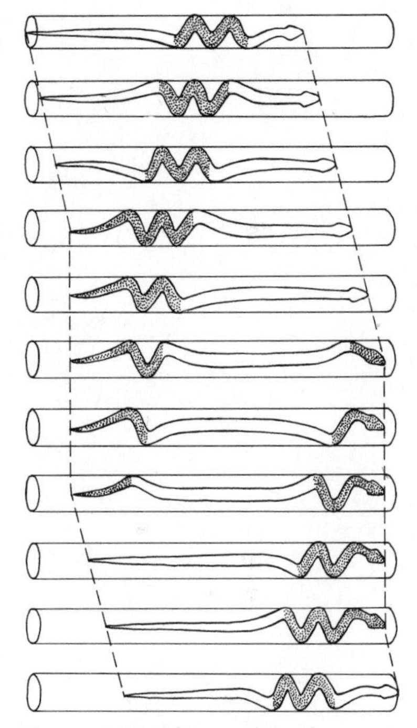

Figure 3-19 This snake, shown in successive positions, is moving by concertina through a straight tunnel with parallel sides. The reaction forces are transmitted across the (shaded) zone in which the snake's body is in static contact with the tunnel walls. Both the weight of the static portion and any forces exerted by widening the coils against the tunnel wall contribute to the maximum static friction or static friction reservoir. The weight of the sliding portion produces sliding friction which acts opposed to this.

Yet the critical aspect is not this seeming alternation of activities, but the use of friction to fix the stationary surface through which forces are exerted on the ground. Concertina movement requires that part of the trunk must always be at rest relative to the ground and consequently forces repeated acceleration and deceleration of the animal's mass. Such stop-start shifting of the body is rather inefficient because the momentum of the moving parts is lost each time they stop. Efficiency would require storage or recovery of momentum, neither of which occurs during slow concertina progression.

In general it may be assumed that the moving portions of the animal are lifted out of contact with the ground. In this case there will be no sliding contact; thus the deleterious aspects of friction are internal to the "black box" and do not enter into our analysis. If a snake slides its head forward instead of cantilevering it off the ground or drags its tail, the static-friction "reservoir" at the stationary site will be reduced by that fraction of weight transmitted across a sliding surface, and the unbalanced force tending to accelerate the system will be reduced by the magnitude of the sliding friction. These several components may be visualized on the basis of Figure 3-20. At X the potential accelerating

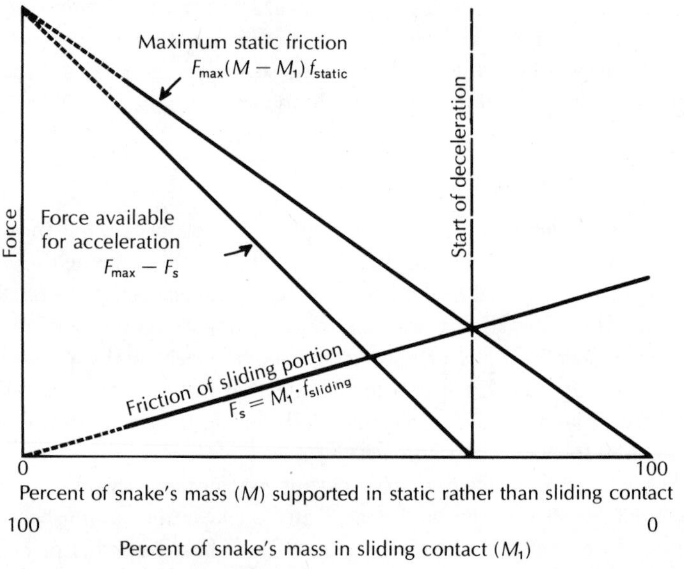

Figure 3-20 As an ever greater portion of the snake is in sliding contact, the force available for maintaining motion or acceleration of the moving part decreases. When the sliding friction becomes equal to the maximum static friction, the moving portion of the animal starts to decelerate unless it shifts more of its mass into stationary contact with the ground. It is important to keep in mind that the snake may cantilever the moving portions of its body, supporting them from the stationary zones. Only the part of the mass in sliding contact with the ground will then produce sliding friction.

force capable of maintaining the system in motion (against friction) has reached zero; somewhat to the right of this line the system would start to decelerate as its momentum decayed.

Static friction may also permit locomotion via lateral forces. As the static-friction reservoir depends upon the force pressing the stationary surfaces together, it may be increased by muscular forces. Gray and Lissmann (1950) showed that snakes use contraction-induced muscular forces to push laterally against the sides of a straight-sided channel; the method then lets various limbless forms traverse smooth tunnels. All of this can be nicely demonstrated by forcing a rat snake, a hognose snake (*Heterodon*), or some other common species to crawl through a transparent plastic tube little wider than its diameter. Since lateral undulation is impossible in such a straight-sided tunnel, the snake will use either concertina or rectilinear movement.

Uropeltid snakes and caecilians use a variant of this method to traverse and widen tunnels (cf. Gans, 1973; Gaymer, 1971). Rather than throwing the whole body into S-shaped curves, they restrict this curvature to the vertebral column which has been freed in anatomically diverse ways to shift within the integumentary envelope. As such curvature tends to widen the trunk, the method may serve simultaneously to form a *point-d'appui* and to facilitate the passage of the remainder of the snake's body (Figure 3-34).

Figure 3-21 This rat snake, *Elaphe obsoleta,* climbs a vertical trunk by a combination of concertina and undulation. Forces applied near the snake's ventral surfaces push its angled side against irregularities in the bark. (Photo by R. G. Zweifel)

Another variant of lateral force application is seen in North American rat snakes (*Elaphe*) and certain other semiarboreal forms. These species climb tree trunks by "stemming" their way between the vertical ridges and projections in the bark. Progression proceeds by a combination of undulation and concertina, and it is curious to see these animals apply themselves to a trunk and climb it with their body thrown into thirty or more short-radius bends. Such species derive an obvious advantage by being able to apply these forces as close to the surface as possible; many of the discontinuities against which they stem do not project far. These snakes (no other tetrapod shows this adaptation) have changed their cross section so that the widest point lies at the ventral surface (Figure 3-14). The wide ventral scutes show a sharp vertical bend on each side, and these lateral keels serve for force application.

Other climbers show a fabulous ability to throw their trunk into multiple, regular, and controlled bends of very short radius. The African file snakes (*Mehelya*) apparently can travel along telephone wires with alternate half-loops hanging respectively over the left and right sides of the wire. In this case we have some clues about the internal mechanisms that make the behavior possible; Bogert (1964) has commented on the curious longitudinally splinted diapophyses that project from the sides of the vertebral centra in these and some other species (cf. Hoffstetter and Gasc, 1969). Although no one has yet studied the associated muscular modifications, it seems plausible to assume that the splints serve as guides, permitting directionally controlled force application by the contracting muscles in a similar way to that seen in the neck of *Dasypeltis* (cf. p. 63).

A very important kind of concertina locomotion involves constriction of the body about a branch or twig so that the forces are medially directed. Two major variants occur. The first, which has already been mentioned, involves mainly those snakes, such as tree boas, that kill their prey by constriction. These can exert considerable medial force and may often be seen to climb by constricting a vertical branch and sending part of their body off from this fixed position to a new stationary site. The capacity to constrict presumably derived its primary selective advantage in prey immobilization. Less clear are the (presumably multiple) origins of the prehensile tails and caudal constricting mechanisms seen in many snakes and even in a variety of truly tetrapod lizards. Some boas and tree vipers manage to lie in a flat coil suspended only by the tail which is wrapped about a horizontal branch. Is this exclusively a hunting posture? What might be the long-term contraction pattern of the caudal muscles of a suspended snake? Are there locking mechanisms in the caudal tendons similar to those described in the leg tendons of some birds (Schaffer, 1903)? Unfortunately even the most obvious questions about this behavior remain to be answered.

Concertina movement or the use of localized stationary zones for transmission of propulsive forces then seems to be an obvious mechanism for limbless vertebrates. Its simplest version using weight-induced forces may be performed by any limbless animal. The ancestral alternating contraction pattern used in lateral undulation is brought into play and only the control needs modification. Consequently it is more than just an alternate method. It may also be combined with undulation so that different portions of the trunk progress by different means. Concertina may be advantageous because it involves no loss due to sliding friction and does not require lateral *points d'appui* or other special configurations of the substratum. Furthermore, the force in the static-friction reservoir may be increased by muscular effort.

The price of concertina locomotion is the loss of momentum in each transition; thus it is relatively inefficient and slow. Both lateral undulation and (unmodified) concertina are particularly inefficient for animals that have become short and stout in response to other selective forces. Sidewinding and rectilinear locomotion, two special uses of static friction, were apparently derived as adaptive solutions that compensate for these disadvantages.

Figure 3-22 Some tree vipers, such as this *Bothrops undulatus* from Oaxaca, Mexico, can suspend their body vertically from a branch to which they are anchored by a coil of the tail. (Photo by C. M. Bogert)

Sidewinding

Certain rattlesnakes of the southwestern United States have long been known to move in a curious and "unsnakelike" fashion. Early settlers referred to them as sidewinders and used the term as an opprobrious epithet for anyone engaging in inappropriate behavior. Walter Mosauer, who conducted some of the fundamental work on the axial muscles of snakes, was the first to characterize the motion (1932). He also noted that it occurred in North African desert vipers as well (Mosauer and Wallis, 1928). Both Mosauer (1932) and Cowles (1956) documented that many other snakes "engage" in a modified sidewinding when forced to escape across flat areas. This has since been confirmed, although behavioral factors and the nature of the ground become most important in establishing the time a particular species shifts to this method of locomotion.

Though sidewinding is often assumed to derive from lateral undulation (cf. Brain, 1960; Gray, 1968), it seems more appropriate to consider it as a special case of concertina. At any given instant at least two portions of the snake's body tend to be in stationary contact with the ground; the remaining portions are either being accelerated forward, shifted between the fixed zones, or pulled to the caudalmost zone. In sidewinding across open terrain, a snake places its neck into stationary contact with the substratum and then rapidly moves the entire trunk, bit by bit, into adjacent stationary positions (Figure 3-23). Thus it appears as if the snake "rolls out its length" along a track in such a way that the position of the tail lies anterior (in the direction of progression) to that of the body and that the body lies anterior to the head. The curious point is that the snake picks up its anterior end to start a new track lateral and somewhat anterior to the first before the old one has been completed, and two or more separate parts of the body form two or more discrete contact zones with the ground. A simple way of visualizing the track pattern is to roll a short, helical spring with open end forward over a surface. Although the actual motion of a sidewinder differs considerably, both will produce a similar trackway. As the snake's body is lifted from track to track, the trackway will consist of a series of discrete, nontouching tracks, and the snake will move at an angle to the line of a single track—hence it winds to the side.

The snake's head remains in the air facing more or less in the direction of travel, and the neck must be bent through an angle to let the body rest on the forward-pointing track. As part of the track is formed by the neck, the track seems to have a forward-open hook at its posterior end. The main part of each individual track is straight, often showing cross marks where the ventral scales rested as well as other signs of lateral slippage. Different species show different tail marks and tracks

Figure 3-23 Tracings from a film showing a sidewinding rattlesnake crossing a layer of smooth sand. Note how the snake's head points generally in the direction of motion, whereas the body and its track lie at an acute angle to this. Also note the way the back edge of the body digs into the soil, building up a windrow of sand on the posterior side.

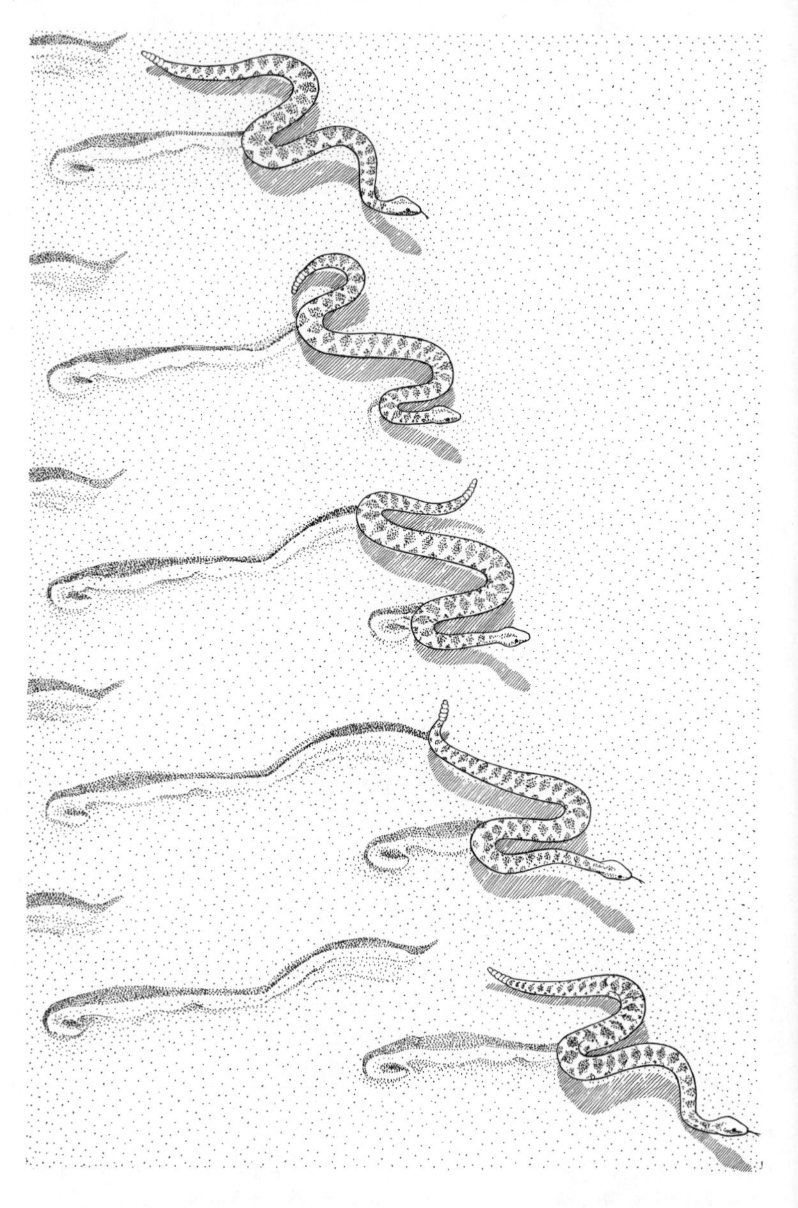

made by snakes sidewinding over sooted paper, or smooth sand may show diverse drag marks of body and tail between tracks (Brain, 1960).

The method described appears to provide an ideal way of crossing flat, low-friction surfaces such as sand dunes, which are prominent features in the habitat of many sidewinders. As the several parts of the trunk rest in stationary contact with the ground, there is a static friction reservoir that permits acceleration, and no forces need to be exerted

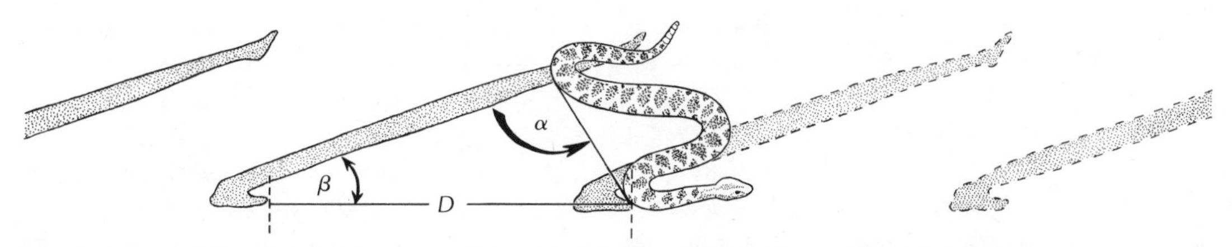

Figure 3-24 The various sidewinders differ in the distance between tracks and in the angle (β) that the tracks make with a line (D) drawn through their anterior or posterior ends. We know that these parameters are affected by behavior, but we still lack enough measurements to make predictions about their magnitude with reasonable probability.

against points projecting from the surface as in lateral undulation. Each bend, whether established as the trunk is laid down or when the body is lifted on the way to a new track, travels posteriorly from segment to segment of the trunk. Hence there may be a continuous wave of muscular contraction permitting an increased velocity. As long as the surface is plane the snake would seem to need few "decisions" about loop placement; minor irregularities of the surface do not affect travel.

Literature reports suggest that sidewinding may be the mechanically most efficient method of snake locomotion; it certainly permits continuous travel of a mile or more at a good velocity. Cowles (1956) suggests that sidewinding is also useful to a desert snake that may have to travel during the day; the lifted body will reduce conductive heat exchange from the surface of the soil, indeed from the high-temperature layer of air adjacent to the surface. A buried snake disturbed by a predator might then sidewind rapidly to a safer location without overheating in the hot sun; both speed and coil lift would presumably assist survival. Is the avoidance of thermal stress alone really a sufficiently critical factor to establish such a method? The diversity of temperate-zone species using (at least a simple version of) sidewinding as an occasional high-speed escape device across smooth surfaces suggests that predation avoidance may also have been a selective factor.

The regular occurrence of miscellaneous drag marks between tracks of a trackway led us to undertake an investigation of sidewinding mechanics, the data of which have as yet been only partly analyzed (Gans and Mendelssohn, 1972). Friction had seemed to be the key to sidewinding as a method for crossing flat surfaces; if so, stationary contact would be advantageous and the slippage suggested by the drag marks to be avoided. Could the published illustrations of extraneous drag marks therefore be misleading and mainly reflect the nature of the surface or some other experimental condition?

An obvious test of such a view was to consider, independently, the contour and the frictional coefficients of the respective substrates to be contacted by the snakes. One might then separate flatness from a low frictional coefficient and an undulant contour from a high coefficient.

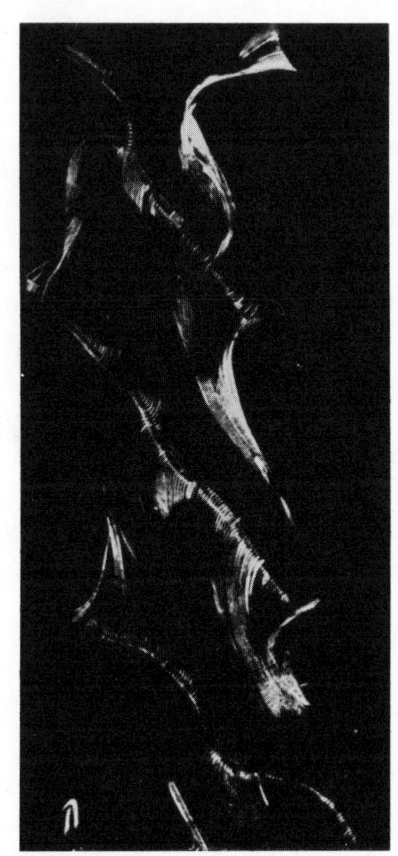

Figure 3-25 Photo of a sidewinder's track across sooted cardboard. The dragmarks between tracks suggest that snakes do not expend much energy lifting portions of their body out of sliding contact on a smooth surface. (Photo from Brain, 1960)

To this end snakes were filmed from two angles while they were traveling across a variety of simple substrata, each marked with a grid of 30-cm spacing. Since the films were taken in bright sunlight, shadow effects indicated the clearance between snake and ground.

In analysis it became very clear that different species moved differently under the identical physical conditions and that the locomotor behavior of even a single species was affected by factors such as time of day, size of the individual, body temperature, and the animal's psychological state. Even geographical races of snakes showed differences in their locomotor patterns; specimens of the saw-scaled viper (*Echis*) from Ceylon sidewind less readily and less effectively than those collected west of India. Each sidewinder generally demonstrated a hierarchy of substratum-specific locomotor patterns. Some surfaces were almost never crossed by sidewinding; for others, some version of sidewinding was the preferred motor pattern. Then there was a whole assemblage of snakes that would only sidewind under stress—and most ineffectively at that (Gans and Mendelssohn, 1972).

The basic difference between good and bad sidewinders seemed to be in the control of loop placement and the response to the frictional coefficient (rather than the contour) of the ground. Ordinarily coils are lifted just high enough to clear the ground. The films show that the snakes are continuously testing for substrate contour; this produces some of the drag marks. They also respond more directly, dragging the loop over surfaces with a very low frictional coefficient; this accounts for the beautiful drag marks obtained on smooth, low-friction, sooted cardboard (Figure 3-24). In contrast to these patterns, all sidewinders swim by lateral undulation.

These statements represent an oversimplification, as a snake in static contact with a high-friction surface can obviously overcome a greater amount of horizontal force (produced by sliding friction) without slipping than can a snake in static contact with a low-friction surface. If the coefficients of static and sliding friction rise uniformly and proportionately over an undulant surface, the movement pattern will not be affected. If the surface is irregular with regard to friction, the animal's strategy seems to represent a compromise between the energy required to lift the loop higher between tracks and the energy required when lower loops might encounter an occasional zone with high coefficients of sliding friction. Irregular contours of the ground will modify the pattern in a slightly different way, since the moving loop may encounter lateral obstructions. Localized lifting then occurs and some films even show snakes shifting briefly and locally into lateral undulation, applying posterolateral forces against the obstructions. Both tactics are affected by the animal's behavioral state. As sidewinders get more excited they tend to lift their loops higher, perhaps expending more energy for the sake of greater and more sustained speed with reduced chances of encountering high sliding-friction or lateral obstructions.

Sidewinders have another tactic for increasing the actual forces required to keep them from slipping (lateral-force reservoir) above the theoretical limit. Natural surfaces are irregular. The number of irregularities encountered will be a function of the area swept out by the snake's body, and the edge markings from the ventral scales document that sidewinding snakes tend to slip sideways. The inefficiency resulting from a temporary shift to sliding friction is reduced as the probability becomes higher that increasingly more local irregularities will be "found" as the body slips sideways along the ground. Since each such irregularity increases the drag, the slippage is self-limiting and the system must automatically shift back into a stationary state. Both regular and occasional sidewinders appear to capitalize on this effect by twisting[2] the portion of the trunk that is in contact with the ground; the backward-facing side is depressed, forming an edge that tends to dig in as it touches the ground.

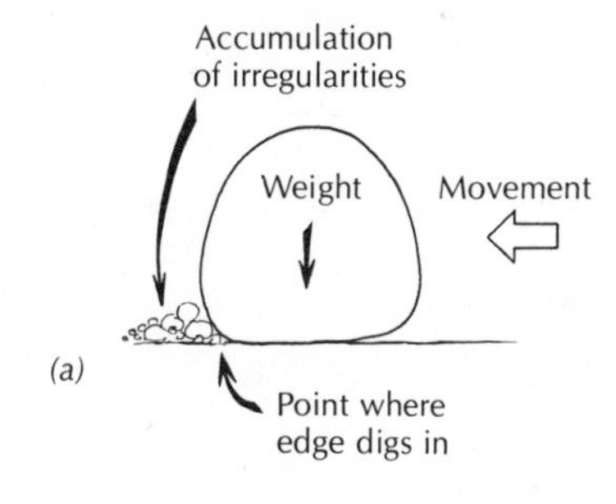

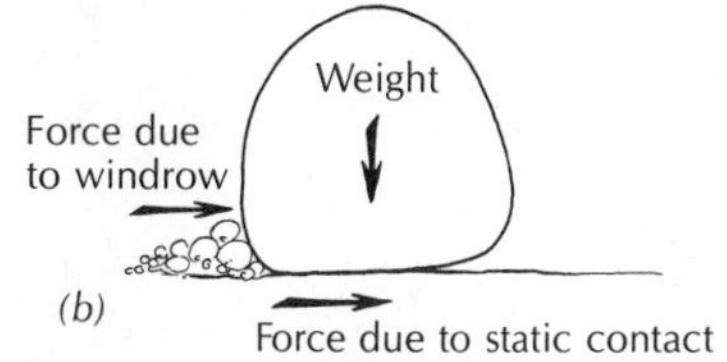

Figure 3-26 These sketches show how the body of a sidewinder contacts the ground and the forces exerted by it. The trunk is tilted in the direction of force application. Any slippage then causes the posteriorly facing edge to dig in, accumulating a windrow of particles. This windrow, in turn, increases the resistance to further slippage. When the magnitude of the resistance exceeds the applied force, the trunk stops sliding.

Establishment of a reaction-force reservoir, then, involves more than friction, and some reaction is obtained from forces at right angles to part of the horizontal surface of the ground. Many published photos of sidewinder tracks on sand certainly confirm this; they show a slight windrow of sand on the posterior edge of a track. The digging in is even more nicely shown in the photo of the track of a sidewinder crossing the drying mud of a stream bed (Figure 3-27). A few species continue this digging in along the entire length of each track. The increasing pressure on the ground due to the decreasing contact area (as the body is pushed to the next track and the body's diameter decreases toward the tail) causes the tail to dig into the sand. The tail also gives a last push, propelling itself toward the next track.[3]

Sidewinding is thus seen as an effective method for crossing or searching flat surfaces by the use of static friction combined with simultaneously sought or automatically created irregularities of the surface. In general this version of concertina locomotion may also be assumed to proceed by (left-right) alternating waves of muscular contraction. The method is fast and simultaneously lifts part of the body out of contact with (hot) surfaces. Yet the long distances traversed by this method suggest that the major advantage of fast sidewinding is that it probably incorporates some conservation of momentum. Each part of the snake is obviously at rest some time in each cycle. However, it seems as if the momentum imparted to the head and neck will be partially conserved

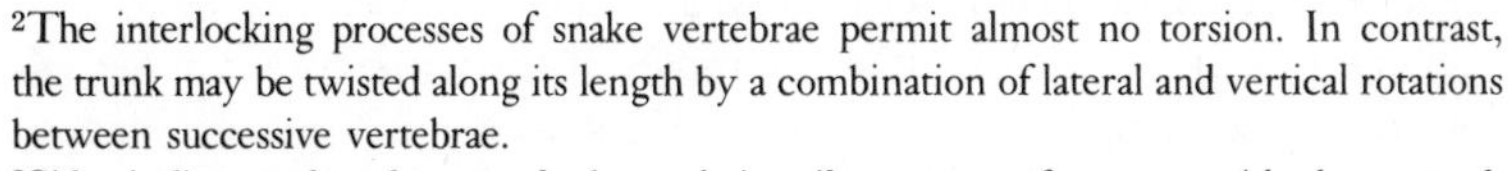

[2]The interlocking processes of snake vertebrae permit almost no torsion. In contrast, the trunk may be twisted along its length by a combination of lateral and vertical rotations between successive vertebrae.

[3]Sidewinding rattlesnakes mostly keep their tails up, out of contact with the ground.

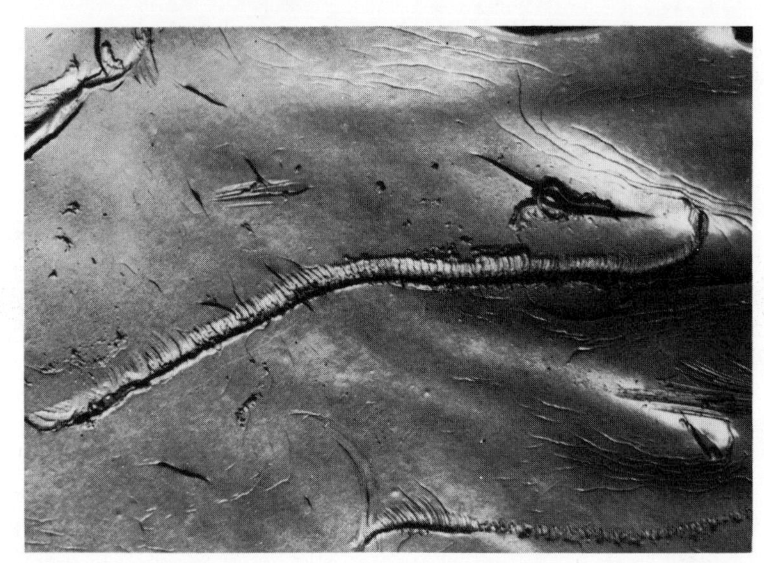

Figure 3-27 Track of a sidewinder (*Crotalus cerastes*) on a muddy river bottom. Note how the body has slipped sideways so that the edges of the ventral scales have left lateral scrape marks and the backward-facing side has dug into the mud. (Photo by W. Porter)

in the rapidly traveling loop in the same way as a loop may progress along a flipped rope or whip. Any such conservation would sharply reduce cost per length traveled, an obvious economy in food gathering. Certainly we have here a locomotor method that becomes more effective as control of loop height increases. We also have a locomotor pattern in which the trunk is bent in more than a single plane.

Escape Saltation

I hope that the preceding discussions have tended to erode the sharp edges between the textbook categories of lateral undulation and concertina and between concertina and sidewinding progression. Definitions of movement categories have considerable didactic value, as the categories describe the events of particular, sometimes brief, time intervals and portions of a snake. No snake, much less any group of snakes or other vertebrates, uses only one of these methods at all times. The general capacity for diverse locomotor methods exists and the differences appear to be in the energetics, in the nuances of control with which loops can be placed, in the forces applied, and in the environmental conditions. This leads us to a variety of specialized locomotor patterns for specialized interactions with prey and with predators.

Some tree snakes are excellent examples of further locomotor specializations for a particular environment. Various slender species, such as the Indian *Ahaetulla nasuta*, use lateral undulation when traveling off the ground. They do not travel continuously, and particularly when hunting, they sway the raised anterior part of the body. Swaying is coordinated with slight wind movements and may have a camouflaging effect. When films of the movement were analyzed, the swaying proved purposive and regular. A swaying sequence consisted of some five or more movements during which the head remained relatively stationary. Each sequence was followed by an advance of the head to a new position. Each of the swaying movements of the sequence involved an anterior shift of the body and the formation of a loop of increasing size into the neck; it was the partial straightening of this loop that advanced the head. Both undulatory and concertina movements could be seen along the trunk, depending only on the number and spacing of available twigs and branches. Concertina was always used when the body showed fewer than three half-loops.

Maximal psychological or physiological stress induces an even more curious series of saltatory (from the Latin *saltare*: to jump about or dance) locomotor patterns in various other snakes. No matter what the preferred method of gradual progression, these various escape reactions share the property of convulsive movement. The animal shifts position in a series of rapid jerks, moving much faster than it progresses. Although such behavior, which is also seen in some tropical worms, has the advantage of deterring some predators, it may also be useful in confusing an unfrightened potential attacker. Most important, we have here a class of motor patterns in which efficiency and nicety of control are sacrificed to gain the capacity of crossing terrain without searching for sites at which optimally efficient forces can be applied.

The first group of such escape saltations is seen in a variety of snakes and has given rise to the comment that such forms may sidewind (Cowles, 1956). Even the boa constrictor, as well as occasional water and rat snakes, will exhibit this behavior when placed on an open zone of hot, unshaded sand.

Other cases that may belong here are the "jumps" described for the colubrids *Psamnodynastes*, *Helicops*, and *Ficimia* (Mertens, 1946). The snake places its neck down somewhat ahead of its starting position and rapidly follows with its body, which is unrolled in a straight line deviating but very slightly (low β, Figure 3-23) from the ultimate direction of progression. The head, neck, and trunk then pick up as in sidewinding, but only after almost all of the snake's body has been stretched out. They move anteriorly at a very sharp angle so that the loop of neck and body passes almost vertical to the resting position before reaching the

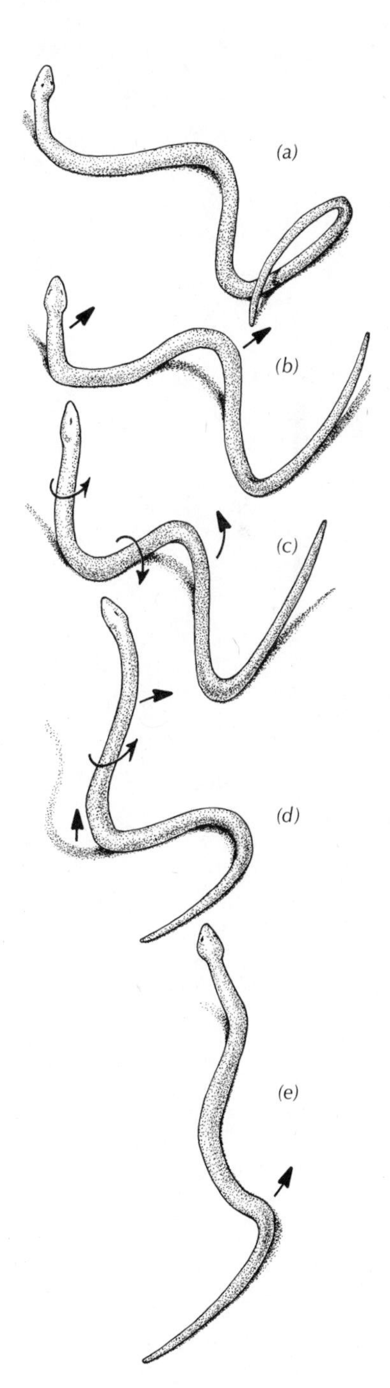

Figure 3-28 A series of tracings (filmed at 24 fps) of an Indian water snake (*Helicops*) rolling on its side. This inefficient escape saltation, in which the whole body twists to a temporary and unstable position over the contact zone, reflects a primitive sidewinding.

forward track. Rather than looping from track to track, the snake almost rolls along.

The upward bend is not produced by pure dorsiflexion of the body. (This would, indeed, produce a "hoop snake"!) Instead, the lifted part of the body bends laterally (with a slight dorsal component) and the snake then rotates the entire portion still in contact with the ground around its long axis. This means that the lowest dorsal scales as well as the lateralmost portions of the ventrals of one side will be in contact with the surface and will determine the frictional coefficient; the roughness of the dorsals makes it likely that their frictional coefficient will be higher than that of the smooth midventrals (acted on by selection for efficient lateral undulation). The increased friction is presumably critical since the resting portion of the snake might otherwise tend to slip forward as the bulk of the snake is being accelerated toward its tail. This method propels the snake in a series of bounds with only one or two portions of the trunk in contact with the ground at any time. Even though the entire animal may retain some contact with the ground, the entire process is sudden enough to justify the term saltation.

Bitis caudalis, the southwest African desert viper, has taken this pattern to its logical conclusion and does indeed jump. Specimens below 23.5 g in weight can accelerate the body into a flat trajectory and, by pushing with and then lifting the tail, travel through the air for about four-fifths the length of their bodies. When the trajectory attempted is too steep, the tail apparently has to propel so much weight that the animal fails to get off the ground. Larger specimens are seemingly incapable of achieving take-off velocity. In simplest terms, this reflects the fact that the force required for upward acceleration is a function of the animal's mass, roughly proportional to the length cubed. The force produced by the animal's muscles will be only a function of the muscle's cross-sectional area, which is proportional to the length squared. (See Thompson, 1951, for a suggestive discussion of the effects of size.)

There is also a very distinct temperature effect on the movements of the *Bitis*. Heat, due to the sun's incident radiation, seems to be the best and almost the only way of eliciting the jumping behavior. Until the snakes we observed had warmed up to 31°C, they would not jump. They were either insufficiently stimulated or their muscles were at suboptimum temperature. Above 37°C their behavior became disorganized, and the jumps lacked the coordination to get all parts of the snake airborne at once.

Even when the conditions are at an optimum, somewhere between 31°C and 37°C, this method certainly appears to involve a remarkable energy cost. Jumps are interspersed between two or more standard sidewinding movements, but can be maintained for only roughly one-tenth the distance that a snake sidewinds with little effort. Does the re-

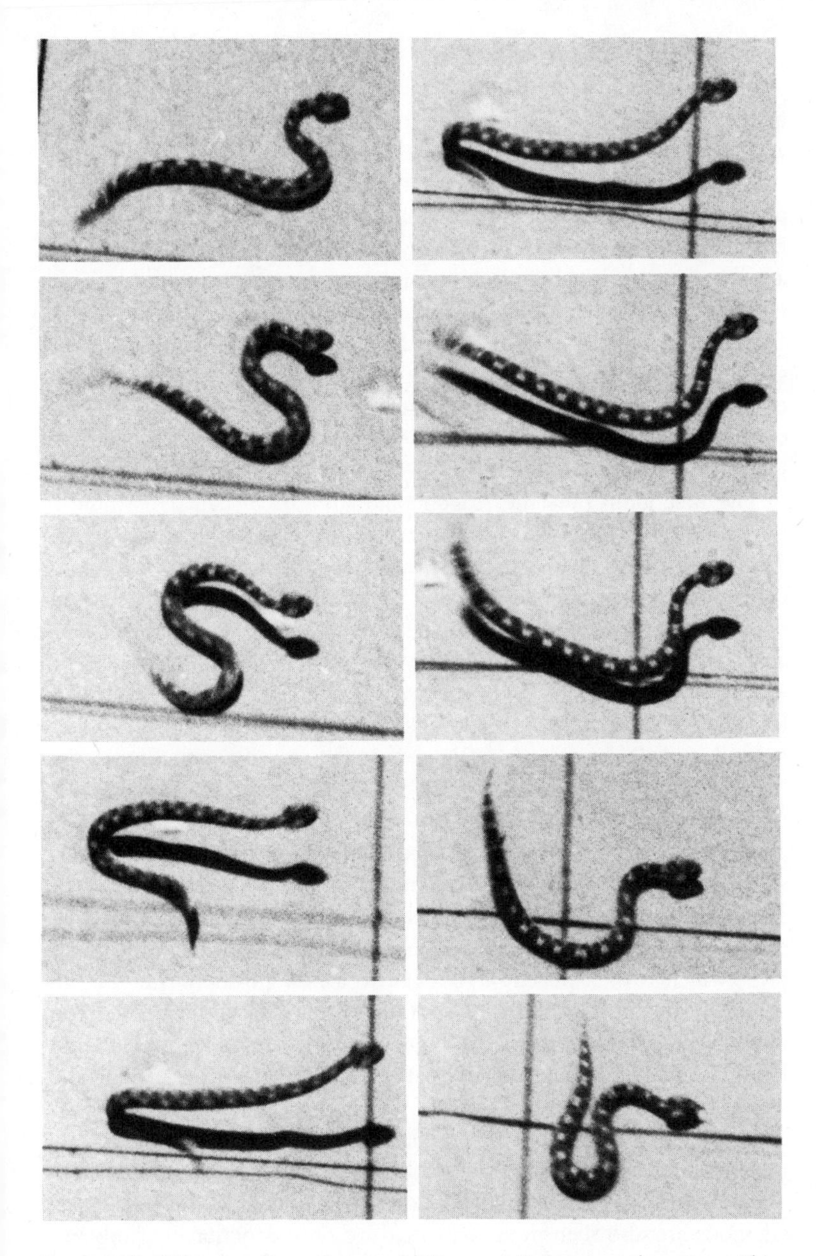

Figure 3-29 When small specimens of *Bitis caudalis* from southwestern Africa become extremely excited, they may utilize escape saltations to propel themselves from track to track, as seen in the successive frames of this film (64 fps). The specimen's shadow shows the height that the individual (photographed at 75 degrees from horizontal) is off the ground. The black lines represent threads used to mark off distances on the experimental plots. (After Gans and Mendelssohn, 1972)

duction in thermal conductance and slight increase in speed justify this, or are we dealing with an escape device of last resort? To answer such questions one should carry out direct measurements of metabolic cost and heat gain in the animal's normal environment.

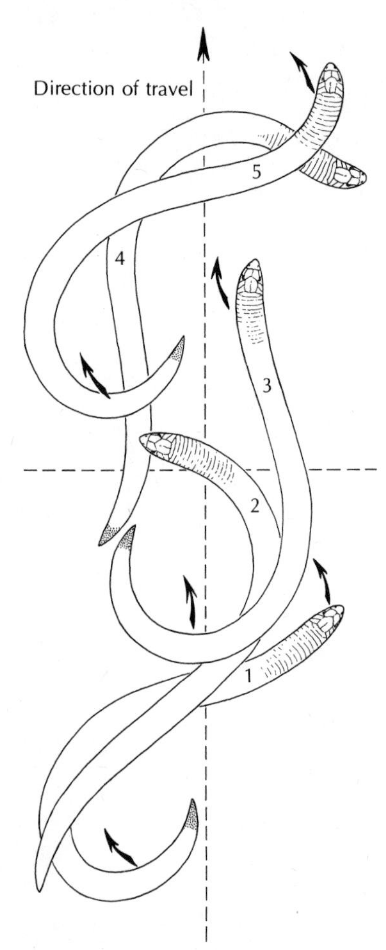

Figure 3-30 Tracings from a film to show the reversing escape saltations of the Arabian trogonophid, *Diplometopon,* when disturbed while away from its burrow. Numbers indicate successive positions at intervals of 1/24 second. The tail is shown shaded where it is dug into the ground and serves the animal as a fulcrum for further propulsion. Each convulsive jump moves the animal some 40 percent of its body's length. The method is clearly fast but has a high energy cost, since the entire trunk is repeatedly flipped from side to side.

A somewhat different escape saltation is seen in the trogonophid amphisbaenians, a group of "worm lizards" found in North Africa and from Somalia to Iran. All of these forms burrow (see Chapter 4). All are also characterized by a lack of caudal autotomy (the ability to break off the tail, sometimes at a cleavage plane) and exhibit a short, pointed, ventrally curved tail. The most modified species are, furthermore, the shortest and become rigidly inactive in various postures when dug out of their tunnels. Upon further disturbance (which under natural circumstances might signal the attack of a predator) the animal starts to jump convulsively (Figure 3-30).

The first jump puts the trogonophid into an S-shaped position on the ground with head and tail curled inward so that the animal almost describes a figure "8." The downward curvature of the tail is then increased and serves as a take-off point as the animal convulsively straightens its body in a second jump. The open, left and right half-loops apparently straighten simultaneously, extending the animal to its full length. The movement is so convulsive that much of the animal's mass loses sliding contact with the ground. The anterior portion lifts off the ground at the start of each cycle so that the stored momentum causes the animal to fly and slide into a new position showing a mirror image of the previous S.

As the dug-in tail provides a rigid connection to the substratum, it serves as a pivot so that the pure straightening forces do not cause the amphisbaenian merely to flip-flop. Rather, each movement advances the center of gravity, accelerating the animal's mass away from the caudal pivot which is then pulled or dragged to a new position. The tail immediately digs in and the next straightening starts with opposed contraction propelling the animal onward. The successive indentations left by the tail define the animal's path. The movement pattern is somewhat faster than the other ones of which this species is capable—it is certainly more confusing to a predator since only two points of any given resting position will overlap the last track.

The examples given could be expanded, of course. Indeed, escape saltations are also seen in animals that are only functionally limbless—for instance, in some elongate plethodontid salamanders (Bishop, 1941). These several escape saltations clearly represent adaptations to some kinds of stress, whether induced by sunlight or a predator. Brief bursts of speed and creation of confusion may become critical factors in response to such environmental pressures. Mechanical analysis evaluating the efficiency of transit in terms of energy cost per unit of distance must take the environmental circumstances into account. The species is adapted to survive rather than to move most economically under all ecological conditions.

Rectilinear Movement

We have now examined a whole series of progression patterns, some avoiding and others using friction, but all utilizing sinuous lateral bends. For such movements the structural pattern of the animal is not changed; only the control mechanism and sequence need be modified. However, a number of snakes and all amphisbaenians are capable of moving in a straight line by so-called "rectilinear" movement. Many of these are relatively plump species. Several major structural as well as neurological modifications seem to have been needed to effect the rectilinear movement.

Lissmann (1950) provided a careful film analysis of a snake (*Boa constrictor occidentalis*) moving in this pattern. He noted that the back of the snake seemed to progress at a fairly constant velocity with minimal accelerational changes. In contrast, the ventral skin moved by jerks. It would stop, accelerate sharply to a velocity higher than that of the snake (Figure 3-31), and decelerate again to another full stop. He also observed that the accelerating skin would first move upward, sharply but slightly, before "running" to catch up with the rest of the snake. It may have been this upward movement that led to the erroneous idea that rectilinearly moving snakes were walking on their ribs; it certainly contributed to the concept that some snakes moved by vertical undulations. Whatever the internal mechanism, the animal clearly anchors itself at certain sites. Such sites seem continuously to move toward the tail because their anterior edge is lifted out of contact as more skin is placed into contact along the posterior edge of the site.

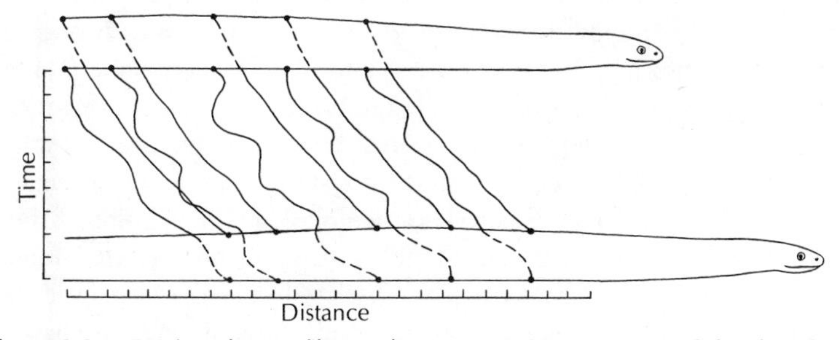

Figure 3-31 Tracings from a film to show successive positions of the dorsal and ventral surfaces of a *Boa constrictor* moving by rectilinear locomotion. The skin of the dorsal surface, which is quite firmly attached to the underlying tissues, progresses smoothly, whereas that over the ventral surface moves intermittently. (After Lissmann, 1950)

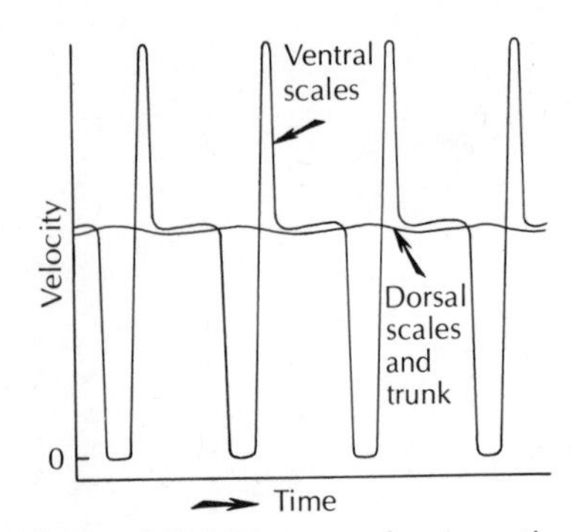

Figure 3-32 Diagram showing the velocity of the dorsal and ventral skin of a snake propelling itself by rectilinear movement.

The critical aspects of rectilinear movement then appear to be (1) that it is symmetrical with both sides operating in parallel, (2) that the skin seems to have remarkable anteroposterior flexibility (it contracts as it goes into contact with the ground and expands as it comes out of contact, like the bellows on a concertina), and (3) that most of the animal's mass continues to travel at a fairly constant rate, thus conserving its momentum (Figure 3-32). There are two patterns of movement. In the first, seen most commonly in short, stout snakes, the skin of the back seems to be fixed with respect to the vertebral column; in any case it moves at a constant rate during progression while the ventral skin oscillates. The second pattern is seen only in amphisbaenians among the vertebrates; here the skin over the entire circumference at one section of the trunk seems to move at the same rate during these movements. Amphisbaenians have their segments arranged in rings, or annuli, and one may see these lengthen and shorten in their entirety as the animal moves along (Figure 4-11).

The snake pattern is obviously one for straight-line travel across a plane surface utilizing its variable friction. The amphisbaenian pattern permits the animal to fix any portion of its circumference; it can "walk" with all of its sides. This would, of course, be advantageous to a burrower that spends its life in irregular tunnels in which the best place to exert forces may not be the bottom surface. Whether on the surface or in a tunnel, rectilinear locomotion has the disadvantage that it must proceed fairly slowly. If the contact zones start to slip, they will shift to sliding friction and exert forces opposed to the direction of motion. Nevertheless, rectilinear movement permits even a relatively short and stout animal to make a slow, momentum-conserving, straight-line traverse of smooth-contoured surfaces and straight-sided tunnels, as well as to make a head-on approach to prey.

We now have a case in which the black-box approach may be taken one step further. The forces acting across the under surface of the box (the snake in rectilinear motion) have first been determined; one can now use this information to gain some information about functions inside the box. To do so we next divide the box into a central (vertebrae, ribs, muscles, and viscera) and a peripheral (skin and intrinsic muscles) part and treat each as a separate unit in analysis. The outer box is still subject to all of the forces imposed on the unit as a whole. Some of these forces may be shown to be transmitted to the inner box via the connections (muscles and connective tissues) disclosed in analysis (dissection); the unbalanced forces will be seen to induce various kinds of changes in the outer box.

Simple manipulation shows that the observed stretching and shortening movements occur in the skin (the outer box). The loose skin slides over a mass of underlying tissues (the second box consisting of vertebrae, ribs, and their muscular sheathing). One can go one step further and document that the ribs do not move during locomotor progression, but stay in regular alignment with their vertebrae and each other. X-ray motion pictures have confirmed this (Oliver, undated); however, Bogert (1947) long ago documented the same process by loosening a small flap in the snake's side and watching the ribs as the snake moved in a rectilinear sequence. Lissmann (1950) allowed his snakes to move by rectilinear motion across a set of moving platforms. He thus measured the place and magnitude where horizontal forces are applied. His results may be interpreted to indicate that such propulsive forces are indeed transmitted where the skin is placed into static contact with the ground.

Dissection of the connections between the two boxes disclosed that snakes have two major muscle systems and one connective-tissue system attaching the skin to the skeletal core. The connective tissues tie the skin fairly tightly along the middle of the back, with many of the fibers concentrating their attachment on the tips of the neural spines. Other loose connective tissues attach the skin to the fascia over the axial muscles. Various muscles have some connection to the skin, but the ones of concern here are (1) a series of inferior costocutaneous muscles that run posterior from the skin to attach near the ventral tips of the ribs and (2) a series of superior costocutaneous muscles that run anteriorly from the skin to insert on the upper third of the ribs. In amphisbaenians a

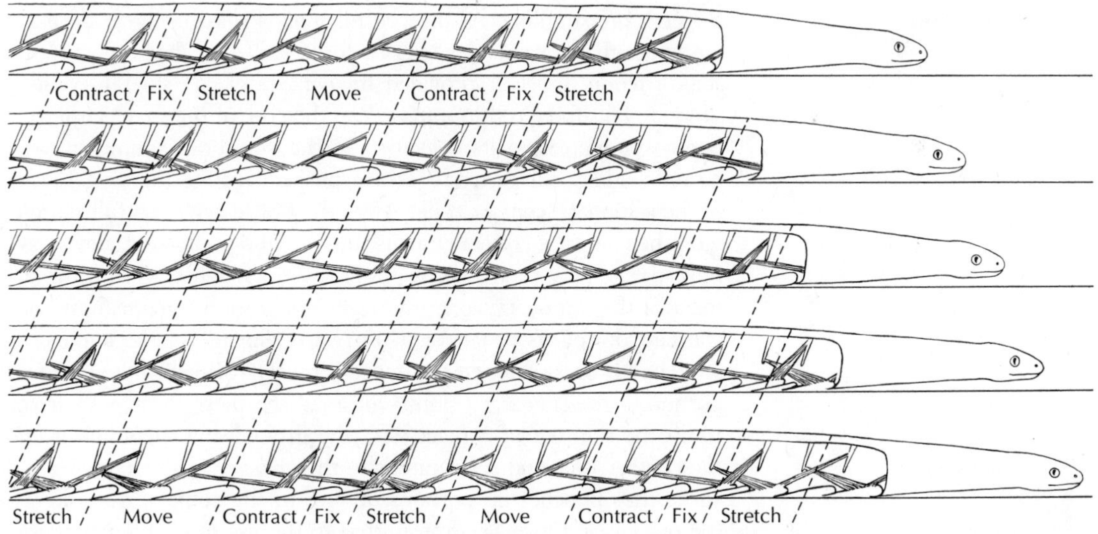

Figure 3-33 These sketches show rectilinear progression of a *Boa constrictor* with only every tenth rib and scale shown. However, the positions of these scales—those shown contracted and fixed and those shown stretched and moving—are to scale. When snakes move in a rectilinear pattern, they give the impression of traveling by vertical undulations with the waves passing from tail to head. Both are optical illusions resulting from the vertical movement of the scales away from the fixed site and the speeding up of these scales on the way to the next site of static contact. (After Lissmann, 1950)

third series of muscles runs posteriorly from the skin to the dorsal portion of each rib. These are apparently required to move the sleevelike skin which is otherwise free not only ventrally but around the entire circumference.

The loosely sliding outer layer (outer box) can be further subdivided into components by recognizing that it consists of a stretchable integument and its associated layer of cutaneous muscles. In snakes this stretching capacity results from the series of scales acting like overlapping plates. In amphisbaenians the skin is divided into annuli that bend to narrow rather than slip over each other (see Chapter 4). The cutaneous muscles interconnect the inner portions of these scales. Contraction of these muscles pulls the scales into maximum overlap or narrows the annuli, and very little force is applied to shorten the tube of skin. The tube is stretched by external forces. All of this suggests that the major propulsive forces do not arise within the outer shell; they must be transmitted to it.

The only significant connection between the external and the more internal "black box" is via the costocutaneous muscles. This means that these muscles must transmit the forces applied to the substratum at the contact zone. Observation and inspection indicate that these connecting muscles not only transmit, but actually induce these forces. Once a ventral scale has been placed in static contact with the ground, its pair of inferior costocutaneous muscles may be seen to contract (in Bogert's preparation). As the ventral shields (gastrosteges) do not slip, we know that the muscles exert a force less than that reflected in the static-friction reservoir provided by contact of the shields against the ground. This force is thus imposed on the core of the snake either maintaining or increasing its velocity. As the vertebral column moves anteriorly, additional gastrosteges are pushed against the posterior edge of the stationary contact zone. After they have achieved full static contact, their inferior costocutaneous muscles also contact. A slow wave of muscular activity then sweeps posteriorly in synchrony with the placement of the stationary zone; the force application remains more or less constant during its progression. Forward slippage of the animal's core stretches those superior costocutaneous muscles which run to the anterior portion of the stationary skin. One may see these muscles contracting to lift the skin out of contact with the ground and the anterior end of the contact zone and accelerate it to the velocity of the axial mass. The snake thus bunches up portions of skin in the contact zones in which the propulsive forces are transmitted to the ground. Other parts of the skin are stretched as they travel unloaded from one contact point to the other.

Examination of the muscles indicates that their size and position reflect their function. The stout inferior costocutaneous muscles of snakes lie closely parallel to the ventral surface of the snake. Consequently the force of contraction lies very close to the direction of motion of the animal as a whole. This maximizes the resultant in the useful direction and is hence advantageous. As the inferior costocutaneous muscles contract when the skin is in contact with the ground, the narrow angle also advantageously avoids lifting forces that would pull the ventral scales from contact with the ground and with this reduce the static-friction reservoir. In contrast, the much more slender superior costocutaneous muscles angle sharply dorsad. Their contraction not only pulls the skin anteriorly but simultaneously lifts it out of contact with the ground. This has the advantage of avoiding sliding friction between ventral scales and ground; such friction would obviously reduce the force available for moving the snake. As the superior muscle acts only on the mass of the skin itself, it can be a slender muscle. The inferior muscles, in contrast, move the snake as a whole and are therefore stout. Amphisbaenians may travel forward or backward (hence their name < L < Gk *amphis*, both ways + *bain* (*ein*), to go). They have diverted a third set of muscles to the skin and both anteriorly and posteriorly directed muscles are consequently quite stout.

Vector analysis at the interface between internal box and the sliding layer shows that the horizontal component of the inferior muscles is equivalent to the force that the snake applies to the ground (if internal friction is corrected for). The force to be applied by the superior muscles cannot be derived by analysis. It is not transmitted to the outside but serves primarily for an internal deformation of the periphery. It must be measured directly (using a strain gauge in series with the muscle), though its order of magnitude may be estimated from the force theoretically required to deform the mass of the outer "black box" at the observed velocity.

The inherent complexities of a comparative motion analysis and the importance of understanding the physical principles may be seen by comparing the rectilinear locomotion of amphisbaenians to the superficially similar concertina pattern of earthworms or caecilians (Figure 3-35). All use static friction by transmitting their propulsive forces to the ground across stationary areas of the skin. In earthworms the trunk is narrowed and widened by hydrostatic forces; the resemblance to amphisbaenians is external only. In the Amphisbaenia the vertebral column remains straight, and the skin is ordinarily bunched up and stays compact; it then stretches to rush to the next site. Most of the trunk, framed by vertebral column and long lateral ribs, moves at a constant velocity, and only the thin skin and cutaneous muscles engage in a stop-start-stop cycle and incur an oscillation of accelerating forces.

In contrast, the caecilians only slightly shorten the skin in the contact zone but significantly shorten the entire body. The vertebral series and

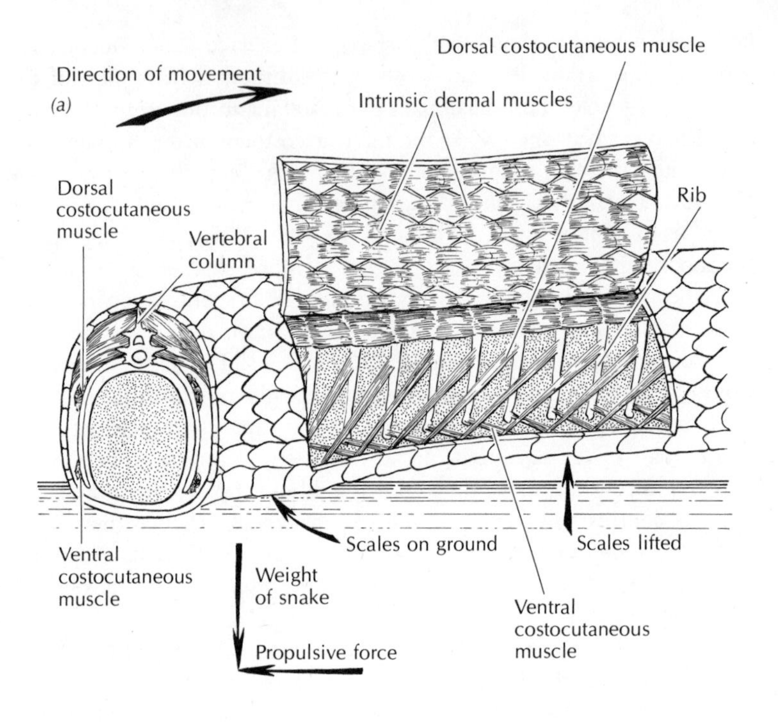

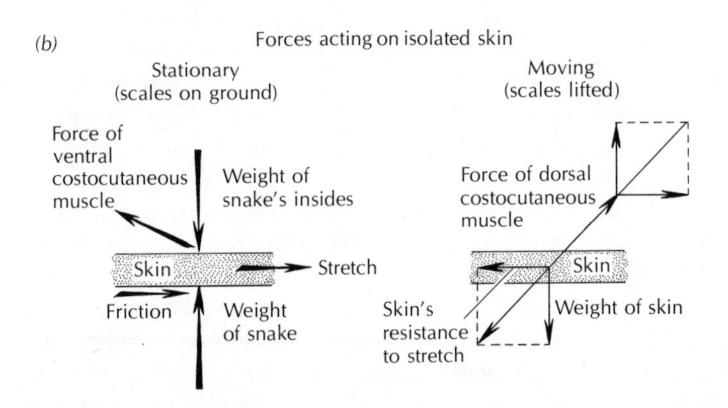

Figure 3-34 Simplified reconstruction (*a*) of the side of a snake in rectilinear motion showing the ribs, the costocutaneous muscles, and the forces transmitted to the ground. In (*b*) the snake has been subdivided, and two portions of ventral skin are each treated as a free body. The vectors here are not to scale but indicate the directions of the imposed forces. They derive from a knowledge of the external forces and the instantaneous positions of the contracting muscles. Given these angles one can carry the analysis one step further and determine the forces supplied by individual components.

surrounding muscles curve, locally thickening the trunk and achieving its contact with the wall of the subterranean tunnel. In doing this they look superficially like a worm that thickens and thins its whole trunk in progression. As in a worm, the entire mass of the caecilian starts and stops during locomotion and undergoes oscillating accelerational forces. The caecilians lack the long lateral rib that supports the vertical dimensions of the trunk in reptiles; hence caecilians cannot slide the entire trunk within the skin. Instead their visceral mass is sheathed and apparently supported by a thick layer of body wall muscles, which do not participate in the internal curvature but are pushed outward when the vertebral column curves. Caecilian locomotion may superficially resemble rectilinear motion, but it actually is a modified concertina motion. Their structural plan does not permit rectilinear movement, but the substitute is less effective as it does not allow the animal to conserve the momentum of its internal mass.

It is particularly helpful that we now have an observational and theoretical hypothesis explaining muscle action in rectilinear movement. Rectilinear locomotion, after all, not only involves such structural changes as a relatively mobile or loose skin and development of costocutaneous muscles, but also demands a unique basis of motor control. Rectilinear is the only method of snake locomotion in which the muscles of the two sides must contract in synchrony rather than alternate. This is particularly interesting because the vertebral column may still be bent, thus maintaining the capacity to undulate. Cases of parallel rather than alternating contractions of muscles are relatively rare in vertebrate locomotion although they do occur, for instance, in some flying birds, galloping reptiles and mammals, swimming whales, and jumping frogs. Rectilinear movement might well allow experiments to ascertain how parallel contractions are established in the central nervous system.

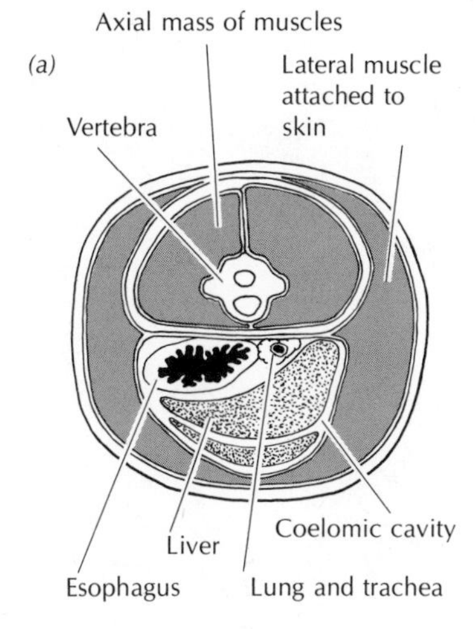

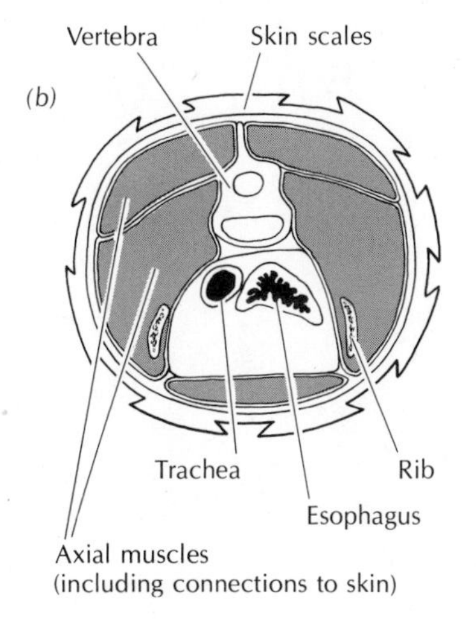

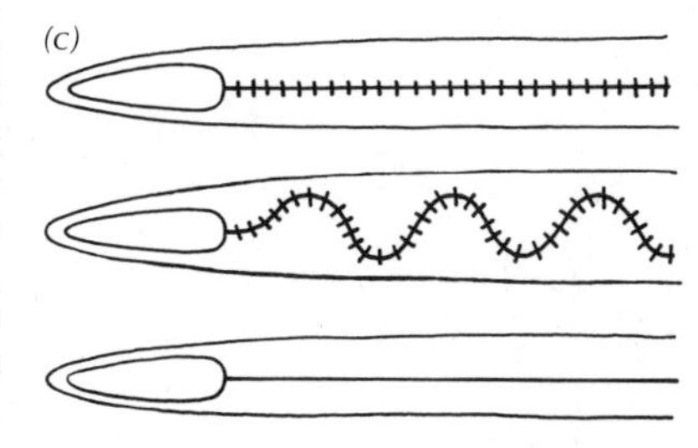

Figure 3-35 These cross sections of a caecilian (*a*) and a uropeltid snake (*b*), taken at places in the anterior part of the body, indicate the fundamental difference in which support and the curvature for concertina movement (*c*) are produced in these two groups. The caecilian's short ribs are often fused, and its visceral cavity is supported by sheets of lateral muscles. Consequently, the turgid body becomes flabby as soon as a caecilian is anesthetized. In uropeltids, the ribs extend around the visceral cavity so that these become an integral part of the axial mass. In caecilians, the curves that widen the trunk and allow it to be pushed against the sides of the tunnel are produced by the vertebrae and by the axial musculature immediately surrounding them; in uropeltid snakes, they are formed by the vertebrae plus the elongate ribs, the muscles surrounding these, and the viscera they enclose. The spaces around these areas clearly define the zones within which motion proceeds. In caecilians, only the axial mass, but not the body cavity, bends; in uropeltids, the visceral mass bends with the axis.

Phylogeny and Problems

There can be no question but that lateral undulation seems to be phylogenetically the oldest method of progression among vertebrates. It was clearly carried into the tetrapod series and appears, sometimes in variant patterns, even in limbed forms—many of which undulate with folded limbs (Daan and Belterman, 1968; Suchanov, 1968). The very efficient method of lateral undulation hence had wide applicability and remained the major pattern up to the evolution of mammals; only in mammals,[4] in which the vertebral column tends to be bent vertically rather than laterally, does it seem to have been replaced (Hildebrand, 1959; Slijper, 1946).

In most cases in which lateral undulation is disadvantageous to a limbless tetrapod, the problem has apparently been solved by a shift to the use of static friction, still retaining the old pattern of progression by bending of the body. Control of the curves and their travel at irregular spacings appear to represent major adaptations for the use of static friction. The various escape saltations provide a good example of the origin of such a system; the shift to utilization of static friction here involves almost no restructuring of the morphology and minimal modification of the behavior. It may well have involved increased metabolic cost as the price of traversing otherwise unsuitable surfaces.

The control sequence of the basic undulatory pattern itself leads to further perfection. This sequence apparently proved to be remarkably adaptable, not only to travel on the ground but to travel above the ground and in water. The adaptability accounts for the limitation of locomotor modifications, which occur mainly in ecologically stressed forms, in species that occupy habitats that could be invaded only by forms that possess a modified locomotor pattern. This may explain the adaptations seen in species of the open ocean, some deserts, and perhaps treetops. One may then suggest that the possession of an effective and also of a plastic locomotor pattern is coevolved with the other traits responsible for the remarkable success of snakes. Caecilians and amphisbaenians, in contrast, incorporate high specialization for only one situation. Perhaps they lacked the developmental flexibility to radiate into the surface environments. The earlier emergence of snakes may also have preoccupied those niches suitable for limbless species.

I hope that this analysis of limbless locomotion has also shown some of the merits of limiting a question, in this case by the study of interfaces. Rather than analyzing forces at all moving points, one studies only those forces and movements acting on the animal's outside. By successive

[4]There are a few exceptions to this statement. Perhaps the most obvious occurs in cryptodiran turtles which withdraw the head into the shell by a vertical curve of the neck. The several kinds of parallel contraction of muscles given above also pertain to this issue.

Table 3-1 Major Locomotor Methods of Limbless Vertebrates

Vertebrate	Locomotor Method: Lateral Undulation	Concertina	Sidewinding	Saltation	Rectilinear
Elongate fishes*	Common	Rare	–	Rare	–
Caecilians	Common	Rare; special variant used	–	–	–
Limbless lizards	Common	Rare?	–	Rare	–
Amphisbaenia	Common; also special variant	Common	–	Special variant	Common
Snakes	Common	Common	Common in some species; environmental influence important	Isolated cases	Common in boids, uropeltids, viperids, some colubrids

*These may also use the dorsal fin and even the paired fins for propulsion.

approximations one next moves closer and closer to the action patterns of internal structures. The story of rectilinear locomotion indicates the nature of the further subdivisions that become possible once the forces imposed upon the first interface have been determined. In general one can say that observation of living animals leads to simple tests to characterize the forces they apply to diverse environments. Film analysis provides a basis for calculating the velocity of parts and components of the whole animal. A combination of operative manipulation and cinefluoroscopy then leads to a preliminary analysis of the forces generated by the most exterior muscles.

Several of the situations discussed in this chapter are, of course, ripe for an extended attack on a variety of levels. In sidewinding we now have records based upon multiple species responding to roughly similar conditions. It may be possible, by numerical analysis, to establish responses of structural components to particular subsets of behavioral conditions (cf. Oxnard, 1973). Where discrete muscle groupings have been identified, it may be possible to observe the actual motor sequences by electromyography. Ultimately one may ask how the contraction sequence of these muscles is programmed, what environmental parameters are monitored by the snake, and how the recognition of certain conditions results in activity at the motor level.

Either through electromyography or respirometry we may finally approach the question of energetics, namely unit energy costs for travel under different speeds and conditions. We now have energetic data for freely moving mammals (Schmidt-Nielsen, 1972; Taylor and Rowntree, 1973) and the first report for small snakes (Chodrow and Taylor, 1973); many of the present hypotheses regarding the relative advantages of the several kinds of limbless locomotion are amenable to testing by similar approaches. Once we have such information we may begin to answer the second question posed in the introduction to this chapter and reevaluate the various current hypotheses of the origin of limblessness.

REFERENCES

Aleev, Y. G. (1969). Function and gross morphology in fish. [Translated from the Russian.] Israel Program for Scientific Translation, Jerusalem, iv + 268pp.

Alexander, R. McN. (1967). Functional design in fishes. Hutchinson Univ. Library, London, 160pp.

*―――――― (1968). Animal mechanics. Sidgwick and Jackson, London, xii + 346pp.

Badoux, D. M. (1964). Friction between feet and ground. Nature, 202(4929): 266–267.

Bernstein, N. (1967). The co-ordination and regulation of movements. Pergamon Press, Oxford, xii + 196pp.

Bishop, S. C. (1941). The salamanders of New York. N.Y. State Mus. Bull. (324):1–365.

Bogert, C. M. (1947). Rectilinear locomotion in snakes. Copeia, 1947(4): 253–254.

―――――― (1964). Snakes of the genera *Diaphorolepis* and *Synophis* and the colubrid sub-family Xenoderminae. Senck. Biol., 45(335):509–531.

Brain, C. K. (1960). Observations on the locomotion of the Southwest African adder, *Bitis peringueyi* (Boulenger), with speculations on the origin of sidewinding. Ann. Transvaal Mus., 24:19–24.

Breder, C. M., Jr. (1926). The locomotion of fishes. Zoologica, 4(5):159–297.

Chodrow, R. E., and C. R. Taylor (1973). Energetic cost of limbless locomotion in snakes. Fed. Proc., 32(1):422.

Cowles, R. B. (1956). Sidewinding locomotion in snakes. Copeia, 1956(4): 211–214.

Daan, S., and Th. Belterman (1968). Lateral bending in locomotion of some lower tetrapods. I. Proc. Roy. Netherl. Acad. Sci., C 71(3):245–266.

Dempster, W. T. (1961). Free-body diagrams as an approach to the mechanics of human posture and motion. In Biomechanical studies of the muscular-skeletal system (F. G. Evans, ed.). C. C. Thomas, Springfield, Ill. (5):81–135.

*Evarts, E. V., ed. (1971). Central control of movement. Neurosciences Res. Prog. Bull., 9(1):1–170, i–vii.

*Gans, C. (1962). Terrestrial locomotion without limbs. Amer. Zool., 2(2):167–182.

———— (1973). Locomotion and burrowing in limbless vertebrates. Nature (London), 242(5397):414–415.

————, and H. Mendelssohn (1972). Sidewinding and jumping progression of vipers. In Toxins of animal and plant origin (A. deVries and E. Kochva, eds.). Gordon and Breach, London, 1:17–38.

Gasc, J. -P. (1967). Introduction a l'etude de la musculature axiale des squamates serpentiformes. Mem. Mus. Nat. Hist. Natur. (N. S.), A 48(2):69–125.

Gaymer, R. (1971). New method of locomotion in limbless terrestrial vertebrates. Nature, 234:150–152.

Golani, I. (1969). The golden jackal. Movement Notation Soc., Tel Aviv, 124pp.

Gray, J. (1933). Studies in animal locomotion. I. The movement of fish with special reference to the eel. J. Exper. Biol., 10:88–104.

———— (1946). The mechanism of locomotion in snakes. J. Exper. Biol., 23:101–120.

*———— (1968). Animal locomotion. The World Naturalist. Weidenfeld and Nicolson, London, xi + 479pp.

————, and H. D. Lissmann (1950). The kinetics of locomotion of the grass snake. J. Exper. Biol., 26:354–367.

Heckrote, C. (1967). Relations of body temperature, size, and crawling speed of the common garter snake, *Thamnophis s. sirtalis*. Copeia, 1967(4): 759–763.

Hildebrand, M. (1959). Motions of the running cheetah and horse. J. Mammal., 40(4):481–495.

Hoffstetter, R., and J.-P. Gasc (1969). Vertebrae and ribs of modern reptiles. In Biology of the reptilia (C. Gans, A. d'A. Bellairs, and T. S. Parsons, eds.). Academic Press, London and New York, 1:201–310.

Hudson, R. C. L. (1969). Polyneuronal innervation of the fast muscles of the marine teleost *Cottus scorpius L.* J. Exper. Biol., 50(1):47–67.

Lissmann, H. W. (1950). Rectilinear locomotion in a snake (*Boa occidentalis*). J. Exper. Biol., 26:268–279.

Mertens, R. (1946). Die Warn und Droh-Reaktionen der Reptilien. Abhandl. Senckenberg. Naturf. Ges., 471:1–108.

———— (1970). Reptilienauge und Umwelt. Natur und Museum, 100(10):435–446.

Mosauer, W. (1932). On the locomotion of snakes. Science, 76:583–585.

———— (1935). The myology of the trunk region of snakes and its significance for ophidian taxonomy and phylogeny. Univ. Calif. Los Angeles, Publ. Biol. Sci., 1:81–120.

————, and K. Wallis (1928). Beiträge zur Kenntnis der Reptilienfauna von Tunisien. Zool. Anz., 97:195–207.

*Oliver, J. A. (undated). The locomotion of snakes. [Motion picture.] New York Zoological Society.

*Oxnard, C. E. (1973). Form and pattern in human evolution. Univ. Chicago Press. Chicago, Ill., ix + 256pp.

Schaffer, J. (1903). Über die Sperrvorrichtung an den Zehen der Vögel. Ein Beitrag zur Mechanik des Vogelfusses und zur Kenntnis der Bindesubstanz. Z. Wiss. Zool., 73:377–428.

Schmidt-Nielsen, K. (1972). Locomotion: Energy cost of swimming, flying, and running. Science, 177:222–228.

Slijper, E. J. (1946). Comparative biologic-anatomical investigations on the vertebral column and spinal musculature of mammals. Verhand. Kon. Nederl. Akad. Wetensch., Naturk., (2)42(5):1–128.

*Suchanov, V. B. (1968). General system of symmetrical locomotion of terrestrial vertebrates and some features of the movement of lower tetrapods. [In Russian.] "Science," Leningrad, 227pp.

Taylor, C. R., and V. J. Rowntree (1973). Running on two or four legs: Which consumes more energy? Science, 179(4069):186–187.

Taylor, G. (1952). Analysis of the swimming of long and narrow animals. Proc. Roy. Soc., 214A:158–183.

*Thompson, D'A. W. (1951). On growth and form, 2nd ed. Cambridge Univ. Press, Cambridge, 2 vols., 1–464 + 465–1116pp.

*Tricker, R. A. R., and B. J. K. Tricker (1967). The science of movement. American Elsevier Publishing Co., New York, xiv + 284pp.

Underwood, G. (1970). The eye. In Biology of the reptilia (C. Gans and T. S. Parsons, eds.). Academic Press, London and New York, 2:1–97.

Van der Vaart, H. R. (1958). Some remarks on the application of engineering science to biology. Arch. Neerl. Zool., 13(Suppl. 1):146–166.

ANALYSIS BY COMPARISON: BURROWING IN AMPHISBAENIANS 4

The Subterranean Habitat

Many reptiles spend much of their life under cover. They are shelterers (Allee et al., 1949:465), a tendency emphasized because they are also crepuscular (active during twilight periods). They hide in holes, tunnels, crevices, and under rocks and logs as adults; such sites also shelter developing young and eggs. The microclimate of a tunnel is buffered by the surrounding soil. The tunnel permits escape from incident solar radiation and the heat sink of the night sky. Cavities in soil often have a higher humidity than the surface; hence water loss from a tunnel's inhabitant is retarded. Air movements are minimal and slow; the scrubbing action of winds is avoided. Tunnels permit escape from the attention of many predators. Fires may pass overhead, but their heat penetrates only a few centimeters.

Reptiles have utilized diverse strategies to invade these subterranean niches (aspects of the environment normally occupied by a particular form). The first reptiles clearly were small; we have evidence that they entered crevices, because many of those preserved as fossils have been found trapped in hollow trees (Carroll, 1970). Among Recent forms one can easily observe an extensive array from crepuscular forms that shelter in leaf litter or under loose bark all the way to those fossorial forms that occupy complex tunnels of their own construction. The great diversity of sheltering forms suggests that a sheltered life is important to many reptiles; the great diversity of morphologically or functionally limbless forms observed by us suggests that its importance is no recent development. For snakes we have evidence of drastic restructuring of eye, ear, and brain. Apparently vision was nearly lost during a past fossorial stage (cf. Bellairs, 1972, for a recent review), and the degenerate eye (and its neural connections—Northcutt and Butler, 1974) became restructured in those snakes that returned to the surface and were again exposed to selection for good vision.

Although most snakes live on the surface and are at most facultatively sheltering, some other groups of reptiles are completely subterranean. Such an assembly is the Amphisbaenia (Box 4-1), an order of squamate, or scaly, reptiles highly modified for a true burrowing existence. These presently fossorial forms provide an excellent assemblage in which to

Box 4-1 Amphisbaenia: Definition and Phylogeny

The language of each country contains vernacular names for its common animals, particularly when these are easily recognized or of economic or folkloric importance. When travelers of the past encountered strange animals, they tended to name these in terms of the familiar. English explorers wrote about horned toads and sea cows, even though neither had any particular affinity to the more familiar animals after which they were named. The further an animal is from the public's normal experience, the more difficult the explanation of a new name by a zoologist. Thus the inappropriate term "worm lizards" for the Amphisbaenia.

The simple diagram suggests the relationships among the several amphisbaenians mentioned in the text. All 130 species of the Amphisbaenia are elongate burrowers. Most of them lack any trace of external limbs and reduced pectoral and pelvic girdles have evolved to the point where it is difficult to homologize these remnants with the elements of limbed forms. Amphisbaenians differ from other reptiles by such characteristics as having the right, rather than the left, lung reduced in size or lost and their much

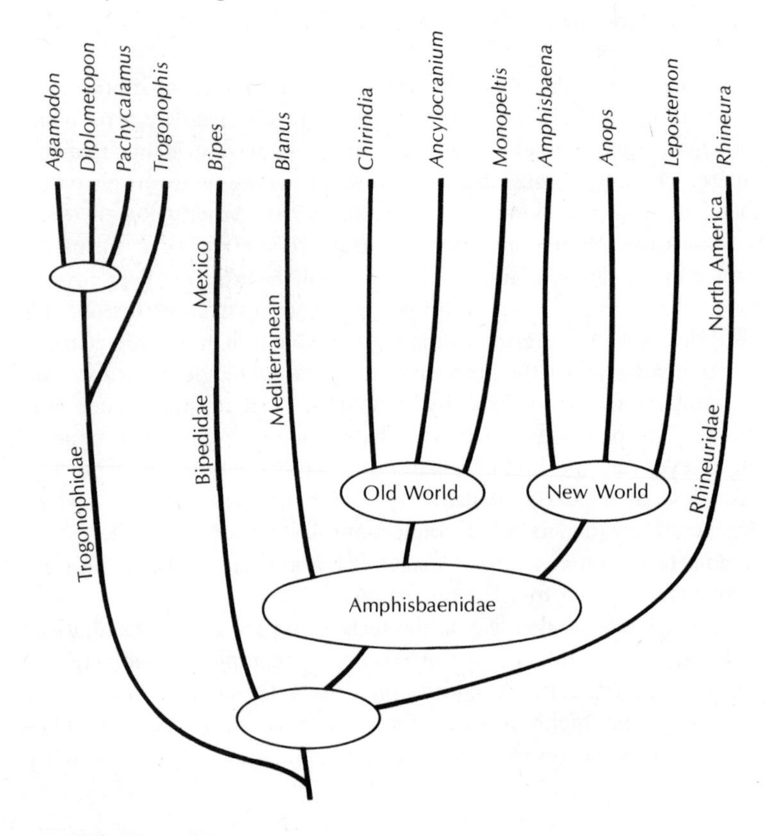

enlarged and medial premaxillary tooth. They have a heavily ossified and extremely solid skull, a marked craniofacial angle (Figure 4-27), and a uniquely enlarged stapes and other modifications of the middle ear (Figure 4-33). Other characteristics almost unique among the Amphisbaenia are a very heavy premaxilla with a prominent facial process, a relatively low number of very large teeth, and a braincase completely surrounded by frontal bones, which in snakes (Figure 2-7) form only the case's dorsal roof. Although amphisbaenians thus differ from snakes and lizards, they also show affinities with one or the other of these groups in the arrangement of the endocrine glands, the architecture of the brain, and the path of some of the blood vessels. There is no indication that they share an unusual number of characteristics with any particular family of lizards or snakes.

The pattern of diffuse similarities in a variety of organs suggests that these three groups have had a long independent history, and the fossil record seems to confirm this. The earliest known Amphisbaenia date back to the Eocene (about 50 million years ago). Yet one has no trouble recognizing that these specimens were true amphisbaenians. This suggests that the three groups separated much earlier, well before the Cretaceous (135 million years ago).

Unfortunately, we lack fossils intermediate between the Amphisbaenia and other groups and can only speculate what their ancestors looked like. Although the earliest fossils seem to be among the largest known amphisbaenians, they were probably less than 3 cm in diameter—and the ancestral forms were unlikely to have been larger. The absence of ancestral forms from the fossil record seems to reflect their small size and perhaps their burrowing habits. After all, a burrower would be unlikely to be trapped and fossilized in a crevice or other cavity; if it did fall in, it could presumably dig its way out! If the transition from protoamphisbaenians to recognizable ones proceeded quickly, perhaps within a million years or so, there would be a very low chance of finding any fossils; in fact, we have fewer than fifty samples to document the history of the Amphisbaenia for the forty or fifty million years since the Eocene.

All we may state with certainty is that the Amphisbaenia belong among the class Reptilia because they share the reptilian characteristics of ectothermy (page 121) and an amniotic or layered egg. Within the Reptilia they should be placed in the superorder Squamata, which also includes snakes and lizards. They show the characteristically scaly skin but also have the double male generative organ or hemipenis, the transverse cloacal opening, and the true egg tooth on the premaxilla. (Hatchling squamates use the forward-projecting egg tooth at birth to cut their way through the tough eggshell. The egg tooth is shed shortly thereafter. In the Amphisbaenia it is replaced by the large median premaxillary tooth.)

study the effect of a major selective force—that of underground existence. Some such studies are discussed here.

What are the corollaries of selection for an underground existence? Are there alternate solutions to each demand of this way of life? Are the solutions to diverse demands congruent? If not, how do they conflict? Why are certain forms more common than others? What limits the range of a particular species, and why do species and major groups replace each other geographically? Similar questions are often discussed on the basis of physiological tolerance of climatic factors for adults or of successful breeding. The present analysis will show situations in which explanations based on biomechanical schemes have significant resolving power for such questions. Rather than deriving principles from the study of a particular species (Chapter 2) or even "typical representatives" of different groups (Chapters 3 and 5), we will try to define questions and derive answers from a simultaneous analysis of many members of a major adaptive radiation. In other words, we will study Amphisbaenia through comparisons of various members of the order.

The various reptiles hide or dig among a wide spectrum of media. The naturally rough bark of certain trees may serve as a simple shelter, as will the forks of twigs and branches, and a diversity of leaf litters. As litter becomes compacted, the price of shelter becomes the ability to displace material and generate a new crevice or to locate an existing one. Compacted litters graduate into more or less organic soils and various clays, packed sands, and gravels, in order of decreasing organic content. These latter media can only be occupied by fossorial (e.g., burrowing) species.

Diverse materials have different properties and hence different advantages to the organism. These advantages must be balanced, in each case, against the cost of living in a particular substratum. The result will always be a compromise; for instance, individuals of the armored South African zonure lizard (*Cordylus giganteus*) occupy short vertical tunnels leading to a horizontal chamber. In some regions the soil is so hard that additional excavations are impossible during the dry season (J. Visser, personal communication), but the soil's hardness also protects against predators on the surface. The lizards' population size then could be limited by the number of tunnels dug in such areas during the irregular rainy periods. We know little about the long-term stability or perturbations of the system. How often is the soil soft enough to permit these lizards to form new burrows? What is the attrition rate of the old burrows? Consideration of the mechanical properties of the soils and of the excavating method combined with demographic analysis may furnish interesting insights.

Opacity, thermal properties, and moisture availability are three aspects of protective shelters that will obviously affect the kinds of organisms that can inhabit them. As important is a shelter's cost to the occupant in terms of the energy required to construct or to occupy it. Let us consider each aspect in turn.

Opacity probably poses the fewest problems, as most solids have this property. The shielding effect of an opaque branch, leaf, or burrow leads to a smaller portion of the animal being exposed; much of the animal's characteristic outline will routinely be shielded. Some predators hide beneath water and sand, media which are effectively opaque and which are useful because emergence offers no problems. Crocodiles lurk with only eyes and nostrils exposed; sand vipers such as *Eristicophis* hide with the body coiled and all but the head's center covered by sand. Both are inconspicuous yet can emerge rapidly to grasp their prey.

The thermal properties of shelters are of particular importance to reptiles, as the reptilian nervous system appears to contain fairly precise temperature indicators. However, although reptiles seem to monitor and to maintain through behavior their physiologically most compatible level, they cannot establish that temperature by metabolic means as do most "warm-blooded" endotherms (the birds and mammals, which mainly use metabolic heat). Reptiles illustrate ectothermy (the use of outside sources of heat). They cycle between heat source and heat sink, picking up heat (from the sun) for some time and then moving to a cooler spot where they can give it up. They thus regulate their internal temperature by behavioral means. Their rate of temperature change will increase, of course, with the difference between the temperature of their body and the environment. Only if they find a site that has the most compatible temperature will they be able to maintain activity without shuttling. For a given set of limits, the cycle time will be established by the animal's thermal capacity. Giant animals, such as elephants or Galapagos tortoises, have limited sources of shade and must spend much of the day in the sun, gradually gaining heat. Since the insolated surface (that receiving the sun's rays) is small relative to their mass, their core temperature rises slowly. In contrast, small animals, such as horned lizards and ground squirrels, operate on a shorter temperature cycle. They expose themselves to incident radiation when searching for food and then use the cooler soil beneath the surface to discharge the heat thus gained before they are able to forage again. A resting animal should ideally select a shelter of a temperature level close to that toward which the animal is controlling. Many reptiles (and a few mammals) simplify their control problem by dropping their body temperature while resting; they warm themselves when they have to become active.

The many different temperatures observed in the environment result from the different thermal properties of its various components and the diverse ways heat is transferred within it. Energy comes from solar radiation, and the amount absorbed depends on the reflectance of the various exposed surfaces. Beyond this, the magnitude of the amount of heat transferred is proportional to the difference between the fourth power of the (absolute) temperatures. Consequently, even a thin leaf will cut the incident radiation, as its underside radiates at a much lower rate than the sun. The situation is complicated by three factors: (1) conduction, or the rate at which thermal energy is transmitted through solids; (2) convection, or the rate at which thermal energy is moved about by the flow of fluids; and (3) evaporation, or the absorption of latent heat in order to transform liquids into gases. Although the rate of heat transfer by radiation has been noted to be a fourth-power function, that by conduction and convection will be close to a first-power function of the absolute temperature differences.[1] The less the thermal conductivity, the greater the insulating properties of the material, or rather the steeper the thermal gradient across the material. Since interfaces between different materials tend to interfere with heat flow, homogenous substances tend to have the greatest thermal conductance.

[1]By the Stefan-Boltzmann Law the quantity of heat flowing (or thermal current), $H_r = \varepsilon\sigma[T_s^4 - T_r^4]$, where ε is the emissivity of the source at an absolute temperature T_s, σ is Stefan's constant, and T_r the temperature of the receiving surface. In contrast the heat flow during conduction $Hc = U(T_s - T_r)$, where U is the thermal conductivity of the material(s). Although the equation for heat flow during convection is much more complex, it again reflects the absolute difference between the temperatures of source and receiver rather than the difference between the fourth powers of their temperatures.

The temperature beneath a rock (which has relatively high thermal conductance) will hence differ less from that of the insolated surface than will the temperature beneath a dry log (the layered structure of which results in a lower coefficient).

The flow of heat is a gradual process. Diurnal cycles of insolation and nightly reradiation are detectable to a limited depth only, and only attenuated variation is noted some 20 to 30 cm beneath the surface of the earth (Allee et al., 1949:219). Seasonal changes penetrate deeper, but even their effects are limited; witness arctic regions with permafrost, a deep layer that remains frozen even in midsummer. The optimum temperatures of tunnel dwellers may be distinct from those of forms living closer to the surface, and some reptiles may change their temperatures by moving back and forth between tunnels adjacent to the surface and deeper ones (Cowles and Bogert, 1944). Such vertical diurnal migrations of amphisbaenians are suggested by dawn-dusk activity patterns of this type recorded in the laboratory (Gans and Bonin, 1963) and were observed for the desert-dwelling amphisbaenian *Agamodon aguliceps* in Somalia (Gans, 1965).

Moisture retention by soils is of extreme importance to a burrower, since it will affect water loss and the access to free water for drinking. Even in deserts with minimal rainfall, the air in crevices remains close to saturation.[2] The absolute percentage of saturation may be affected by the chemistry of the soils; certain salts are hygroscopic (moisture-absorbing) and will depress the water content of the surrounding air. The capillary properties of the soils also influence the life of the burrower in less obvious ways, as they affect the path and rate of downward passage of liquid (rain or flood water). Capillary forces are often sufficient to trap air pockets in small tunnels when the surface is completely flooded. Rain percolating downward will consequently allow air-breathing animals to stay in their tunnels, whereas rising ground water will drive them to the surface.

When amphisbaenians form a burrow in dry sand, the tunnel will collapse as soon as they have passed. When they burrow through wet sand, the tunnels stay open and may continue to do so even after the sand has dried. This resistance to collapse is a complex phenomenon involving the surface tension of the water and its action on the soil's

[2]The term "saturated" implies that the air contains the maximum amount of water vapor (not liquid droplets) physically possible. The absolute amount of liquid water going into and held as vapor increases with the temperature. Thus the nocturnal radiative heat loss of the earth's surface may cool the adjacent air layer to supersaturation. Water droplets will then condense out and moisten the surface. Some reptiles such as the Australian *Moloch* lizard have developed special mechanisms for drinking such dew and manage to survive in very dry areas (Bentley and Blumer, 1962). (Other reptiles manage to survive by obtaining their water as a by-product of metabolizing their food.)

particles. Actually sands and soils always show a mixture of many different particle sizes. This may be expressed as the percentage of particles passing each of several sieves of different sizes (thus 6 percent of the grains are smaller than 0.002 mm, 10.5 percent smaller than 0.05 mm, and 100 percent smaller than 2.0 mm in a typical sand). The particle distribution results both from mixing and because any grinding process will break rocks irregularly, yielding a spectrum of particle sizes (Dallavalle, 1943). Only sedimentation by water and wind action stratify particles by size.

The smaller the particles the greater the surface area per unit of weight of soil and the greater the amount of water it takes to wet this surface. The surface tension of the water will tend to bind the particles together and thus increase the resistance to deformation (or collapse of a tunnel). The size of particles is also important because it affects the packing and, with this, the size of the intergrain spaces that may hold air or moisture. As the soil's moisture contains diverse solutes, it may not only etch the soil particles but also will precipitate out various salts as it dries. Such precipitates can cause particles to stick together and change the mechanical properties of the soils. This also explains the temporary patency of tunnels in formerly wet and now dry sand.

Finally we must remember that the thickness of the soil layer may be critical in determining the size of the inhabitable volume and, indirectly, the number of individuals and perhaps species that may survive in an area. There is clearly a minimum depth of soil, as members of even a single burrowing species must be able to shift their position during the day in order to hide from the changes in heat and moisture levels produced by sun, wind, and rain adjacent to the surface. The lower limit is established by strata that cannot be readily penetrated (rock or clay) as well as by the ground-water level and its fluctuations. If the ground-water level is fairly deep, there will be a greater chance that a thick layer of suitable soil may be able to include more than one vertically spaced subterranean niche. The nature of this zone, its geographic extent, and the temporal constancy of conditions obviously influence the number of species that may occupy it.

Mechanisms of Tunneling

An important fact in determining the kind of shelter to be produced is the energy cost for constructing and maintaining it. In general this cost is some function of the diameter of the tunnel as well as of the penetrability and compressibility of the soil. The latter parameters depend primarily on moisture content, size and kind of particles, and fibrous components (roots and litter), as well as on the packing. Any gardener can document the importance of moisture content; it softens some soils but causes the particles of other soils to cohere in clods. Cars get stuck in dry sand but are supported by wet sands, unless these contain so much water that the particles float as quicksand. Finely particulate clays are denser and more difficult to deform or compress than is humus. Fibrous material may decrease penetrability, as everyone knows who has tried to dig a hole through an extensive root system.

Tunnels may be formed by excavation or by compacting the soil into their sides. (The process of an earthworm eating its way through the soil presumably does not leave much of a tunnel, though such animals do transport soil to the surface as well.) In excavation the soil is scratched or scraped off the end of the tunnel and transported to the surface; here penetrability of the soil at the tunnel's end is of prime importance. Excavation is of use mainly to species that establish fixed tunnel systems or to those that are too large to use the compacting method. It presupposes a permanent connection to the surface and requires that the excavated materials be carried out. Abandoned tunnels may also be used to store excavated material, but not during initial excavation of a tunnel system.

In contrast, the compaction method, which concerns us here, establishes the animal's independence of the surface. Compaction to the sides entails both the penetrability and the compressibility of the medium. Some mammals will use their claws to scrape dirt off the tunnel's end and then force it into the wall with pushing movements of their heads. In contrast, various reptiles, such as the amphisbaenians, have developed ramming heads that can penetrate the soil to form and then widen the tunnel. Movements of these penetrating heads ideally should produce minimal compression and compaction of the soil immediately ahead of the animal. The shape of the head will determine the degree to which this anterior material becomes compacted rather than shifted to the sides, and hence the amount of energy that must be invested in the next unit of progression.

A variety of head shapes occurs among different burrowing reptiles, and even the scale arrangement will vary on heads of roughly similar shape (Figure 4-2). We consequently have to look more closely at the physics of burrowing to understand how it relates to head shape. Assuming that a head is being pushed by a given amount of energy (or equal momentum), it will penetrate for a maximum distance if it has the shape of a pointed cone with a minimal included angle. Yet the strength of such a pointed cone or the absolute load (force) it may carry without failure is a function of the stress level, which is the force per unit cross-

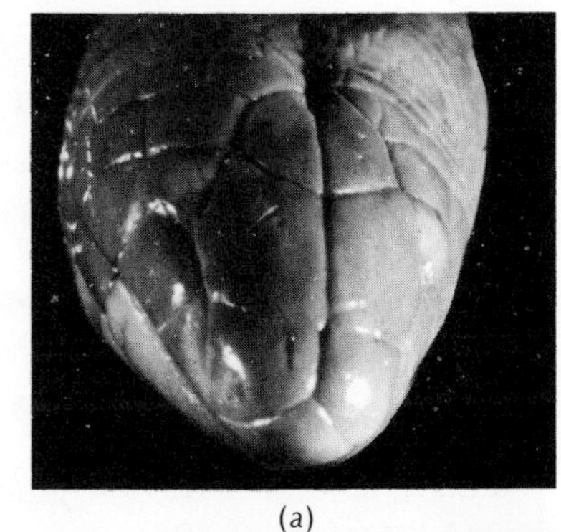

(*a*)

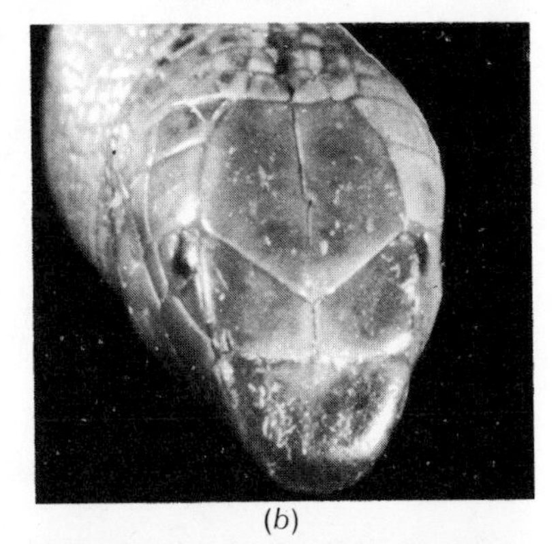

(*b*)

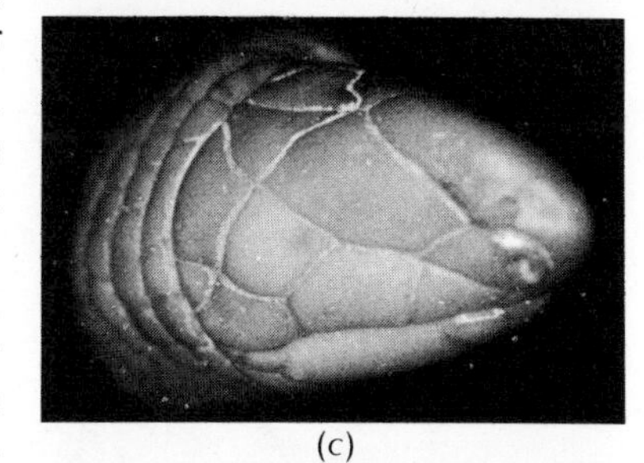

(*c*)

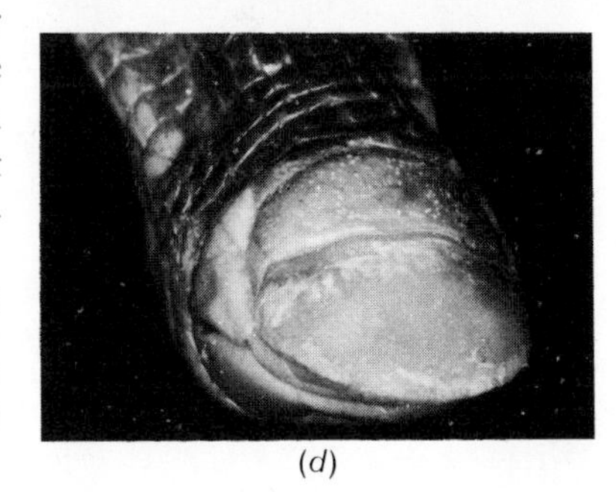

(*d*)

Figure 4-1 The heads of different amphisbaenians have drastically differing appearances. Here we see a round-headed *Amphisbaena darwini* (*a*), an advanced trogonophid, Diplometopon zarudnyi (*b*), a keel-headed *Mesobaena huebneri* (*c*), and a shovel-snouted *Monopeltis anchietae* (*d*). What might be the selective influences that have established such diverse patterns?

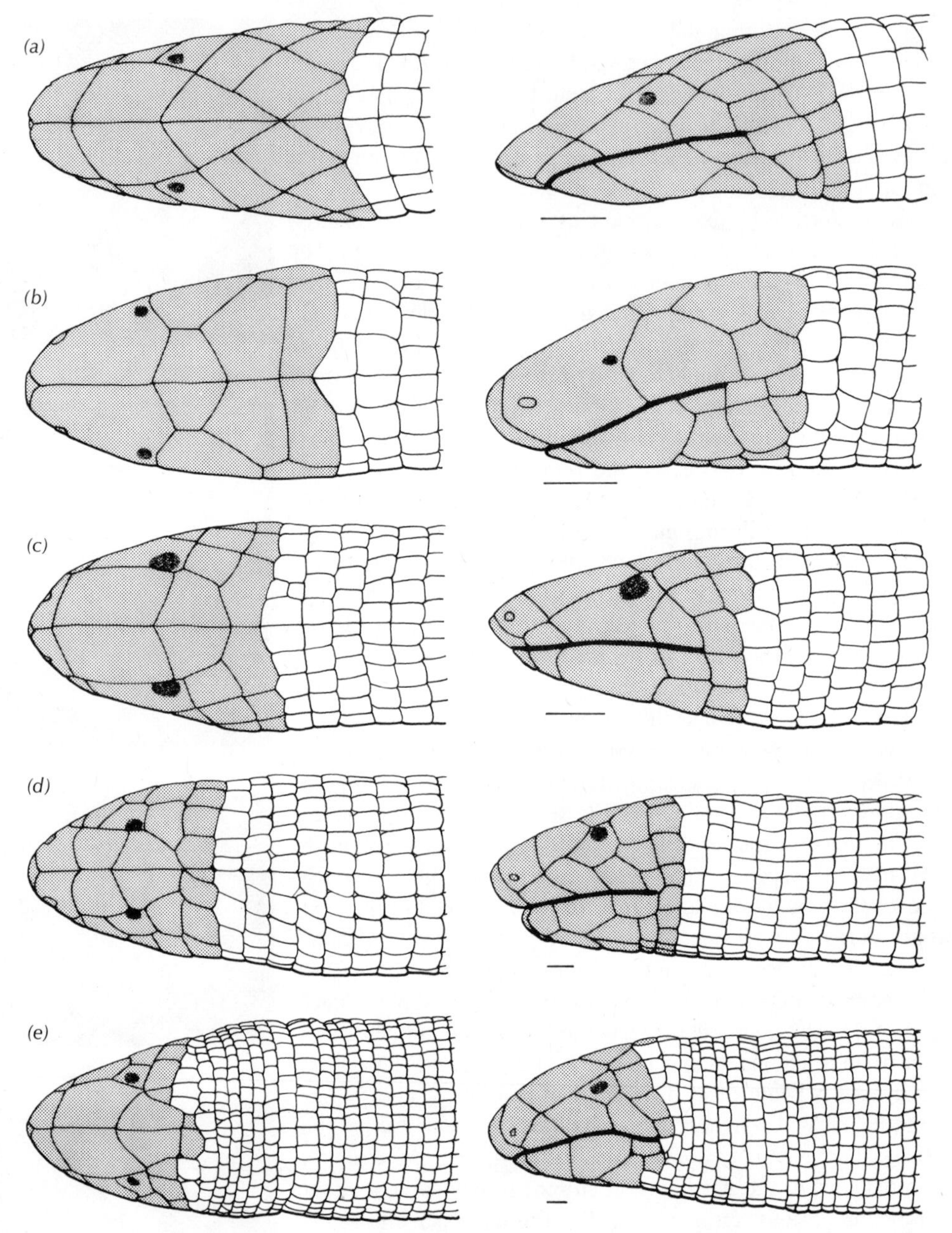

Figure 4-2 Even among the more or less round-headed amphisbaenians one finds drastic differences in segment arrangements, in particular in the size of the zone (stippled) covered with enlarged shields. The relative size of these shields may be seen to shrink as one progresses through the series. The specimens of *Amphisbaena heathi* (a), *Chirindia rondoense* (b), *Amphisbaena cubana* (c), *Amphisbaena vermicularis* (*d*), and *Amphisbaena darwini trachura* (e) are drawn to different scales (each line equals 1 mm), suggesting that the shift in segment proportions involves more than absolute size. These species represent a morphological, rather than phylogenetic, sequence.

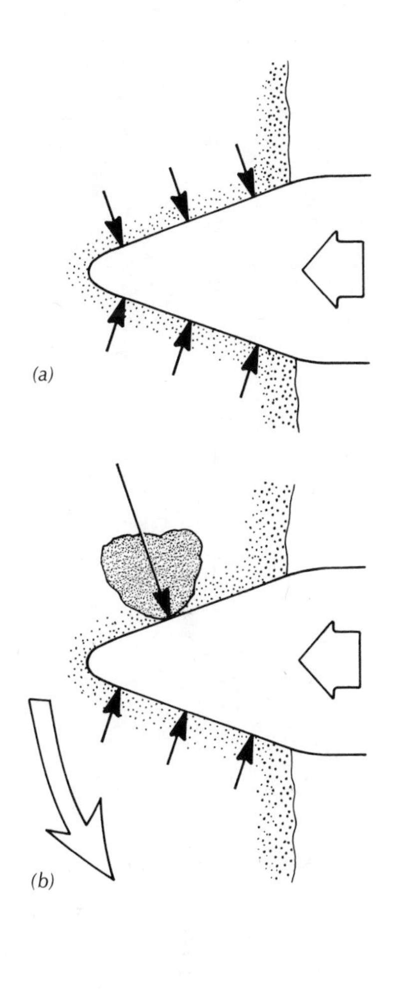

Figure 4-3 When a conical head is driven into a homogeneous substratum, the resistance forces (solid arrows in *a*) act more or less equally and normal to the surface. When the side of the head encounters a rock or similar object (*b*), this concentrates the forces to the contact zone. Depending on the object's site, it may or may not be movable within the soil. Increased and localized resistance on one side may cause a moment to be formed that will force the head in the opposite direction.

sectional area (Frost, 1967). This remains true whether we define failure by the yield point at which the material suffers irreversible deformation or as the ultimate strength at which rupture occurs (Box 2-3).

When the surface of the penetrating cone is in contact with a homogenous material, the stress during penetration will be equal throughout any section of the cone. However, if the cone's side or tip should make contact with a harder substance, such as a small stone that can itself be moved only by compressing a given volume of soil ahead of it, there will be local stress concentration (in this case tending to bend the head). The local stress concentration will approach infinity if the small stone is hit head on by the point of a cone approaching zero cross section. One way to avoid a ruptured snout is to change the cylindrical cone into a simple wedge, which makes for a more rapid increase in area with penetration. Another is a rounded point; and yet another is a covering of the surface with resilient material that resumes its original shape once the deforming forces are no longer present. All of these compromises serve to increase the capacity of the tip to support localized and eccentric loads without undue stress concentration (which exceeds the yield point), and each can be seen in some animal. There is also the option of hardening the cutting tip (or edge) so that it may bear greater stress levels than the tip as a whole. Not only the hardness of the covering but the elasticity of connection with the underlying skull are critical.

The last point suggests the utility of a simple force analysis of such a penetration process. Let us start with the assumption that there will be no friction between head and soil and that the driving force is applied along the axis of the cone. We may then draw Figure 4-5, showing the head of a uropeltid snake in which the forces on the soil are evenly distributed, normal to the penetrating surface. One next resolves these surface forces into a longitudinal component promoting penetration and a set of lateral components promoting widening (Figure 4-5*b*). In order to continue in motion, the driving force will have to equal the

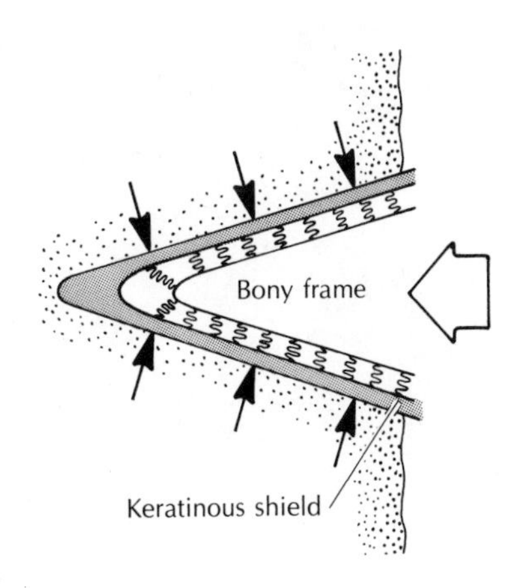

Figure 4-4 A keratinous shield may shape or harden the penetrating edge. The keratin will wear on its outer surface and grow by accretion on its underlying surface. The soft tissues connecting the keratinous sheet to the bony frame not only facilitate growth of the shield (and repair of minor damage), but also assist in shock absorption and shear resistance, as imposed loads will be distributed more widely onto the underlying bony frame.

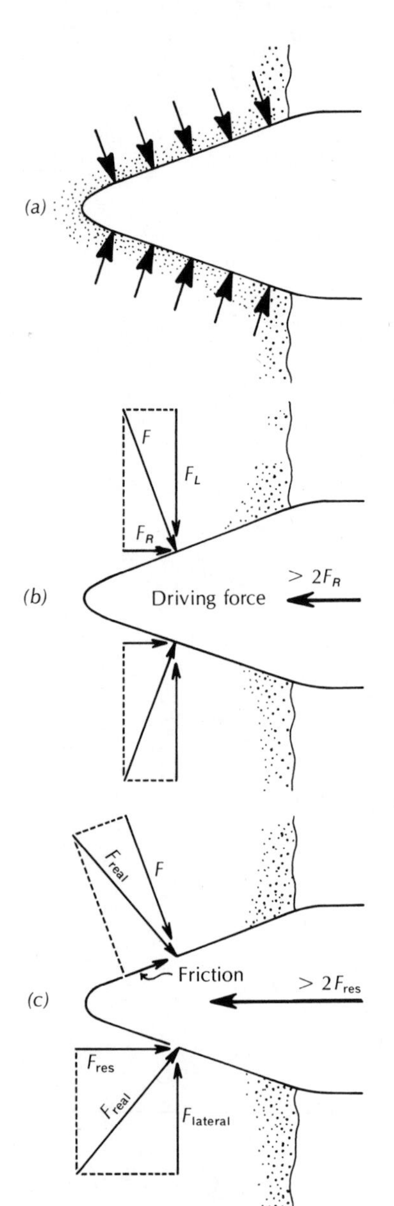

Figure 4-5 These sketches show the reaction-force pattern imposed when a burrower's conical head penetrates a homogeneous soil. The forces are then distributed uniformly and act normal to the shield (*a*). They may be summed into two equal forces *F*, in this case those acting on the left and right sides (*b*). Each summed force may be resolved into a lateral component F_L, which balances the other, and a resistance component F_R. Penetration will take place only if the driving force is greater than $2F_R$. Friction acts parallel to the penetrating surface (c) and opposed to the direction of movement. Since friction increases the resistance component of the real opposed force, F_{real}, penetration (for a given momentum) is less if an animal digs in high-friction rather than low-friction soils. Friction is again disadvantageous.

sum of the penetrating components. The forces at right angles to the direction of motion compact the soil and will balance out to zero as long as the system is symmetrical.

The contact surfaces of the cone-shaped head and the soil pass each other in opposite directions; a nonideal system will show frictional contact along the surface (Figure 4-5*c*). Friction will cause the resultant reaction force to shift so that the resistance to penetration becomes less than that to compaction, leaving less force available for the latter, or requiring more force to penetrate a soil with high than one with low frictional characteristics.

Such considerations permit us to state several limiting conditions in tunnel formation. The diameter of the cone and hence the tunnel diameter will directly affect the force required for penetration. Either the need for penetration or that for compaction will establish the absolute force required to drive the tunnel. Finally, as there are frictional forces parallel to the inclined planes forming the cone, and since surfaces will be exposed to shear forces that will increase with their coefficient of friction, a decrease of surface irregularities will increase the shear's magnitude. This provides a physical basis for the observation that those surfaces of squamates exposed to wear or sliding show fusion or reduction from many small scales to a few large shields (Vanzolini, 1951).

Some Special Cases

Some half-dozen radiations of ectotherms live in the subterranean biotope and may be considered truly fossorial burrowers. They differ from surface dwellers that use the ground merely as a temporary shelter (cf. Allee et al., 1949). Fossorial species tend to construct their own tunnel systems and obtain their maintenance requirements underground, though they may be driven to the surface by a catastrophe such as a flood or ant attack. I will not consider here the two groups of worm snakes (Typhlopidae and Leptotyphlopidae) because these are highly specialized for preying on and living commensally with diverse kinds of ants and termites (Gehlbach et al., 1968). Many of them do only limited burrowing since they occupy the tunnel systems of their prey. Thus they represent a pattern of life history completely distinct from that to be considered here.

One other adaptive pattern deserves special mention—namely, sand swimming. Although sand swimmers are often referred to as burrowers, they are specialized for life in dry sands, a medium of unique characteristics. The surface layers of dry sands may show little adhesion between grains, so that an animal moving through such a medium forms only an instantaneous tunnel and the displaced overburden collapses just as soon as the animal moves on. This keeps these forms from having to use concertina or rectilinear progression as do the inhabitants of straight-sided tunnels (p. 89). The sides of the "tunnel" are at any instant defined by the sides of the animal. Consequently, the medium may be considered fluid, and the animal progresses by lateral undulation involving the customary waves that move posteriorly faster than the animal does anteriorly (p. 81). This kind of swimming occurs most easily in loose, unsettled sands. Such a progression causes minimal compacting; indeed, the overlying zone instead may become looser and well up with the animal's passage, like the ground above the subsurface tunnels of moles.

Sand swimming is utilized by many skinks as well as other lizards and a variety of snakes, and it also seems to involve some special problems. Norris and Kavanaugh (1966) noted that most sand swimmers seemed to be ventrally concave in anterior cross section and also had their external nares directed ventrally. The nostrils thus open into the concavity beneath the head. This zone will be sand-free, since the sand will not flow upward, as would water—a true fluid, no matter what the pressure of the overburden. Just before entry into the nostrils, the air will flow through this sand-free zone of significantly greater cross section than that of the nasal passage. For a given mass flow of air (g/sec)

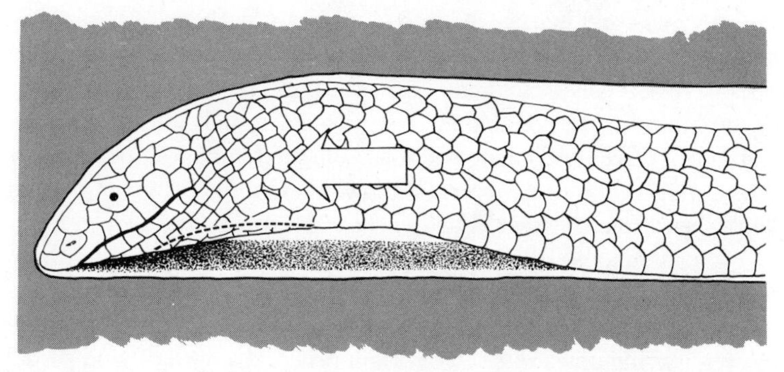

Figure 4-6 The head of small sand swimmers is often dished and ventrally hollow. Such animals, furthermore, push through the soil with the snout slightly below the line of the rest of the body. They thus generate a sand-free space that facilitates breathing. (After Norris and Kavanaugh, 1966)

the cross-sectional area determines the velocity (cm/sec), and the velocity determines the stream's capacity to float particles of sand. The effect is striking as the velocity (cm/sec) varies with the fourth power of the zone's radius; a small change in diameter may then have a major effect. Seemingly, the larger cross-sectional area results in a reduced velocity for entering air, causing fewer grains of sand to be picked up and reducing the need for internal filters and cleaning devices (Stebbins, 1943).

Norris and Kavanaugh (1966) also reported that the snake *Chionactis* showed an 18-hour cycle of immersion in a dry-sand habitat. They attempted to explain this with the suggestion that the pressure of the overburden forced the animal to leave the sand at intervals. However, the general concavity of the snake's ventral surface may also facilitate breathing movements. There does not seem much need for a subterranean air reservoir since this zone of dry sand apparently has enough air diffusing through it so that the animals are not subject to oxygen stress (Pough, 1969). In snakes (Rosenberg, 1973) inhalation requires rib movements; these might conflict with the positioning of the ribs in resisting the overburden. On the other hand, a very elongate animal might need so little lateral rib displacement for breathing that it could operate in loose soils for extended periods. The actual magnitude of overburden-induced load remains to be established, as very few environments consist entirely of the kinds of loose sands used in these experiments. Relief from the load of the overburden does not necessarily require emergence. The root stock of any small plant, a moist area, or a rock (or log) might provide a shelter, easing the load and permitting the animal to extend its stay underground.

The use of rocks to relieve overburden deserves a brief digression as it shows how many factors have to be considered in order to understand seemingly simple phenomena. The aggregation of numerous animals beneath rocks and logs is well known to every collector. One is almost tempted to write a distinct essay on the peculiar properties of the undersides of rocks. Any small body such as a rock, root, or log represents a discontinuity in the soil and, as a consequence, will intersect the path and force a directional change of otherwise randomly positioned tunnels. The critical thing is that the rock allows these tunnels to stay open and keeps them from collapse. Whether the soil is wet or dry, clay or humus, its components will have significant compressive strength to withstand pressure of the overburden transmitted by the rock. Interaction between the soil's particles will be sufficient to keep them from slipping laterally or even moving upward. Even the compressive strength of sand particles is perhaps a thousand times greater than the weight of a large rock; thus a rock will remain supported even if much of its undersurface has been excavated.

Rocks also protect underlying tunnels from wind erosion and the impact of raindrops, as well as minor local flooding of the tunnel systems. If the rock has projecting edges, these form excellent protected entries into tunnels. Many tunnels intersect, and one often finds large tunnel chambers under logs and rocks. These chambers provide an assembly area for subterranean creatures and are rivaled in size only by those constructed by ants and termites. Here sheltering and truly fossorial species meet in a microenvironment determined by the roofing object's size and position. If this object is a rock, deeply recessed and located at a shaded site, there will be little transmission of heat to the underlying zone. If it is a large log in the process of decay, the underlying chamber will lie at a special interface; the log will maintain humidity and may leach nutrients into the surrounding zone. During the day the temperature of the chamber will be buffered by evaporation, and the heat of decay may keep it raised at night. Diurnal vertical migration of burrowers may then end here without exposing the animals to the predators of the surface. The relative permanence of such chambers makes them preferred egg depositories for various invertebrates, and the locally raised humidity may promote germination of seeds incidentally introduced and attract the roots of adjacent seedlings. This combination of circumstances creates a chamber offering access to the surface and to the soil, and offering many life forms a different microclimate than that beneath an equal depth of soil. Consequently, it promotes the aggregation of small organisms and in turn attracts their predators.

The complexity of seemingly simple situations is further emphasized by noting that rocks in tropical rainforests generally lack such aggregations. The kinds of species encountered under rocks generally seem much more limited in tropical than in temperate zones, presumably reflecting the extreme predation pressure in the tropics. I have also observed that relatively few truly fossorial reptiles are commonly encountered under large rocks, even in subtropical zones; collecting records suggest that they may visit such sites, but do not rest there.

General Amphisbaenian Specializations

Setting aside sand swimmers and worm snakes leaves two main fossorial groups of amphibians and reptiles, the caecilians and the amphisbaenians.[3] Caecilians include some 160 species of limbless amphibians, the taxonomy and anatomy of which have received far more attention than has their biology (cf. Tannes, 1971). To the best of our present knowledge, they are an order of elongate, limbless predators that occupy fairly moist (sometimes aquatic) habitats, although they can survive dry

[3]There is also a small family of rough-tailed snakes, Uropeltidae. They will be referred to later when their curious distribution is contrasted with that of amphisbaenians.

Figure 4-7 This caecilian, *Siphonops annulatus,* from South America shows the characteristically smooth, moist skin, the mucous glands of which facilitate low-friction passage through the soils in which it tunnels.

periods in tunnel chambers. They feed on various kinds of worms and arthropods, but the mechanics of their food-intake mechanism remain to be described. As far as is known, they progress mainly by lateral undulation and a special concertina motion (Gans, 1973b; Gaymer, 1971); their mucus-covered, low-friction surface makes lateral undulation an effective method of locomotion. They burrow by driving variously rounded and wedge-shaped heads into the soil; some movements are possible at the neck, and hence the wedge can be shifted to widen the tunnel and also to reposition the head to avoid pebbles or similar objects. In spite of vulnerability to dessication, this curious group is widely distributed, ranging across the moist tropics of Africa on to the Seychelles, from India and Sri Lanka into China and Indonesia, and from Middle America south to Argentina (Figure 4-8).

The amphisbaenians are their reptilian counterparts, but these range far out of the humid tropics into some relatively xeric environments. They dwell in permanent tunnels of their own construction but regularly change the tunnel pattern. Amphisbaenians[4] have undergone major modifications in their locomotor and burrowing systems. The order

[4]There is a nomenclatural distinction that might be mentioned here. All of these animals belong to the *order* Amphisbaenia and would then be called amphisbaenians. However, most of the species here discussed also belong to the *family* Amphisbaenidae and one must then call these amphisbaenids, as opposed, for instance, to trogonophids, members of the family Trogonophidae. Finally there is the *genus Amphisbaena*.

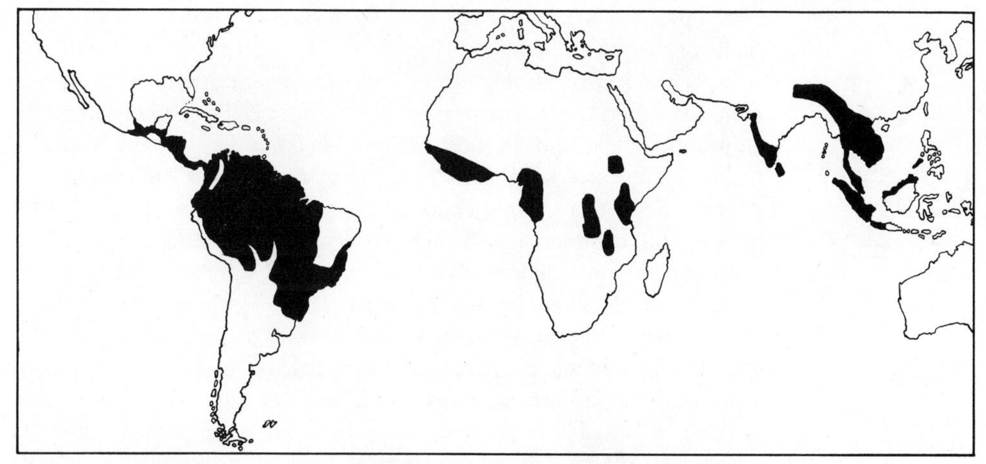

Figure 4-8 Sketch map to show the distribution of caecilians (compare with Figure 4-9).

ranges from Patagonia to Panama in the Americas, with some further species occurring on the Antilles, in Mexico, and in Florida. In the Old World, they range from the northern Cape Province of South Africa, north to Somalia on the east and Senegal on the west coast, though only to Zaire in central Africa; other forms occur on Socotra, across the Arabian peninsula to Iran, from Turkey to Lebanon, and from Morocco to Portugal and Spain. The various amphisbaenians have distinct head shapes and squamation, but differ even more in the shape and construction of their skulls. The overall modification is so extreme in some species

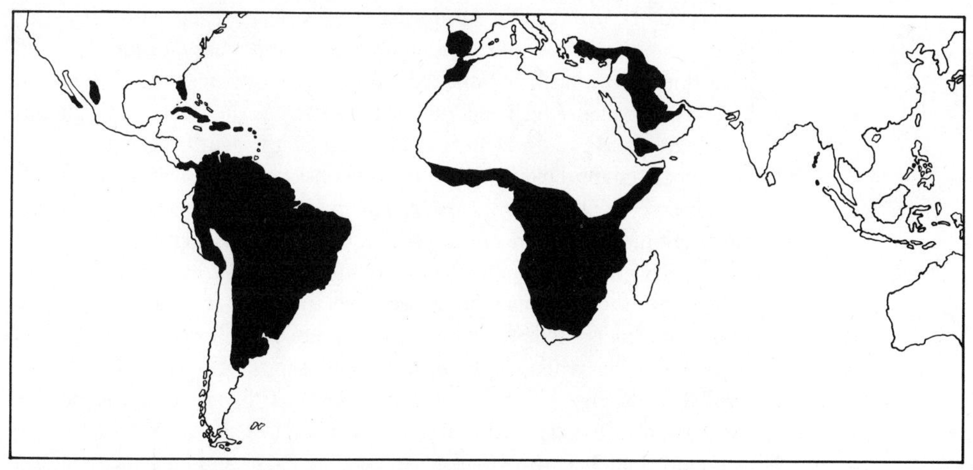

Figure 4-9 Sketch map to show the distribution of amphisbaenians. Note that they extend much farther into the temperate zone than do the caecilians. They have an extensive distribution across tropical Africa and in South America, but curiously limited ranges elsewhere.

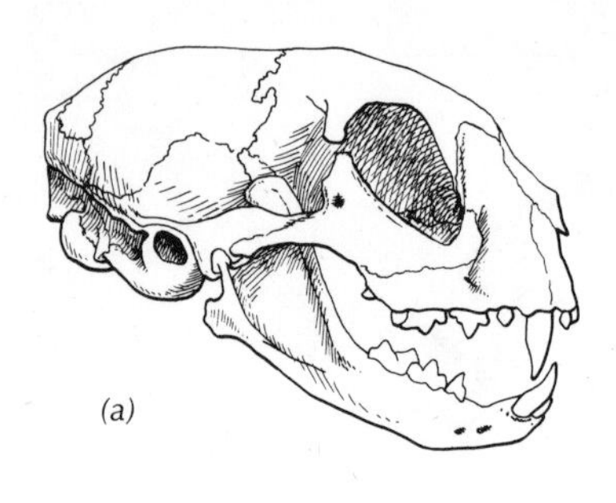

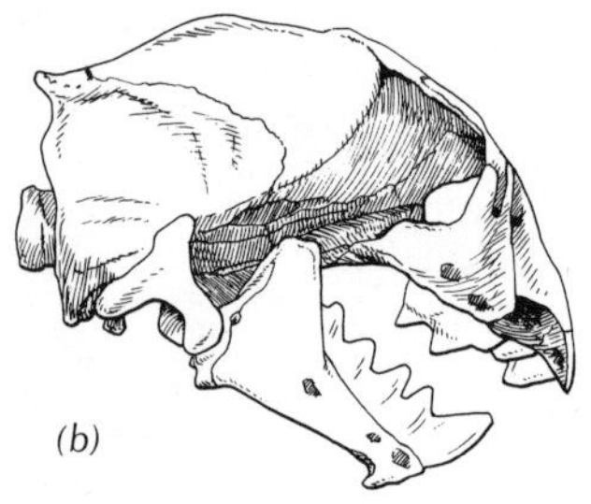

Figure 4-10 A comparison between the skull of a cat (*a*) and that of the trogonophid amphisbaenian *Agamodon* (*b*) documents the superficial similarity which once led to remarks that the skulls of amphisbaenians most resembled those of mammals. Yet the two skulls differ in everything except their solidity and the most superficial proportions.

that some workers have compared their skulls to the stout and solid skulls of mammals (Zangerl, 1944).

Some interesting general amphisbaenian modifications jointly mark their adaptation to a burrowing existence. First and most obvious is the annulation of the skin. In most species this organ has become liberated from the trunk around its entire circumference. The integumentary tube is subdivided into a series of annuli by deep folds to which attach annular muscles, the contraction of which narrows the full surface of each annular segment into an externally convex curve. The annulated integument is connected on each side to the axial mass by three sets of muscles per vertebra (rather than two as in most snakes). Amphisbaenians hence are able to use rectilinear locomotion forward and backward in a tunnel, with fixed contact sites set anywhere along the trunk or its circumference (p. 105). The segmentation of the skin is not uniform but its proportions change around the circumference as well as along the animal's length (cf. Zug, 1970). Minor differences in segment size and angulation, which are not particularly useful in the definition of species and hence tend to be omitted in taxonomic diagnoses, may relate to variants of the pattern of force application. This specialization does not preclude the use of lateral undulatory and concertina movements either in the tunnel or on the surface. The escape saltation used by some species has already been described in the previous chapter (p. 101).

Liberation of the skin from tight junction to the underlying tissues permits amphisbaenians to avoid friction between animal and wall when moving the head and trunk forward in a ramming movement. Such friction would obviously consume energy and consequently reduce the force of the driving stroke, and with this the depth to which the head could penetrate. To avoid it the Amphisbaenia accelerate their trunk (in other words, build up momentum for the burrowing push) inside the skin. The skin of the neck is bunched up and fixed against the tunnel wall before the stroke. This fixed zone is then used as the base from which to accelerate the head and internal portions of the trunk (vertebrae, ribs and associated musculature, viscera) by contraction of the costocutaneous muscles (see p. 107). Only the relatively minor and unavoidable internal friction is hence encountered when the head is driven forward.

The dry, but loosely flexible, skin suggested to some earlier investigators that these animals might achieve their water intake by absorption from moist soil, as do some amphibians. Desiccated specimens gained weight when covered by moist sands (Bogert and Cowles, 1947). Different species of amphisbaenians showed different rates of overall water loss in flowing dry air (Krakauer et al., 1968). These rates generally matched the environmental conditions under which the animals are found in nature. Such species as *Diplometopon zarudnyi* collected in desert areas (Dharan, Saudi Arabia) had very low rates of

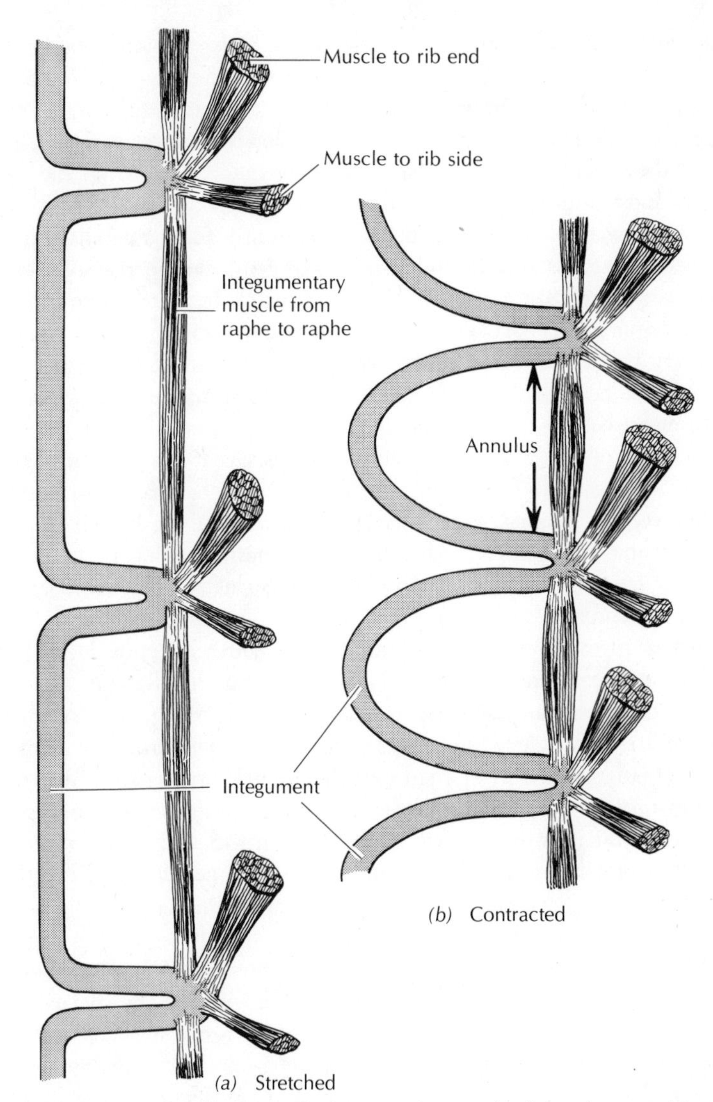

Figure 4-11 Motility of the amphisbaenian skin results from the stretching (*a*) and narrowing (*b*) of the annuli ringing the body. The intrinsic dermal muscles run from raphe to raphe and their contraction shortens the dermal tube but lacks power to propel the animal. Propulsion occurs by contraction of the muscular connections between locally fixed skin and the ribs.

water loss, whereas forms of equivalent sizes, such as *Amphisbaena innocens* from humid areas in Hispaniola, had rates of loss greater by two orders of magnitude. The latter is well within the range for frogs!

Did these high values for water loss imply a converse capacity to take in water via the integument? To test this we restrained animals of several species in a two-compartmented system, body and tail in one compartment and head in the other. Even when the head was in saturated air and the trunk immersed in water so that integumentary water passage was nullified, the animals continued slowly to lose weight. Only when they were allowed to drink did they quickly make up the loss. It was most interesting to note that these animals could even take up interstitial water from wet sand; they apparently form a capillary plate between their lips and are thus adapted to drink within the soil. Thus there is a significant benefit to the good match between environment and integumentary permeability; the drastic relaxation of the integumentary barrier to water loss in some Caribbean species suggests that there must be a considerable cost for the maintenance of this barrier in drier areas.

Analysis of the individual records suggests that water is probably one of the critical factors in explaining amphisbaenian distribution. Wherever the amphisbaenians occupy areas out of the humid tropics, they inhabit microclimates that have high humidity levels. In desert and dry savannas they occur adjacent to seasonal (or occasional) river courses, commonly among the root stocks of shrubs and bushes. They do not seem to move far, and some species seem to occupy local areas sometimes surrounded by wide belts of unoccupied habitat. Certain species do engage in nocturnal migrations which bring them near the surface at night, when heat stress is least and dew formation is most likely (Gans, 1968). More than this, the animals move toward the surface in response to even slight precipitation; at other times they may aestivate (rest in a torpid state) very deep in the soil. Farmers report that the animals are observed commonly only when it rains; in other seasons they are seen but rarely, most often when the farmers dig down a meter or more.

The pigment densities and patterns on the skin of amphisbaenians also correlate with habit. In general the fewer visits to the surface, the less pigmented the animal and the higher the probability of uniform coloration. Wherever two species are truly sympatric (occupying the same geographical area), the one that lives more deeply in the ground will be less densely pigmented; indeed, in Somalia the surface-associated form *Agamodon anguliceps* has a yellow color (that matches the tone of the surface sand) with disruptive brown blotches, whereas the sympatric and deeply burrowing *Agamodon compressus* lacks even traces of integumentary pigment. These observations suggest that the pigmentation reflects the probability of being encountered at the surface; there appears to be a relative advantage for cryptic coloration, but we do not know whether this imposes a developmental cost.

As a different kind of defensive device, the tails of many amphisbaenians may show head mimicry; such species move the tail and hide

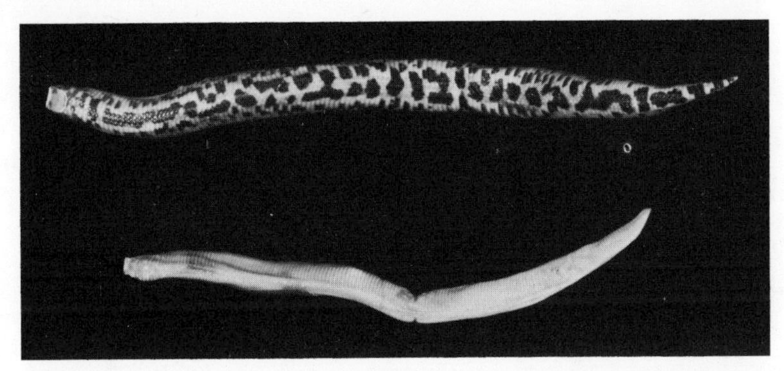

Figure 4-12 Two sympatric species of *Agamodon* occur in the sandy scrub forests of Somalia. *Agamodon anguliceps* commonly occurs in the immediate subsurface region. It shows dark brown blotches on a bright yellow background, which matches the color tone of the local sands. In comparison, one has to dig down a meter or more to find *Agamodon compressus*. This species lacks all dermal pigment, and living individuals appear pink from the color of the underlying blood.

the head (Greene, 1973; Wickler, 1968). Head and tail generally have pigments different from those on the trunk. In some species practicing head mimicry, the tail is short, blunt, and may be supported internally by a set of vertebrae fused into a caudal cup. Such tails often show scars, some of which probably document past and unsuccessful attacks by predators. In other species the tail is longer and bears an autotomy (self-amputating) annulus that is almost always marked by a pigmented constriction coinciding with an autotomy vertebra. Even though the fracture takes place intravertebrally (Alexander, 1966), the Amphisbaenia are in the minority among readily autotomizing forms in that they lack any regenerative capacity. The broken caudal tip heals into a configuration very similar to that seen in the short-tailed species, and some early herpetologists argued that such a perfect tail could not be just a healed stump, so the stump-tailed forms must be a different species! In some species the tail's distal tip is covered with tubercles that may catch sand grains and provide a terminal shield within the tunnel. Although it would seem to be of obvious advantage to an animal to retain such a specialized mechanism, some tuberculate species do autotomize without regeneration.

The presumed advantage of autotomy is clear. When it is grabbed by a predator, the end of the tail breaks off. By some still undescribed neural system (perhaps similar to that of male praying mantises that can produce copulatory movements after decapitation), the isolated tail tip flips back and forth. It is much more active than the rest of the animal and presumably occupies the predator's attention. The advantage of the system may be seen in a series of Antillean species of *Amphisbaena*,

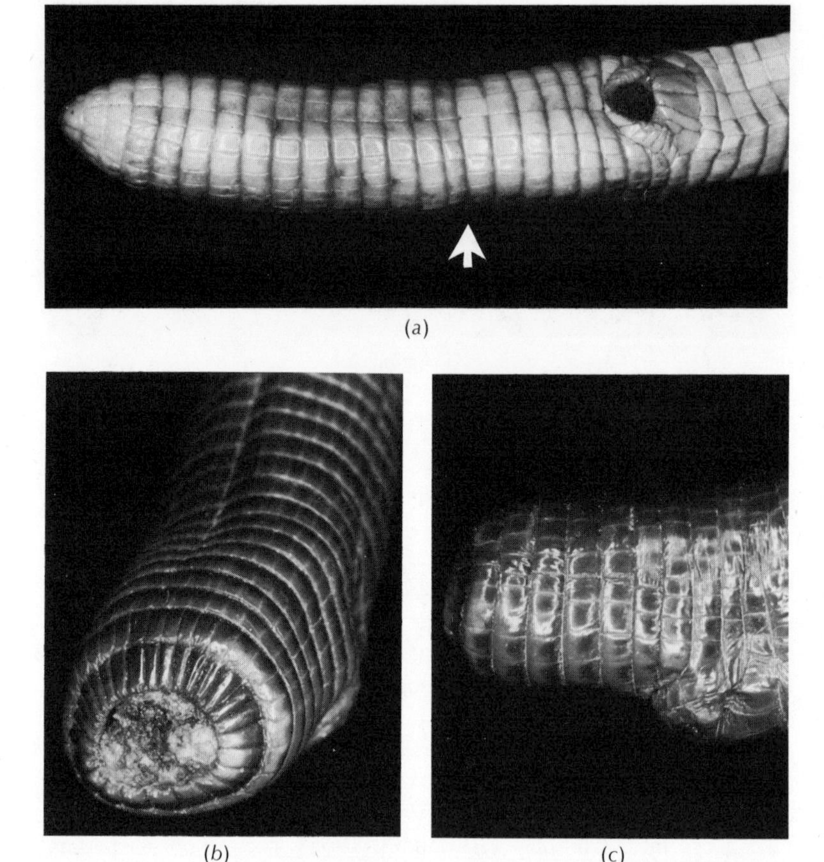

Figure 4-13 The amphisbaenian tail shows an autotomy constriction (*a*, arrow) at which the bones, blood vessels, and skin separate easily when grabbed by a predator. Connective-tissue septa close off the stump (*b*, *c*) so that it will not bleed and will heal rapidly, with the scales growing medially to cover the caudal tip. Curiously enough, the tail does not regenerate.

some of which autotomize their tails with varying degrees of readiness. The frequency of entire, but scarred, tails is higher in species that do not autotomize than in those that do so readily. One can even make the assumption that the frequency of autotomized plus that of entire, but scarred, tails should be constant in a given species and reflect the frequency of major attacks on the animal. Therefore, if a nonautotomizing species shows a lower percentage of attacked tails than an autotomizing one, it might be assumed that fewer members of the nonautotomizing species survived attack and that the difference (all else being equal) is an index to the degree of success under predation and represents the magnitude of the selective pressure. The most that may be said in summing up is that more smaller amphisbaenian species show autotomy than do larger ones. Similar situations have been discussed by Werner (1968) for geckos and Wake and Dresner (1967) for salamanders; both sets of data demonstrate that there are many degrees of autotomy.

Box 4-2 The Bony Skeleton

Most biological materials more readily transmit tensile than compressive forces. Collagen fibers, in the forms of straplike tendons and sheetlike aponeuroses, transmit tensile forces generated by muscles to the supportive elements. Support is provided by skeletal structures capable of withstanding compression. In vertebrates such skeletons are made of notochordal tissue, cartilage, and bone, the latter being a uniquely vertebrate tissue and clearly the most important in resisting compression.

Bones consist of a primary network of collagen fibers onto which crystals of calcium apatite are deposited in regular patterns by specialized bone-forming cells, or osteoblasts. (In certain vertebrate bones the general shape of the tissue is first preformed in cartilage. However, the cartilage breaks down before bone formation so that the action of osteoblasts is not affected.)

The osteoblasts generally surround themselves with osseous tissue and then remain as osteocytes enclosed in small spaces (or lacunae) within the matrix. In larger bones these lacunae are arranged in a regular geometry relative to the bones' surface. When blood vessels pass within the matrix, the lacunae are generally placed in a concentric shell around their path. Very fine channels (or canaliculi) perforate the matrix and connect the central canal to the several layers of osteocytes. Gas exchange, as well as nutrient and waste flow, must proceed through these canaliculi since the bone matrix does not permit diffusion.

The system can grow only by apposition or the deposit of new tissue along the edge. Internal growth (swelling) is obviously impossible because it would require a shifting of the calcium crystals upon each other; this is specifically precluded by the proteinaceous cross connections between the crystals. Growth, then, involves bone formation by osteoblasts on free surfaces and the deposition of additional layers here. If growth is seasonal, one may then expect bone to show growth rings similar to those in the heartwood of trees.

Since change of shape may also involve the loss of bone (for instance, increase of the diameter of the shaft of a long bone involves increase in the diameter of its marrow cavity), a second set of cells, the osteoclasts, can dissolve bone, thereby returning the calcium into solution. Even without such architectural rearrangement, bone serves as a reservoir of calcium salts for other metabolic events, and the amount of bound salt is always in equilibrium with the calcium concentration within the fluids of the body. We may observe changes in this bound calcium at times when lactation or embryo formation imposes major demands on the calcium levels of the maternal circulation.

The two-component, collagen-fiber/calcium-crystal system allows the tissue to resist deformation due to several kinds of imposed loads. The crystalline structure offers significant resistance to compression and thus gives load-carrying capacity. The bonding of crystals onto a collagenous network similarly assures tension resistance as well as the capacity to withstand bending (which may ultimately be resolved into tensile and compressive stresses). A large body of evidence, furthermore, suggests that the proteinaceous matrix orients itself more or less in parallel with the directions of stresses imposed before calcium deposition. Since the orientation of these fibers inevitably determines the orientation of the calcium crystals formed on them, this orientation will maximize the element's carrying capacity. Furthermore, observations document that bony tissue responds both to reduced and excessive stress levels by resorption, whereas intermediate levels serve either to maintain the status quo or stimulate further deposition (see diagram; after Kummer, 1966).

This built-in feedback explains some medical conditions such as the loss of calcium from the heel

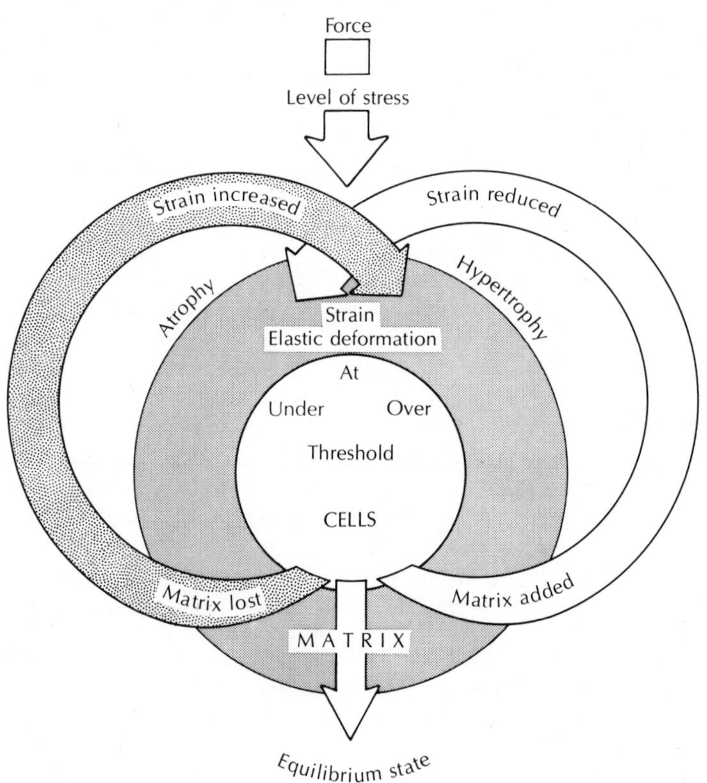

bones of astronauts. It also permits one to make some deductions regarding the direction and general magnitude of imposed stress levels from the architecture of bony elements.

The organization of the plates and surfaces on which the osteocytes are arranged markedly increases the resistance to shear and to fatigue (gradual and progressive failure due to repeated cycles of submaximal loadings). Apparently, by some as yet unclear mechanism, the fiber direction and axis of crystal deposition changes between successive layers of osteocytes. There are thus no obvious slippage planes at which shear resistance might be less. Fatigue effects are observed much more readily in dead bone than in living animals. Apparently the initial crevices are quickly arrested when they intersect planes at right angles to them. This reduces the loading and may give the living tissue a chance to repair itself.

Bone, then, is a uniquely vertebrate tissue that is highly responsive to environmental influences. Not only does it provide a reservoir of calcium salts and may once have served as a dermal armor, it also permits support during growth. Skeletons may simultaneously support the animal and be rebuilt. They respond adaptively to variations on the loadings imposed during the animal's life.

Simple Burrowing and Entry into the Soil

The preceding specializations of skin characteristics, water balance, and autotomy represent patterns apparently common to almost all amphisbaenians. We have already noted that different amphisbaenians show marked differences in head shape and skin segmentation. Moreover, particular shapes and segment arrangements differ in the frequency of species possessing them, in the geographical ranges these species occupy, and perhaps in the density of individuals within the occupied ranges. Could these interpopulational differences relate to burrowing mechanics?

Some 130 species of amphisbaenians are now recognized. Of these, 87 have a head of more or less rounded or pointed shape, 26 have horizontally spatulate heads, and 11 have vertically keeled ones. The six species placed in the family Trogonophidae share various degrees of a fourth pattern, which tends toward a steeply wedge-shaped snout with prominent lateral canthi (edges). Four trogonophids range from Somalia to Iran. One occurs on Socotra and another in Morocco and Algeria. This trogonophid zone is peripheral to the range of the rest of the amphisbaenians, which is well defined by the range of the species with rounded heads (Figure 4-14). Up to three round-headed species are sympatric in some zones. Spade-snouted species tend almost always to be sympatric to round-headed and allopatric to keel-headed ones, though

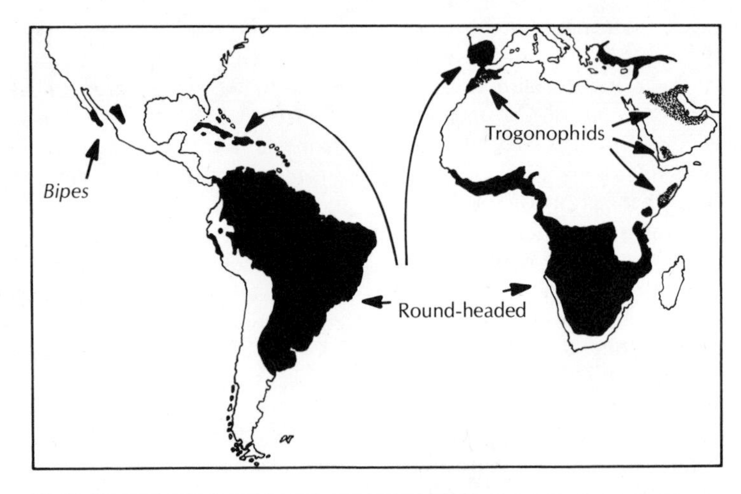

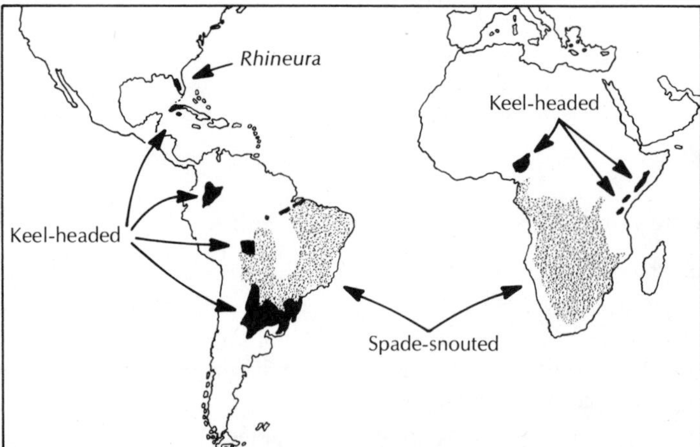

Figure 4-14 Sketch maps to show the distribution of the different subgroups of amphisbaenians. The round-headed forms clearly occupy the greatest zone, with smaller ranges for the several spade-snouted and keel-headed groupings.

there are numerous instances of sympatry between as many as four spade-snouted forms. The composite range of the spade-snouted and keel-headed species overlays the center of the composite range of the round-headed forms. Only the spade-snouted species *Rhineura floridana* occurs allopatric to any round-headed form.[5]

Analysis of shape distributions proceeds best from that of digging patterns. The initial clues that these patterns differ were provided by specimens of some of the most highly modified species, which had been

[5]This species represents a relict of a distribution that once extended across the North American continent (Gans, 1967). Since all of these specimens are spade-snouted, they probably represent a single and independent invasion of the continent.

accidentally preserved with some of the neck muscles in the contracted state. Those specimens with horizontally spatulate heads had the tip of the snout lifted to or above the level of the dorsal surface; the keel-snouted forms showed the head pulled sharply to the left or right. Observations of living specimens confirmed the meaning of these positions, and also showed a much greater diversity of patterns than could have been deduced from the preserved specimens alone.

The various round-headed forms with more or less pointed snouts were seen to dig by forcing their heads into the soil in a series of forward pushes. Reaction forces for this push are developed by contact of the body with the wall of the tunnel. If the tunnel is curved, a bend of the trunk can exert sufficient push against the side to keep driving the head; if the tunnel is straight, the animal uses concertina or rectilinear locomotion. The magnitude of the force that may be exerted in driving the head is then a function of the skin-soil frictional characteristics. These determine the force that may be exerted across the skin-soil boundary without slippage (p. 91). The animal may, of course, build up momentum (a function of its mass and velocity) and then expend it in penetration; this avoids the continuous exertion of a penetrating force. Such momentum may be built up over a greater time or along a greater distance of the animal's trunk than is involved in expending it. Like a runner attempting a broad jump, the animal reduces the muscular force that must be exerted at any instant and stores up momentum to be expended during a brief interval.

The head region of these round-headed species shows the effect and direction of the driving forces. Those zones involved in the penetration (and hence exposed to the surface shear detailed earlier—p. 128) show fusion of head shields. In general the dorsal surface of the head bears the largest shields, and the tip and the sides of the lower jaw are similarly clad. In some of the African genera, the temporal region bears the largest shields and the zone of enlargement extends farther posteriorly (relative to angle of the mouth) as the absolute size of the species decreases. The greatest zone of enlargement is seen in the threadlike *Chirindia* of East Africa, but even in this case the shields stop anterior to the level of the head joint.

The interaction of the need for a regular pattern of annuli on the trunk with that for friction-reducing smoothness along the head produces some interesting effects where the zones of influence overlap. The segments of the gular, or throat, region are subdivided more complexly or, rather, less regularly geometrically than are those of either the snout or trunk; both segment size and arrangement vary. Although the segmental arrangement of the head shields is generally constant within species or subspecies, it is usual to find the architecture of the intersection zone varying greatly within populations. The location of

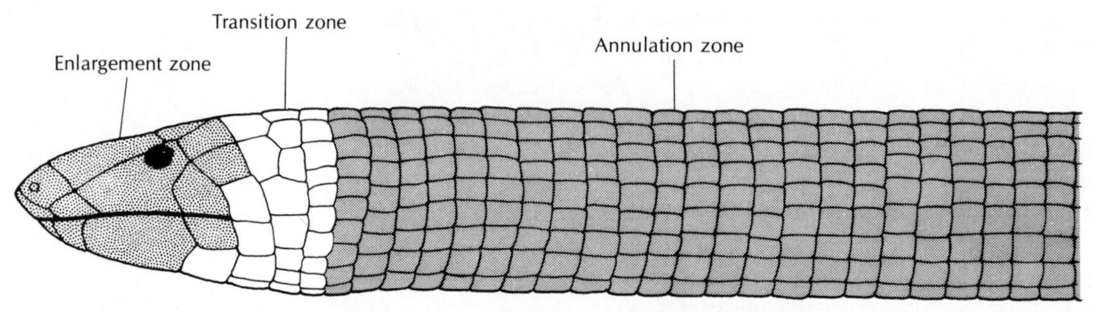

Figure 4-15 Lateral view of a species of *Amphisbaena* to show the segmentation pattern in the different regions of the body. The penetration zone reflects the friction-reducing fusion of head segments. The annulation zone shows the regular pattern of body annuli useful in locomotion. The transition zone from one to the other also allows some flexibility, permitting the mouth to open and the head to swing on the head joint.

this transition zone also differs among species; its limits are clearly defined by the burrowing mechanics. This local variation suggests that selection is for the capacity to fold and stretch the skin in this region, rather than for specific segmental geometry.

Shield differences are most logically explained when comparisons are made between adults. The segmental arrangement of the skin of squamate reptiles is fixed before birth and constant throughout development. Therefore, juveniles apparently must adapt their burrowing behavior to the scale pattern of the adult even though this may be most effective at a larger size. However, the proportions of the head and of its covering shields do change with age. This is due to allometric growth, which is a different growth rate in different directions and for different structures. Allometric growth provides a way of compensating for the different functional needs to which an organism may be exposed at different ages. In the present case, the allometric growth may provide some functional compensation for the disparity between size and segment pattern. Certainly the juveniles of all species have a more sharply pointed head; the diameter of the shielded portion increases more rapidly during growth than does its length, so that the snout becomes more blunt. Juveniles of species with horn-covered head shields regularly increase the relative area of the keratinized zone with growth. All of these allometric changes suggest that there is a conflict between direct functional requirement during the juvenile period and the need for a system that can grow to meet adult needs (cf. p. 12).

Burrowing has thus far been considered as the extension of an existing tunnel. Entrance into the soil from the surface poses a separate and distinct set of problems, in that burrowing animals are forced to start a tunnel without formed tunnel walls to absorb reaction forces. Several strategies are observed here. These generally serve either (1) to reduce the vertical component of force necessary for the initial drive (reaction for the horizontal component may be obtained from the static friction of the posterior trunk) or (2) to increase the capacity for exerting vertical forces; sometimes both approaches are used in combination. The vertical component necessary for penetration is generally reduced by driving the initial divot more or less horizontally into the side of a crevice or a natural rise. The crevices left by decaying stems and roots may facilitate such entry into hard soils. After it has insinuated its head for some distance, the animal can shift the direction of tunneling to a more steeply vertical one. The vertical force available is amplified by two methods. First, burrowers may brace their bodies against roots, twigs, logs, or similar devices having a projecting edge. Even if there is no preexisting chamber under a rock, a rock's mass may be used as a *point-d'appui* for vertically directed forces (thus providing yet another advantage to its "underside"). The second method is the use of the tail as a reaction point. If it is dug in, the whole body may be stiffened, increasing the overall anteriad force and with it the magnitude of the vertical component that serves to form the tunnel.

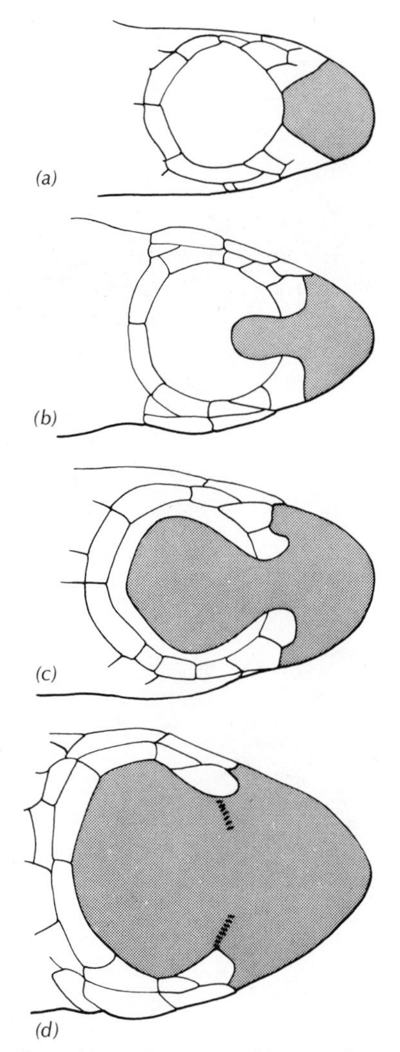

Figure 4-16 As *Monopeltis guentheri* from western Africa grows, the heavily keratinized zone (shaded) extends farther over the head, obliterating sutures between head shields.

Two variants of the above general methods are seen in some round-headed amphisbaenians. The first, which evades the issue completely, is seen in three species of *Bipes* (members of a presumably distinct amphisbaenian family from Mexico); all have retained a pair of short forelimbs. These limbs are well developed and their phalangeal formula (3-3-3-3-3) is greater on the fifth digit than the supposedly primitive reptilian 3-4-5-5-2.[6] The forelimbs are normally pressed against the sides of the trunk, being beautifully braced and constructed to retain maximum strength for minimum thickness. When the limb is swung anteriorly, the animal's width increases only slightly because each joint folds. Although these "hands," placed so far forward on the trunk that the natives considered them to be ears, might be used to extend an underground chamber, they do not seem to serve this function. Rather, the animals use them to climb over surface irregularities (and supposedly even into palm trees—C. E. Shaw, personal communication) and to generate the initial cavity for further burrowing by other methods. In creating the cavity, the animal swings the head to one side and the arm opposite the bend reaches past the face to scrape the surface (Figure 4-17). Even mere contact of the fifth digit with the surface requires lengthening, and this presumably accounts for the hyperphalangy, which is a rather uncommon phenomenon. The scraping method, reminiscent of the excavating strokes of moles, is effective mainly in fairly soft soils.

The second special method of initial penetration is used by diverse

[6]The numbers refer to the phalanges per digit. It is interesting to note that loss of a phalange—indeed, loss of whole digits or other structures—appears to have been a common event in phylogenetic history. The rise of new structures is much less common, hence the emphasis on the extra phalange here.

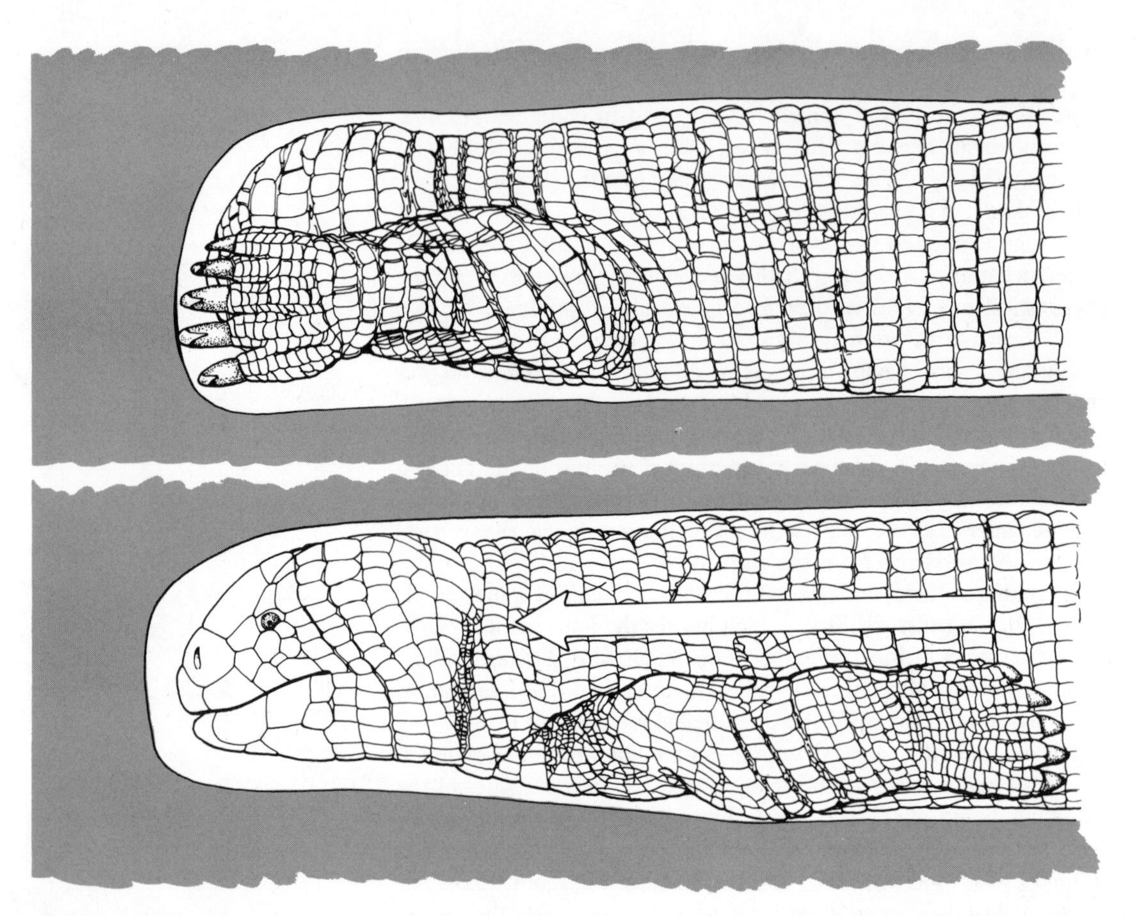

Figure 4-17 *Bipes* can form an initial entry or continue a tunnel through the soil by scraping with its long claws across the tunnel's end. The head need only compact the freed material by pushing it into the sides.

amphisbaenians with small and often pointed heads. The body is shortened into a series of narrow curves of short radius. The tail and posterior body are then pushed against the ground as is the tip of the snout; action and reaction forces are exerted to snout and tail by the straightening of the anterior trunk. The body may even be arched so that the snout points farther downward, increasing the vertical vector to include much of the animal's weight. The anterior part of this arch is then vibrated vigorously from left to right by alternating contractions of the lateral musculature. This oscillates the tip of the snout, displacing soil particles until an opening is created. During burrowing, the same method will also shift the head from a position at which a rock may be directly encountered to a more suitable angle of attack; the snout may slip along the rock's surface or even slide to a new site. It has not been determined if the vibratory frequency matches a natural frequency of the soil or, in other words, whether or not vibrations set up in the soil enhance penetration.

Complex Burrowing Styles

Many species use a two-cycle digging movement. The ramming strokes are each followed by a general lifting or seemingly random movement of the snout's tip. This vertical or sideways movement slightly widens the anterior end of the tunnel. Since this two-cycle movement occurs in all of the spade-snouted and keel-headed forms, it must have some major advantage and consequently deserves consideration.

It has been noted earlier that the force required to extend a tunnel is a function of the tunnel diameter. Driving a conical snout into the end of a tunnel will then require a force that increases with penetration (cf. Gans, 1960). In other words, the momentum required of the driving system to sink fully a pointed cone of height h into a flat surface is less than one-half that which would be needed subsequently to drive it the same distance h through the substratum.[7] If after each stroke the new portion of tunnel will be fully widened to the diameter of the cone's base, the animal needs to build up less momentum in effecting the next ramming movement. Thus even partial widening of the tunnel would be advantageous because the animal needs to exert less force against the tunnel walls to compensate for the reaction. This observation suggests the nature of the first evolutionary steps leading to mechanically more complex systems for separating the penetrating and widening functions.

Both the spade-snouted and keel-headed amphisbaenians can now be seen to have taken this system to its logical conclusion. Random movements of the head may exert only limited force to widen the tunnel; the muscle attachments lie much closer to the fulcrum (here the head joint) around which rotation will be induced than do the forces the snout imposes on the wall. As long as the potential for building up momentum for the driving stroke is limited (by the potential slippage between the body's skin and the tunnel wall), there will be an advantage to keeping momentum requirements low by separating the penetrating from the widening systems. The shape of the penetrating wedge may then be designed for maximum penetration and the widening handled as a separate function.

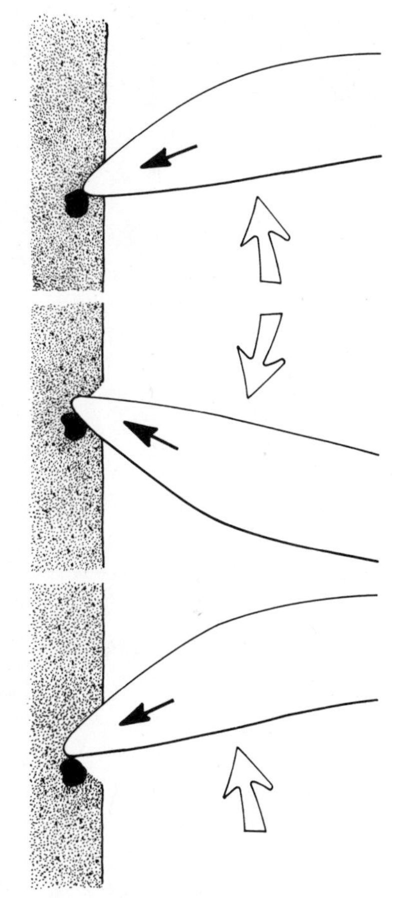

Figure 4-18 Some amphisbaenians can flip the whole body from side to side so that the head swings back and forth around the rostral point. This point then penetrates by displacing particles laterally and gradually developing an entry to the soil.

[7]The force to drive a cone is at any instant some function of the maximum diameter in contact at a time, as well as the friction of the cone and the soil type. The momentum required to sink the cone to a particular depth thus reflects the sum of those forces required to sink the cone at any point. Since the diameter increases as the cone penetrates, the average momentum is less than half of that required to continue driving the cone for a unit distance once it has penetrated to its maximum diameter, namely that of its base.

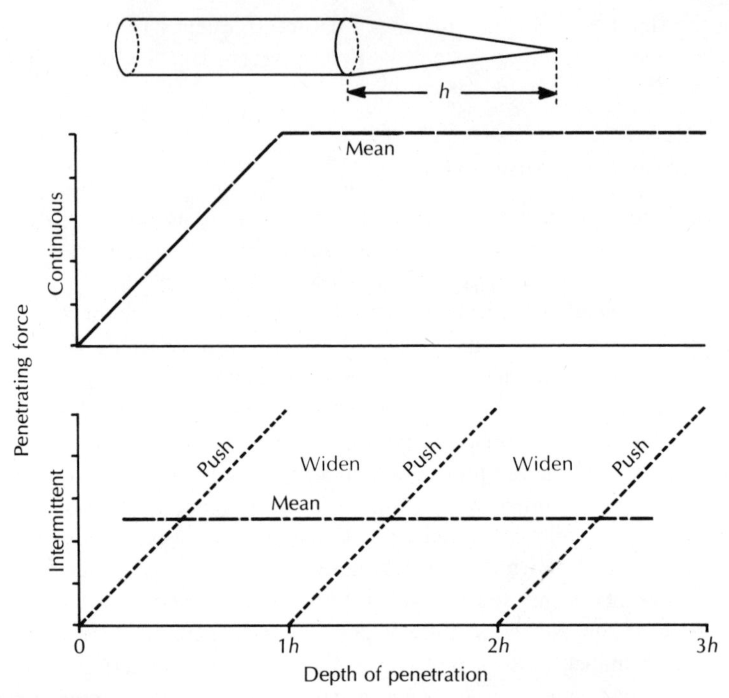

Figure 4-19 The resistance force opposing continuous penetration into the soil of a rod tipped with a cone (of height *h*) rises until the cone has penetrated completely. It then remains constant (*a*). If the cone is pushed forward in a series of motions (each time to the depth *h*) and between drives the tunnel is widened to the cone's maximum (or basal) diameter, the penetrating force for the next push will again start at zero. Consequently, the mean force or mean momentum required for such intermittent (push and widen) penetration will be one-half or less of that needed for a continuous drive.

In both cases it is the entire axial musculature of the anterior trunk, rather than the musculature of the neck alone, that provides for the widening aspect. This musculature had a limited cross-sectional area, since an increase in the overall diameter of the amphisbaenian would force an exponential increase in the digging work. Natural selection has accomplished this through arrangement of the muscle fibers in a feathered or pinnate, rather than a parallel-fibered, pattern (cf. Gans and Bock, 1965). Parallel-fibered muscles stretch in a more or less straight line between origin and insertion. As the force of contraction of the fiber will be some function of its diameter for a given speed and percentage of shortening, parallel packing would imply increased width or more limited force. In a pinnate arrangement, the aponeuroses (flat sheets of connective tissue) running respectively from the origin and the insertion site are parallel to each other and parallel to the line from origin to insertion. The fibers of a pinnate muscle run between these tissue

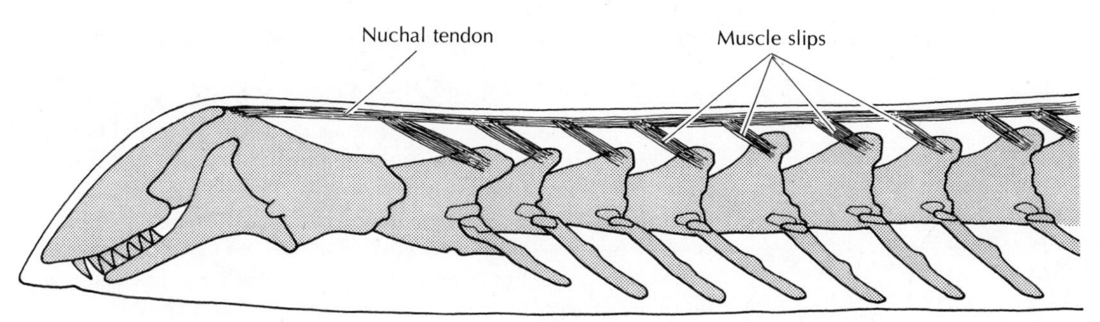

Figure 4-20 The muscles that pull the head up, when an amphisbaenid widens a tunnel, insert on the nuchal tendon from their origins on the neural arches of the anterior vertebrae. Not only does this allow the animal to generate the force far back along the body, but it also makes the muscle pinnate, since its fibers lie at an angle to the line of force that they produce.

sheets and hence lie at an angle to the straight line between origin and insertion. Where two or more aponeuroses leave one site, the fibers appear feathered in cross section; hence the name (pinnate < L. *pinnàt(us)*: feathered) (Boxes 4-3, 4-4).

Each fiber will pull at an angle to the line of action of the entire muscle, and its contractile force can then be resolved into two components, each generally smaller than the force of the muscle as a whole. There is a useful component along the line connecting the two sites, and a wasteful component at right angles to it. Since only the useful components are summed, each fiber adds less to the total "force at tendon" than does each fiber in a parallel-fibered muscle. On the other hand, by increasing the muscle's length one can add fibers, increasing the force without simultaneously increasing the contracted muscle's diameter. Each fiber will, of course, increase in diameter as it shortens (fiber volume being essentially constant), but the overall diameter of the muscle will not increase during contraction. Thus the force to be exerted in lifting the head can be increased by extending the system posteriorly (attaching to more and more vertebrae) without increasing the trunk's diameter as the muscles contract during tunnel-widening operations. Thus neither the penetration nor the widening need depend on the nuchal region ("nape of the neck") alone; the great selective advantage to this use of a pinnate muscle is that it permits an animal to generate forces along the trunk far from the site of the widening action.

In many spade-snouted species the initial penetration of the tunnel end is made with the snout's edge *depressed* to a level near the ventral edge of the trunk. Since the lower edge of the mouth is countersunk and the nostrils depressed, this position minimizes the risk that shear along the ventral surface might wedge open the lower jaw. The shear resistance is concentrated along the dorsal surface of the spade. Its shields are variously enlarged, and they, as well as the edge of the spade, are

Box 4-3 Muscle Fibers and Their Contraction

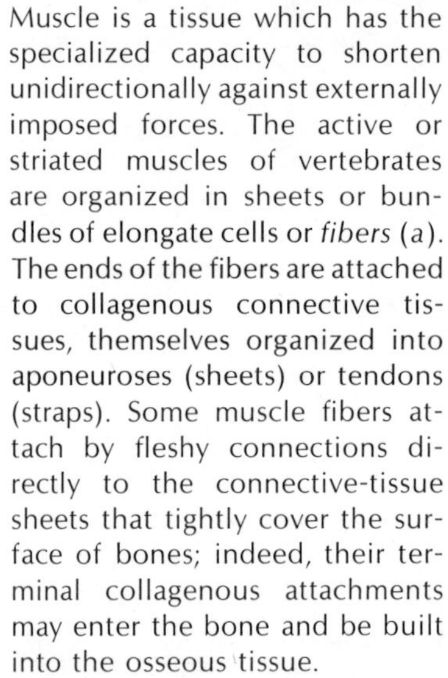

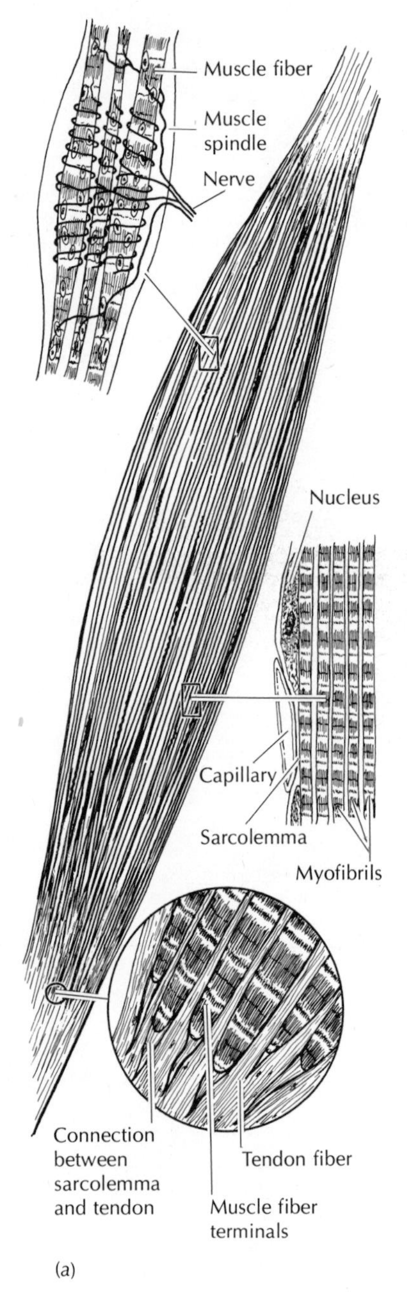

(a)

Muscle is a tissue which has the specialized capacity to shorten unidirectionally against externally imposed forces. The active or striated muscles of vertebrates are organized in sheets or bundles of elongate cells or *fibers* (*a*). The ends of the fibers are attached to collagenous connective tissues, themselves organized into aponeuroses (sheets) or tendons (straps). Some muscle fibers attach by fleshy connections directly to the connective-tissue sheets that tightly cover the surface of bones; indeed, their terminal collagenous attachments may enter the bone and be built into the osseous tissue.

Because at least one end of a muscle will move when the fibers contract, the nerve and blood supply have to be able to follow such excursions without rupture. Two options occur. First is entry into one end of the muscle in parallel with the tendinous attachment. The second is an entry into the muscle's belly in a loose coil that avoids strain when the muscle shortens. Once they have entered the muscle, nerves and blood vessels subdivide and course down the septa between the fiber bundles and ultimately between the individual fibers. Some of the neurons derive from muscle spindles that monitor the muscle's state of contraction.

Each muscle fiber consists of an elongate, cylindrical, multinucleate cell. Its inside is packed with striated fibrils that generally run the full length of the cylinder, though one does observe cells with some fibrils terminating short of the end with cross connections to the cell membrane. The membrane is externally coated by a matrix containing a network of fine collagenous fibers. This assemblage explains some of the elastic response when muscle fibers are stretched. It also furnishes continuity with the collagen fibers onto which tension is exerted.

The force generated by a stimulated muscle depends on its length at time of stimulation. The maximum contractive force is observed near the length which the resting muscle occupies in the body. This observation was made during the last century and numerous papers commented on the way in which such an arrangement might be established during development. In 1940 Ramsey and Street dissected out a single muscle fiber, floated it on a pool of mercury, and stimulated it while measuring the tension generated. They demonstrated that the length-tension curve of such a single fiber was most similar to that of an entire muscle (*b*). The various individual points generated by their careful and difficult determinations suggested that the tension gradually increased from a minimum corresponding to 60 percent of the length at which maximum tension was generated. After reaching maximum, the tension decreased again to zero somewhat above 150 percent of the resting length. They noted that muscles allowed to shorten to less than 60 percent

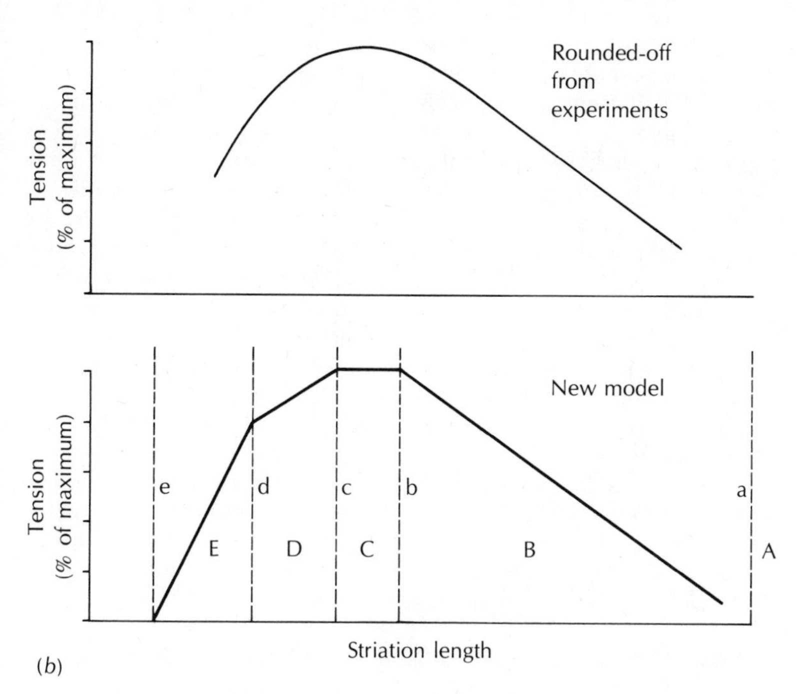

(*b*) The length-tension curve smoothed out from the actual data points obtained by Ramsay and Street (top) and the series of straight lines that describe the actual behavior of the fiber (bottom). Compare the real lines with the interactions of the myosin (heavy) and actin (thin) rods in the sliding filament model of Gordon et al. (1966). Refer to (c) below.

of resting length suffered a kind of irreversible change.

Twenty-six years after the first report, the shape of the length-tension curve was elegantly explained by a combination of ultrastructural and physiological work (Gordon, Huxley, and Julian, 1966). It had been known for some time that a striated bundle or fibril, which gave this muscle its name, was actually composed of an arrangement of interdigitating filaments consisting of the two proteins, actin and myosin. The actin filaments are hexagonally spaced upon so-called Z-disks (often referred to as Z-bands, as histologists sometimes describe the world in two-dimensional terms). Myosin filaments are regularly spaced between the actin filament by myosin bridges. The center of each myosin rod lacked the capacity to form these.

Gordon, Huxley, and Julian were able to measure the force generated by fibers restrained at different degrees of shortening. Their analysis confirmed that the tension generated is proportional to the number of myosin-to-actin bridges and thus to the amount of overlap between the two sets of filaments (c). When the fiber is stretched so that the actin and myosin filaments do not overlap, no bridges will be formed and no tension generated (A). From there until the overlap reaches the bridge-free region in the center of the myosin fiber, the tension increases linearly (B). Further shortening and overlap does not increase the number of contacts and one apparently reaches a zone of constant tension (C) fol-

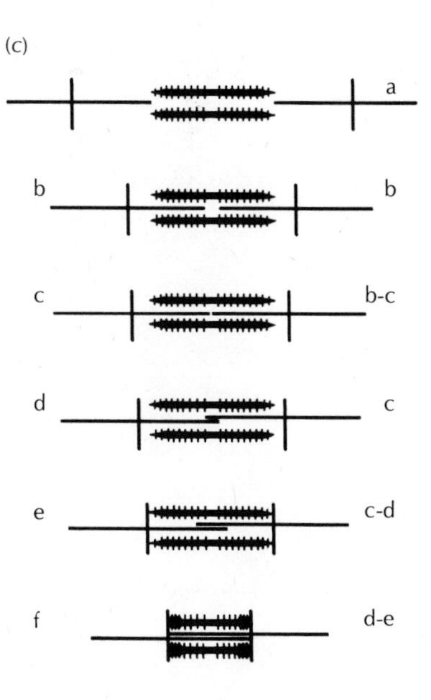

lowed by two kinds (D, E) of overlap between the actin filaments of adjacent disks, each resulting in a further decrease of the tension generated. Once the muscle has contracted to 60 percent of resting length the actin filaments of one disk contact the opposite one. Further shortening will then kink them and lead to the irreversible deformation.

The several results leave us with a general and a more specialized message. The general message is to note the risk that a seemingly mathematically perfect curve may mask a series of diverse events which cannot be separated if their magnitude lets them be "averaged" away with the experimental errors. The second one is that the position of an animal's muscles must sometimes compensate for the fiber's tendency toward decreased force production with increased shortening. Muscles should not be compared in their "resting" position but in all the positions through which they move as they shorten. Certain kinds of retroarticular processes compensate by increasing the effective lever arm of (and with this the moment produced by) a muscle as it shortens. The force produced needs to be evaluated in terms of the biological role that a particular muscle plays in the life of the organism.

often heavily keratinized. Only in several species of the South American genus *Leposternon* is the keratin thin and the snout surface variously underlaid with spongy tissue sheathing the bone. The several species differ in the size and placement of the zones of enlarged scales along the sides of the head. Apparently these reflect slight differences in the digging motions employed.

Following the penetration, the entire head swings upward, rotating about the occipital condyle. This rotates the dorsal surface of the spade into horizontal position, simultaneously compressing the overburden into the tunnel roof. The forward pressure is apparently continued during this upward stroke; not only does this move the spade's tip vertically rather than in a backward curve, but the ventral reaction forces are transmitted to the opposite surface of the tunnel by the pectoral skin. The upward movement of the head is powered by contraction of the pinnate nuchal muscles which, furthermore, cause the anteriormost part of the vertebral column to bend dorsally.

Some of the freedom to move the head dorsad (as well as to open the mouth) is produced by a deep "gular" fold in the skin at the back of the lower jaw; however, the pectoral part of the skin transmitting the ventral forces simultaneously slips anteriorly, smoothing and further compressing the bottom of the tunnel. This explains why the segments of this pectoral skin have fused into elongate pectoral shields in various species of the genera *Monopeltis*, *Dalophia*, and *Leposternon*. A reduced need

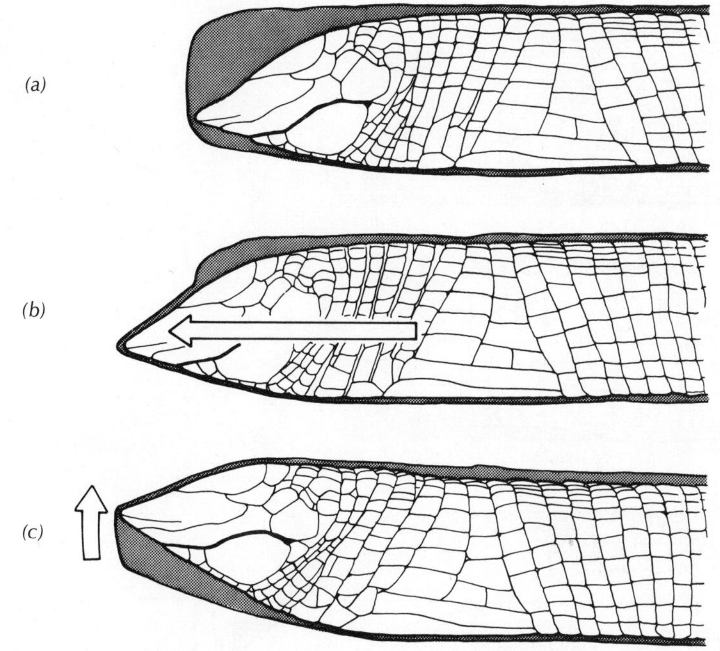

Figure 4-21 The spade-snouted amphisbaenians extend their tunnels by two-cycle movements. The snout first drives a penetrating divot at the bottom level of the tunnel (*a*). Once all of the snout has penetrated (*b*), the forces applied by the nuchal muscles lift it, compressing the overburden and widening the tunnel (*c*) for the next penetrating stroke.

for sliding the skin forward in the largest species, or the increased force that would be needed thus to widen the largest tunnels, may explain why such large (diameter > 2 cm) forms as the Cameroons species *Monopeltis jugularis* and *Leposternon infraorbitale* from Mato Grosso have little or no pectoral-shield enlargement. The smaller species, in contrast, have spectacularly regular, parallel, and smooth pectoral shields.

Excavation by keel-headed species is similar, except that the ramming stroke tends to occur with the head in medial position, so that the widening of the tunnel may proceed by movements to either side. Consequently, one might expect some very slight tendency for enlargement of the segments along the sides of the head. The reasons why such enlargement does not occur are probably the relatively smaller angle of bend in keel-headed than in spade-snouted species and the common use of a different widening system. In this alternate system, the tip is rammed into the soil, but retains its central position. The anterior body is then driven left and right, bending the head from side to side around the rostral tip which forms the rotational fulcrum. The wedge-shaped snout then remains at the center of a wedge-shaped tunnel end, the

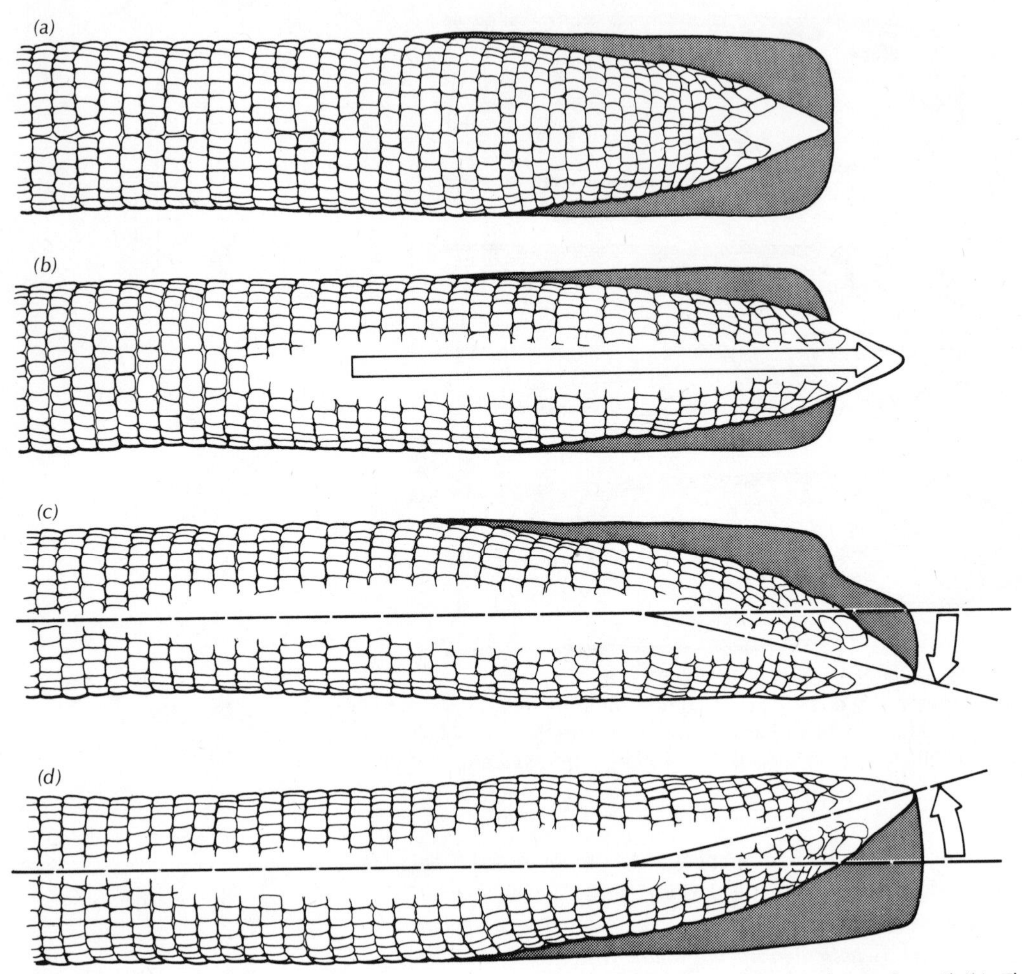

Figure 4-22 Keel-headed amphisbaenians extend their tunnels (*a*) by driving the head into the soil (*b*). They then widen the tunnel by shifting the head first to one side (*c*) and then to the other (*d*). This is followed by the next penetrating stroke.

included angle of which is greater than its own. The next penetrating stroke will again proceed with gradually increasing lateral contact until the head has fully penetrated and the next widening cycle starts.

In most of this discussion I have omitted mention of the Trogonophidae because their pattern really does not fit any of these schemes. The most advanced members of the group occur in Somalia and Arabia. They have a slightly concave, sharp-edged spatulate head roughly reminiscent of that of the spade-snouted amphisbaenians. This almost led me to the prediction, primarily from morphological data, that they must excavate their burrows in a similar manner. Yet they lack any trace of

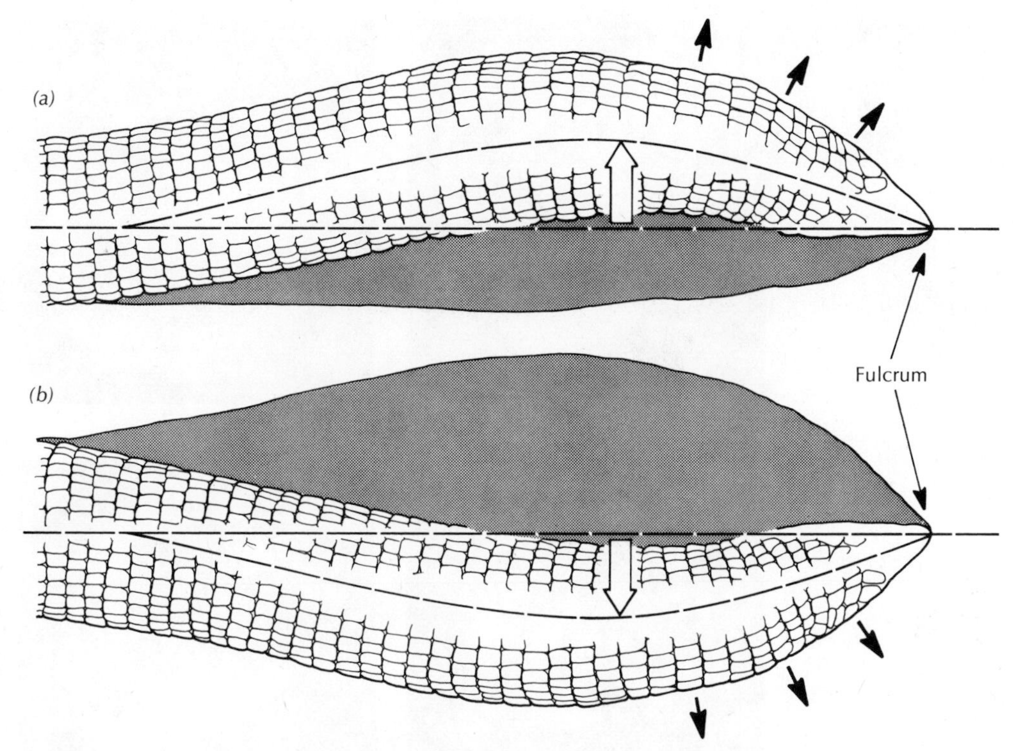

Figure 4-23 When the tip of the keel becomes wedged in a particularly resistant medium, a keel-headed animal may widen the posterior portion of the tunnel by curving head and neck first to one side (*a*) and then to the other (*b*). This again is followed by another penetrating stroke which drives the keel-shaped head farther.

enlarged pectoral shields or of nuchal tendons and muscles such as seen in the spade-snouted forms; they are unique among amphisbaenians in having a deep ventral sulcus and a reverse U-shaped (ventrally concave), rather than cylindrical, cross section. They also have only a semiannular segmentation restricted to the sides of the trunk. The anatomy of the group is obviously distinct, and its members share a short, conical, downward-pointing tail, as well as a spectrum of internal differences affecting their teeth, their cervical vertebrae, and even their endocrine pattern.

Although I obtained some specimens of the African *Trogonophis wiegmanni*, the most primitive member of the group, and even took motion pictures of their digging, I was unable to discern the key to their specialization. Only some three years later, when the first living members of the advanced *Agamodon anguliceps* came to hand from Somalia, did the burrowing mechanism of the Trogonophidae immediately become obvious, and the functional basis of the modifications

Figure 4-24 Frontal views of the heads of *Trogonophis wiegmanni* (a), *Pachycalamus brevis* (b), *Agamodon anguliceps* (c), and *A. compressum* (d) show some of the diversity of the trogonophid amphisbaenians.

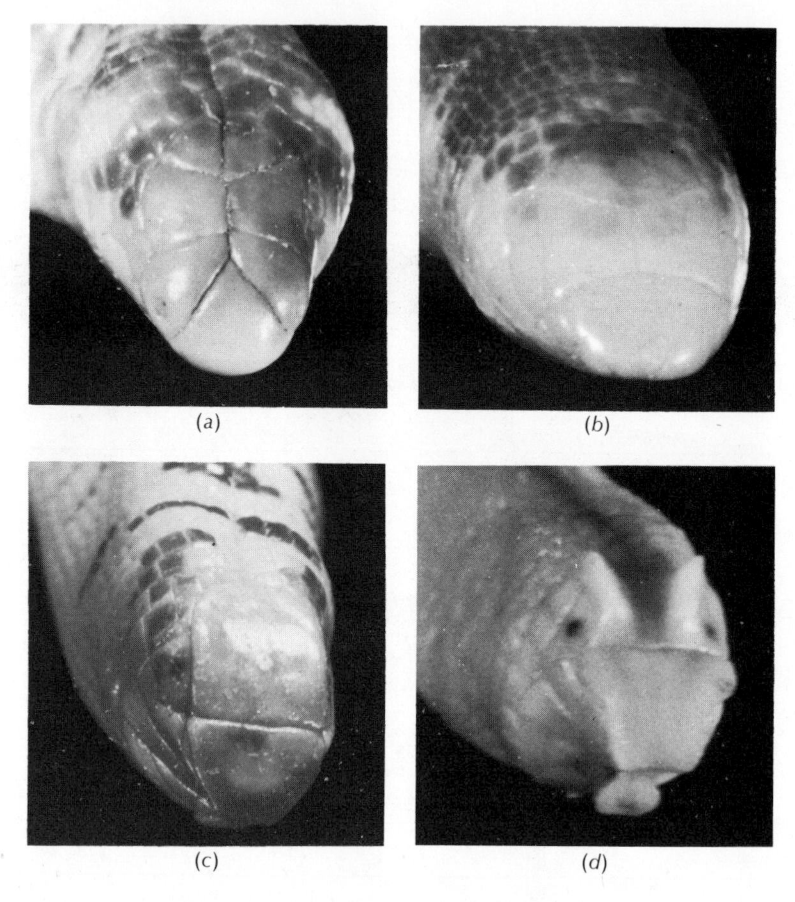

explained.[8] Reexamination of the films of *Trogonophis* then showed indications of an early stage of the trogonophid pattern. This experience supported the philosophy that the extrapolation of function from structure is risky. Furthermore, adaptive patterns are much more easily recognized in the advanced than in the primitive members of a group.

How does *Agamodon* burrow? The movement is oscillatory. The animal rotates its head about its long axis and then reverses immediately (Figure 4-26). The head is simultaneously twisted so that the resulting motion scrapes the sharp-edged lateral canthus across the tunnel end, shaving off sand grains. The loosened material is then compacted into the walls by the head's side. The oscillating motion not only explains the function of the sharp lateral edges but also explains why the trunk

[8]I must admit that this signal discovery occurred in quite unscientific surroundings. The shipment arrived late at night by air freight. Rather than waiting until morning, I emptied the can of animals and sand onto the kitchen table and witnessed a totally unexpected movement.

Figure 4-25 When *Agamodon* is disturbed by being suddenly uncovered near the surface, the animal freezes in a defense posture and only later rights itself quickly to crawl away. This view of an upside-down animal demonstrates the three major trogonophid specializations: the downward-pointed edge of the snout; the triangular and ventrally hollowed (noncircular) cross section of the trunk; and the short, pointed, and downward-curved tail.

of these animals is not circular or cylindrical. The ventrolateral edges of the inverted U, beneath the rib ends, clearly dig into the ground to balance the torsional reaction forces. An ordinary amphisbaenid with a cylindrical trunk would tend to spin in the tunnel when oscillating; only the friction between skin and wall would facilitate penetration forces. The oscillating pattern seems to be adapted to burrowing in sandy soils; the shaving action of the canthi reduces direct compression to a minimum and shakes up the grains while compacting them. This presumably repositions them and, with minimum work, packs the grains into a tight formation.

The most extreme adaptation to this oscillating pattern is seen in the unpigmented *Agamodon compressum* (Figure 4-12), a species sympatric in central Somalia with the blotched *A. anguliceps*. The skull is laterally narrowed as is the entire trunk, which is three times higher than wide. The keratinous canthal edges of the frontal shields project over the face like pairs of bony plates. The animal moves through the soil in a ribbon-

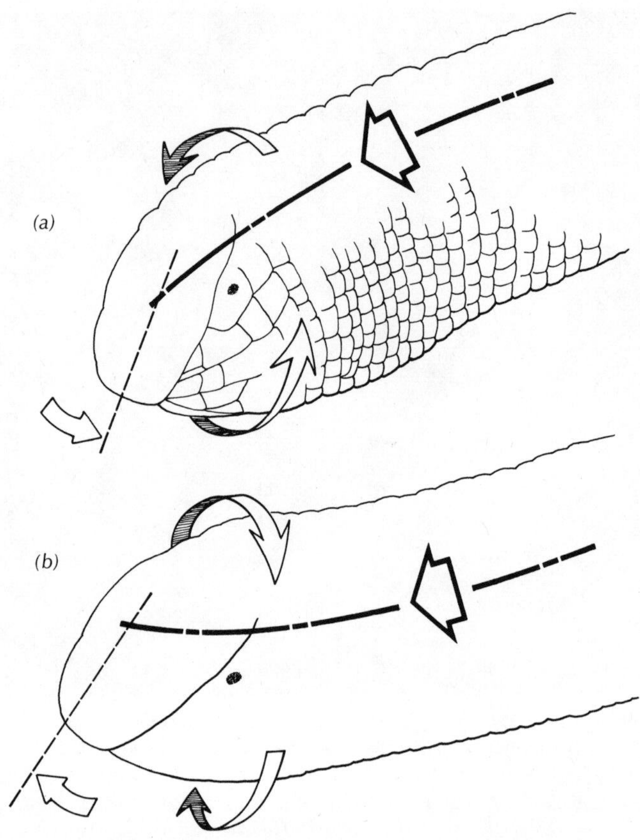

Figure 4-26 Stages in the oscillatory movement of a trogonophid amphisbaenian. Although the two movements rotate and shift the head in opposite directions, they are otherwise symmetrical. Note that the movement combines rotation about two axes, with translatory (forward) movement along the long axis.

like tunnel, and the trunk's shape provides maximal stability for balancing the reaction forces incurred when penetrating deeper, denser soils.

Individuals of *A. compressum*, adapted as they are to the deep layers, do have great difficulty when moving along the surface; they wobble from side to side and they have difficulty in staying upright and in reentering the soil. In contrast, the dished head shield of the intermediate species of the family is advantageous for the initial penetration of the soil. Its curve seems to dig slightly into the surface no matter how little force is applied to it. It proved most difficult to photograph the Arabian species *Diplometopon* on sand with the snout exposed. Even the slightest movement caused grains to pile up and cover the anterior end of the spade. Consequently, these animals can traverse a sandy surface in concertina

Box 4-4 Muscle Architecture

Muscle fibers in animal bodies show a variety of arrangements. Often they lie adjacent to each other in sheets or bundles. Their lines of action will then be parallel to each other and to the tendons or aponeuroses connecting them to the structures they move. Yet contraction changes the architecture. As the center fibers of a bundle shorten, they swell, as there is almost no change in the volume of a fiber during contraction. This deflects the peripheral fibers laterally, away from a parallel pattern; they will then curve from origin to insertion. The thicker a strap-shaped muscle, the more its center will bulge upon contraction (witness the human biceps), and the wider the curve of the peripheral fibers.

Such generally parallel fiber arrangements also pose packing problems. For a given degree of shortening the force at the tendon is likely to be a function of the number of fibers (fibrils and filaments) acting in parallel. This suggests that muscles that pro-

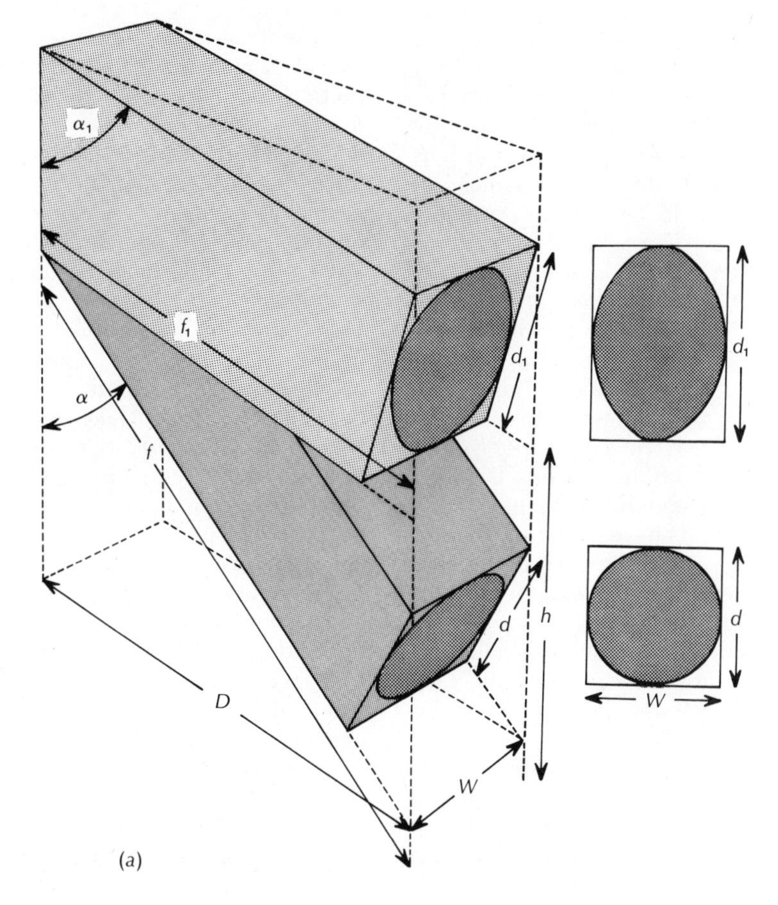

(*a*) This sketch shows how the contraction of a pinnate muscle is accommodated without a swelling of the diameter (distance *d*). The sketch also shows that the shortening of each fiber, $f - f_1$, will be less than the vertical excursion of the tendon (*h*).

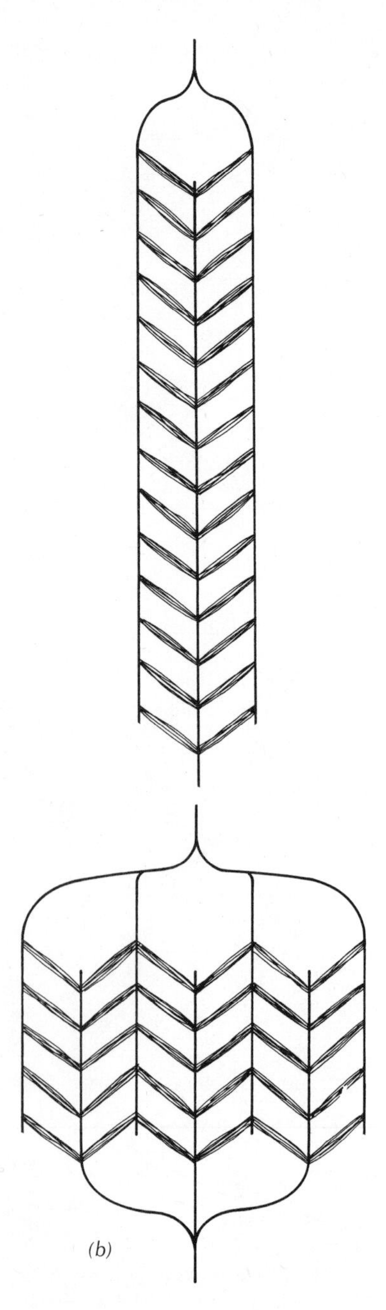

(*b*) These two muscles theoretically exert the same force for the same excursion, but the packing shows drastic differences.

duce more force should have a greater diameter. Since animals are selected for a particular configuration as much as for the ability to generate force at a particular tendon, there will then be an obvious selective advantage to particular packing arrangements. Obviously the biting muscles of an amphisbaenid must not bulge beyond the diameter of its trunk or they will establish the need for an increased tunnel diameter. Pinnate or feathered positions of muscle fibers are more suitable for such sites, as they increase the flexibility of packing arrangements. When muscles are pinnate the tendons of origin and of insertion do not lie on the same but on parallel planes, and the muscles cross at an angle between them (*a*). This angle α obviously changes as the muscle shortens. Since multiple sheets may be placed in parallel, one can discover very complex patterns, even in small animals (*b*).

Pinnate muscles also provide the opportunity for attaching a large number of fibers to a relatively restricted site. These fibers may be arranged into a volume that may be either elongate or wide without affecting the force at the tendon (*b*). A pinnate muscle shifts its mass during contraction within the column in which fibers are arranged rather than by lateral bulge of the sides of the muscle. It is easy to visualize this if one remembers that the sides of each muscle fiber lie at an angle rather than parallel to the sides of the entire muscle.

There is obviously a cost to this pinnate arrangement. This cost results from the shift of the vectors of shortening away from the line of the tendon. These forces must then be resolved into useful (parallel to tendon) and wasteful (at right angles to tendon) components. The reduction of the useful component over the absolute force generated is a function of mean angle of pinnation. Pinnate muscles incorporate an intrinsic compensation since angulation of the muscle fibers causes the tendon to move a greater distance than each fiber will shorten. We know (Box 4-3) that the less a fiber shortens the greater the force generated (below resting length), so we can see that the fibers of a pinnate muscle may remain in the high-tension portion of the length-tension diagram. Each fiber will then produce a slightly larger contractile force than its counterpart in a parallel arrangement. Depending on the angle of pinnation the force at the tendon may be only slightly smaller for the same number of fibers.

The architecture of muscle obviously reflects more than these fairly simple mechanical parameters. There is no reason, for instance, that the motor units of a pinnate muscle need be distributed regularly across the muscle. If the tendons are fan-shaped rather than merely parallel, differential contraction of the left or right side of the fan may impose a lateral component to the movement of the element. Pinnate arrangements, then, represent selective responses to different problems of force application.

or rectilinear progression and push their snout into the soil with just the frictional force of the body. The downward-pointed tail, which in the last chapter was shown to serve in escape saltation (p. 104), digs in and significantly increases the force that can be exerted anteriorly. The need for rigidity in transmitting forces presumably also provides the functional reason for the observed shortening of the trogonophid trunk, in a sequence from those species least specialized mechanically to those most highly modified (Gans, 1960; Gans and Pandit, 1965).

What Shapes the Skull?

The evolution of the burrowing mechanisms described above has affected the shape of the animal's head. The shape will later provide a basis for explaining the differential distribution of the several forms. However, the head is also involved in other functions, and serves the organism in different ways. What structural aspects of the amphisbaenian head can be identified as involving the performance of particular functions? Or, to be more specific, what shapes the amphisbaenian skull?

The four primary and three secondary functions of any vertebrate skull may be summarized as follows:

1. The skull contains and protects the anterior end of the nervous system; this demand obviously became important with a shift toward bilateral symmetry and cephalization in higher vertebrates.
2. The skull contains the entrance of the respiratory gas exchanger.
3. The skull contains the anterior end of the alimentary canal and suspends the jaws in a pattern suitable for food catching, manipulation, and mastication.
4. The skull contains the nasal, optic, and otic capsules.

Beyond these, there are at least three secondary functions of the vertebrate skull:

5. The skull supports appurtenances for secondary sexual characteristics; here belong tusks, horns, antlers, and flaps.
6. The skull supports the external aspects of the sensory receptors.
7. The skull maintains the external configuration of the animal's head.

Each of these functions imposes demands in one of the four following categories:

1. *Spatial*: Each structure will have an optimum size.
2. *Positional*: The jaws must open forward and the sensory systems must be placed so as to intercept the sensory input.
3. *Associational*: The reception of feeding cues must relate to the line

Figure 4-27 The skulls of amphisbaenians document the diversity of burrowing methods: (*a*) the small, round-headed *Amphisbaena bakeri* from Puerto Rico; (*b*) the keel-headed *Ancylocranium ionidesi* from Tanzania; (*c*) the inch-thick *Monopeltis jugularis* from Cameroon; and (*d*) the deep-digging *Agamodon compressum* from Somalia. Each excavates its tunnel system using a distinct specialization.

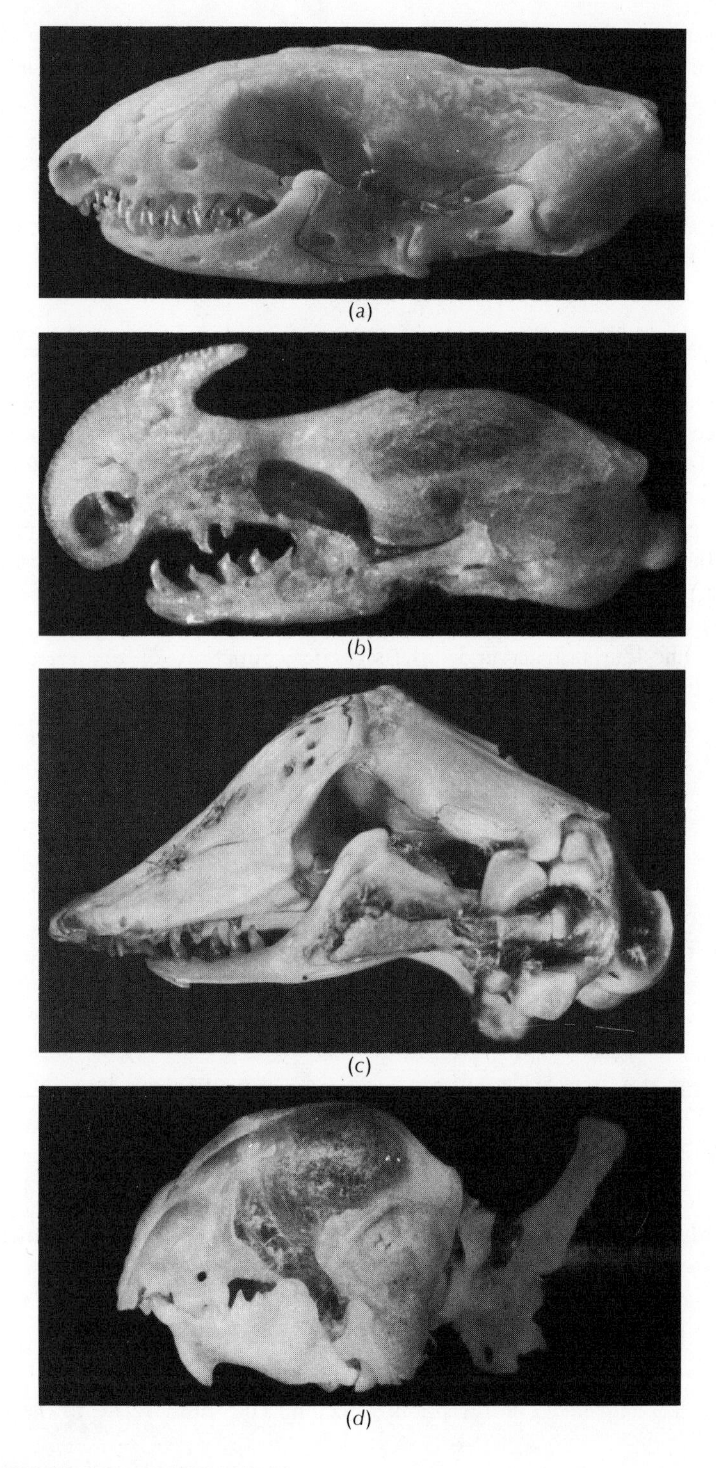

of bite. The distance between left and right sensory input must be fixed.
4. *Structural*: The integrity of the brain and sense organs must be protected.

The demands of different functions overlap. Since multiple functional components affect each mechanical unit (pp. 4, 9), conflict will be inherent in the system and compromise must occur. Masticatory efficiency increases with a forward shift of (for instance) muscle placement, whereas the biting of prey may require a wide gape. The most effective chewing muscle would connect nose to chin; this is hardly a position that will facilitate the initial bite. If the absolute dimensions of the skull were unrestricted, the components might be spaced to reduce the amount of overlap between the various demands. In this case, the gaps might be filled with structures such as pneumatizations, sinuses, and other links that serve mainly to connect the primary elements. Maintaining the several functions becomes ever more difficult as the external dimensions are constrained, and individual elements are utilized for multiple functions. One example of such compression of demands is seen in aquatic forms, in which the head must be streamlined.

The amphisbaenian skull offers an instructive case of complex compromises to meet conflicting demands. The animals bearing it are clearly predators on fossorial prey (such as spiders, termites, grubs, and worms), most of which only rarely emerge from the soil. Amphisbaenians locate and catch prey encountered during tunneling or which have entered their tunnels. Consequently, one may state that the farther an amphisbaenian tunnels, the more prey it is likely to meet (assuming other factors remain constant). We have seen that the work needed to form a length of tunnel varies with the tunnel's diameter. Consequently, it is selectively advantageous for a burrowing predator to elongate its body; though the mass remains the same, it can produce and travel through a tunnel of smaller diameter. It would be even more advantageous to a burrower to be able to perceive prey some distance away in the soil. This would increase the "effective diameter" of the tunnel, and thus the volume within which prey may be found, without increasing the digging work.

Reduction of the body's diameter mechanically stresses internal organs, particularly the skull. If selection would act to reduce the linear dimensions of a roughly spherical skull in parallel with the body's diameter, three things would happen. The jaw would shorten (thus reducing the absolute gape), the volumetric ratio of sensory capsules and brain to body would decrease, and the semicircular canals of the ear would become less effective as the absolute diameter of their curves decreased. Hence the reduction of the skull must be allometric; in order to maintain the functionally useful volume of the skull, at least part of the skull must become more elongate as the body becomes more slender.

In amphisbaenians such allometric reproportioning must do more than compensate for the simple elongation that will occur if the skull's volume becomes "repackaged" to a different transverse diameter. Since the digging stresses are applied near the tip of the snout and the muscular forces near its rear, an elongated skull is subject to extreme bending torsion. Asymmetric stresses encountered while driving the snout into the soil become more important with increased length. The tendency toward buckling cannot be counteracted by fusion of cranial components early in life, as the skull must enlarge during ontogeny by appositional growth upon the free surfaces and edges of the various cranial bones.

The amphisbaenians have overcome all of the foregoing problems with three specializations general to the order. (1) The digging unit is formed by having the facial portion of the skull anterior to the nasal capsule; this reduces the need for direct deformation of other sensory zones. (The eye lies hidden beneath the skin immediately lateral and posterior to the facial portion; it has been independently reduced and hence poses no packing problem.) (2) The brain, lying in the zone between digging unit and occipital condyle, is completely encapsulated by interlocking, sandwiched bony sheaths arranged into a mechanically strong but narrow braincase. (3) The vibration detector, which serves for prey detection, has been shifted anteriorly from the cranial to the facial portion of the skull by the development of an elongate extracolumellar process.

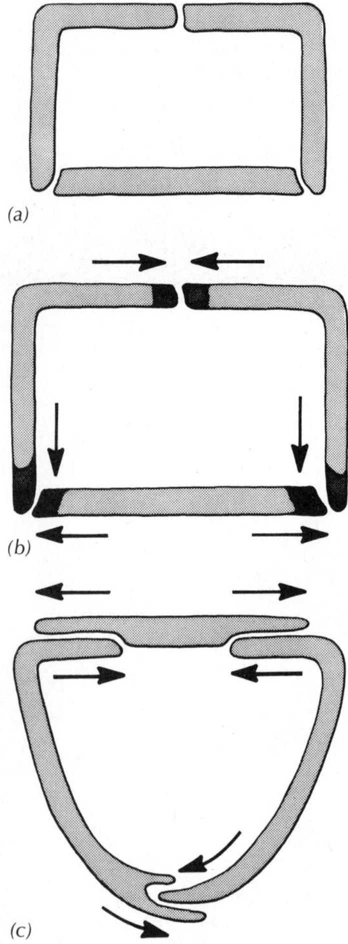

Figure 4-28 Bone grows by accretion; thus only the outside surfaces of a bone may continue to extend. The diameter of a closed bony shell, like a braincase (*a*), may only increase by the deposition of further layers along the sutural regions (*b*). This limits strength of the skull during growth but may be compensated for by convolute sutures. In the amphisbaenian skull (*c*), the bones overlap each other singly or doubly. Growth (arrows) still proceeds along the ends and the sutures remain open, but the layered construction increases the strength.

The general specializations are important enough to suggest that we study them further. The penetrating surface, whether keel-shaped or spatulate, is formed by a bony sheet covered by a soft or keratized integument; the arrangement obviously needs internal reinforcement to avoid deflection or fracture as it is rammed into the ground. Since this penetrating unit contains the nasal passages and anchors the upper jaw, there are additional demands for rigidity. The integrity of the skull is indeed further assured by having a keratinous surface layer thickest at the rostral edge, shielding the closely adjacent bony supports to which it is tied by fibrous connections.

There is, on the other hand, no need for this bony support to be formed by any specific elements. The contribution of a particular bone (for instance, nasal, premaxilla, or frontal) varies from form to form; selection seems to have been for the rigidity of the overall assemblage. The patterns seen in Recent forms perhaps depended on the nature of developmental flexibility within the ancestral population and on the effect of the possible architectural schemes on the arrangement of nasal passages or dental supports. This has led to a situation in which one may differentiate between parallelism and convergence. In parallelism, equivalent selective forces acted upon organisms with equivalent genetic potential; they produced structures that are both homologous and analogous (Box 1-1). In convergence, equivalent selective forces, such as those set up in burrowing, acted upon organisms with different genetic potential. It is unlikely that selection for other activities, such as breathing or biting, would accidentally be similarly equivalent; hence convergence tends to produce analogous, but nonhomologous, structures. There are, for example, three superficially similar groups of spade-snouted amphisbaenians, but the genera *Leposternon*, *Monopeltis*, and *Rhineura* each have a drastically different arrangement of the underlying bones.

The second requirement of burrowing specialization is for the transmission of forces from occipital condyle to anterior wedge. These forces arise when the penetrating surface is driven into the soil; different forces are imposed when the surface is shifted laterally or dorsally by rocking the skull about the condyle or rostral edge. The portion of the skull connecting spatula and condyle also serves as the braincase; its contents are particularly susceptible to pressure and deformation. Deformation is obviously disadvantageous; consequently, the posterior portion of the skull should be capable of withstanding high compressive stress and high bending moments with minimal deflection. Lateral to the braincase lies the visual system and associated glands; next to these run the attachments of the muscles that power the bite and control the attitude of the skull.

Since the diameter of the skull's periphery influences the energy cost per unit length of tunnel, one may interpret the structure in terms of advantage to a thin but strong frame which can grow from juvenile to adult while supporting burrowing stresses. Amphisbaenians have achieved this by joining their cranial elements in mortise rather than butt joints. For instance, the braincase of the largest African fossil, *Lystromycter leakeyii*, has the posterior end of the parietal divided into an internal and external flange that sandwich between them the anterior edge of the supraoccipital. The three sheets of bone do not lie just over each other but their facing surfaces bear matching grooves and rills, curved to prevent sliding. This strengthened braincase forms a tubular strut that supports the anterior penetrating plate but remains narrow enough so that the masticatory muscles in their postocular position lateral to the skull will not significantly increase the diameter of the head.

The digging forces are transmitted to the amphisbaenian skull in several ways. The penetrating strokes are powered via the vertebral column. They induce compressive stresses in the braincase, as well as downward and lateral bending stresses in the penetrating wedge. The muscles attaching to the occipital surface of the skull facilitate these movements by stabilizing the skull on the vertebrae composing the head joint. However, the various kinds of rocking movements serving for tunnel widen-

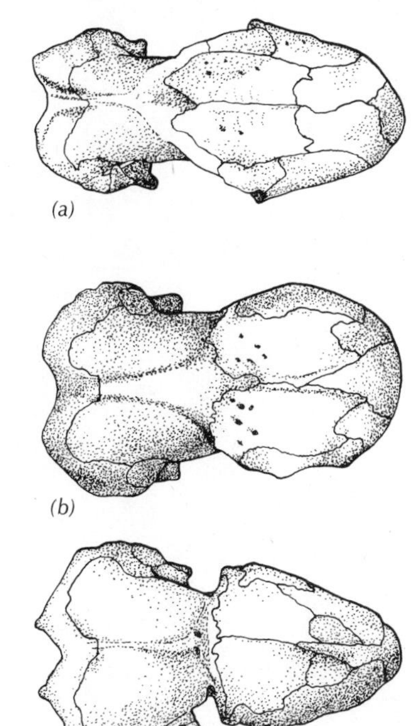

Figure 4-29 *Rhineura* (*a*), *Monopeltis* (*b*), and *Leposternon* (*c*) each show a spade-shaped rostrum. Although the skulls indicate superficial similarity, this applies only to their shape and not to the component bones. The extent occupied by the azygous premaxilla which forms the anterior tip provides a good example and supports the idea that we are dealing with a superficial convergence of three lines.

ing, rather than penetration, require more extensive muscular attachments. As rotation of the skull around the occipital condyle requires moments, there is a selective advantage to inserting the muscles a maximum distance from the instant center of rotation about the condyle. A series of neck vertebrae is normally involved in twisting the head, actually allowing the instant center to move some distance away from the condyle itself. The insertion of the bending muscles on the skull tends to shift anteriorly in the more specialized keel-headed and spade-snouted species; muscle attachment is to the posterior edge of the penetrating shield in spade-snouted forms and to the often-protruding top of the keel's bony base in keel-headed species. The neck vertebrae of these more specialized forms also have considerable capacity to bend on each other. When this occurs, the ventral tendon bands of the head-tilting muscles depart from the arc (formed by the vertebral series) to approach its chord, constrained mainly by ligaments and the skin (Figure 4-30). This departure from the arc shifts the muscle's line of action away from the condyle, and increases the moment arm on the head as bending

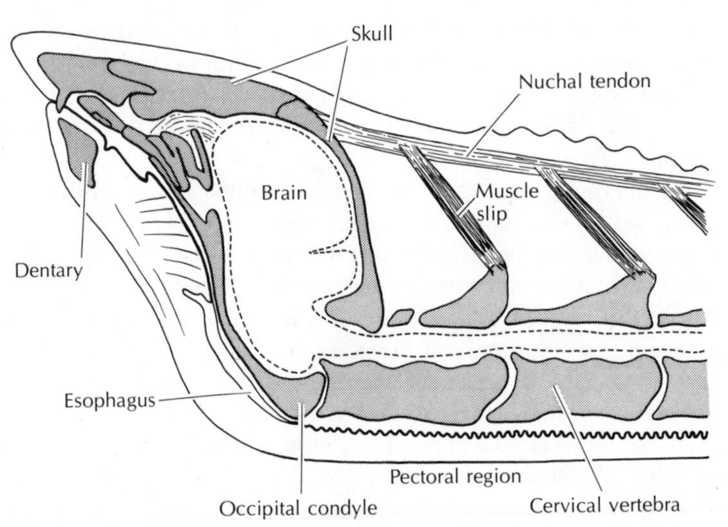

Figure 4-30 The sagittal section of an African spade-snouted amphisbaenian, *Monopeltis*, shows the position of the nuchal tendon when the head pushes against the top of the tunnel. As the head rotates about the occipital condyle, the tendon's attachment to the dorsal edge of the facial shield increases its moment arm; the distance from condyle to the tendon becomes greater. During the movement of the head, the enlarged pectoral shields (which lie ventral to the cervical vertebrae) must slide forward, as the tip of the snout does not extend as far forward in the dorsalmost as in the ventralmost position of the widening movement. This, of course, accounts for the selective advantage of the polished pectoral shield. Also interesting is the sharp bend of the spinal cord at the head joint and the compression of esophagus and trachea between vertebrae and skin.

proceeds; the gradually increased moment arm compensates for the reduction in the force produced by each muscle fiber as it shortens (cf. Gans, 1966).

Sensory and Feeding Influences

Amphisbaenian eyes retain high sensitivity but are of little use in the deeper tunnels (Gans and Bonin, 1963). Olfaction alerts these animals to the existence of prey in their immediate vicinity, but is of little use in detecting more distant prey. Hearing, as well as touch and the reception of vibration, is found to have achieved major importance (Gans, 1960).

Review of numerous series of skulls suggests that the otic capsules represent a limiting factor in terms of reduction of skull width (Figure 4-31, cross section of *Bipes*). In each instance the skull is widest at the level of the semicircular canals, and the inward-bulging otic capsules are restricted by the brain medially. This regular feature apparently reflects the fact that the response of semicircular canals to movements of the head will, for a given sensory acuity, be a function of the size of the canal and its maximum distance from the center of the animal (Jones and Spells, 1963). Since the system facilitates balance (Trendelenburg and Kühn, 1908), maintenance of its bulk should be important to a subterranean predator unable to utilize visual cues in determining direction.

Other aspects of the skull similarly seem to reflect modification in response to sensory demands associated with the burrowing habitus. The stapes is large, indeed enormous, in all amphisbaenian species. It is joined by an anterior element or extracolumella which, unlike that of lizards (Wever and Gans, 1972), is homologous to the cartilaginous epihyal. This epihyal cartilage is present as a lateral element of the hyoid apparatus in *Bipes* and *Blanus*, which consequently lack an extracolumella. In all other genera it has lost contact with the hyoid and become articulated with the stapedial tip, from which it runs anteriorly to terminate along the side of the face. This extracolumellar apparatus has different configurations, ranging from a simple slender cartilaginous rod (*Amphisbaena*–Figure 4-33) to an enormous mass of cartilage and connective tissue (*Monopeltis*–Kritzinger, 1946) or a large bony plate

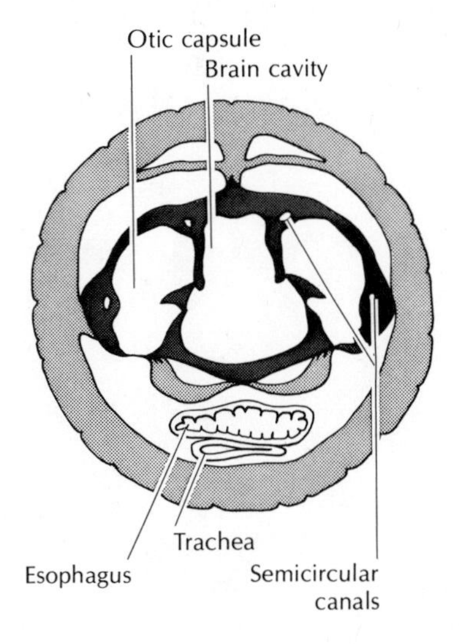

Figure 4-31 This cross section through the back of the skull of *Bipes biporus* shows that the otic capsules occupy the lateral extremes of the skull. Although modification in this genus does not allow the extreme movements seen in the tunneling of *Monopeltis* (Figure 4-30), the skull here fills most of the animal's width and a significant portion of the height. The remainder of the section reflects the position of the thick layers of connective tissues and of the dorsal muscles; esophagus and trachea occupy but a small, restricted ventral zone.

Figure 4-32 The hyoid of *Bipes* (*a*) still retains epihyal horns, and this genus lacks the extracolumella. In *Amphisbaena* (*b*) and most other genera the epihyal wing has become separated from the hyoid. Attached to the stapedial bone of the middle ear, it is now called an extracolumella and its connection produces a sensitive zone on the side of the face.

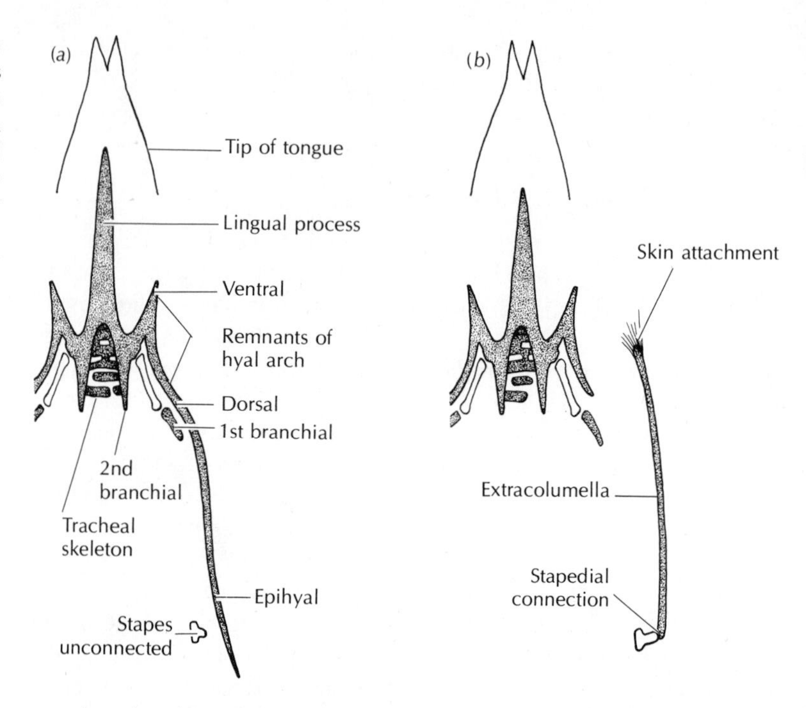

covering the side of the face *lateral* to the quadrate (*Diplometopon*—Gans, 1960).

The hearing system has been investigated both histologically and by recording cochlear potentials (Gans and Wever, 1972; Wever and Gans, 1973).[9] These experiments showed that the amphisbaenian extracolumella serves as an amplifying system which (1) provides for the pickup of airborne sounds on the side of the face (except in *Bipes* where maximum sensitivity is in the region of the ear), (2) transmits the vibration to the inner ear via the extracolumellar-stapedial linkage, and (3) amplifies the vibrations in the process. Exposure of the extracolumella by removal of the superficial soft tissues has only a minor effect on the system; disruption by removing even a small section of the extracolumella causes an immediate loss of sensitivity of 20 to 40 decibels (Figure 4-34*a*). These specialized vibration receptors have significant directional discrimination; an asymmetrical sound source produces asymmetrical cochlear potentials (Figure 4-34*b*).

[9]Cochlear potentials are recorded by placing a needle electrode on the round window or, in this case, into the sacculus; both sites are adjacent to the sites where the animal presumably transduces mechanical to electrical energy. Sounds are then directed at specific portions of the skin and electrical output of these electrodes is compared to those attained by two indifferent electrodes in the soft tissues of the head. The potentials thus measured reflect the electrical response of the otic system to imposed vibrations. One may then test for reaction to aerial and mechanical vibrations applied at different sites and frequencies.

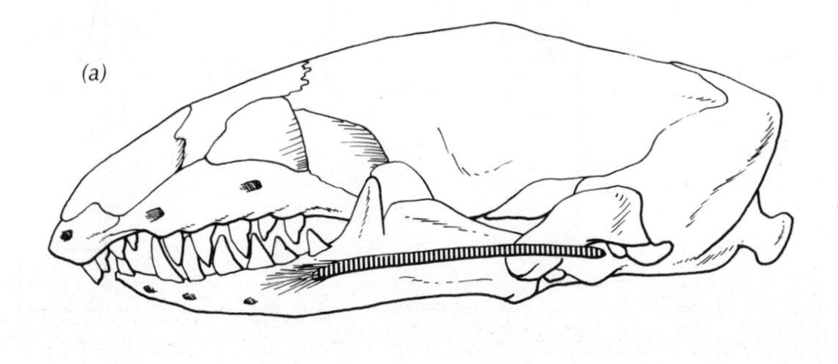

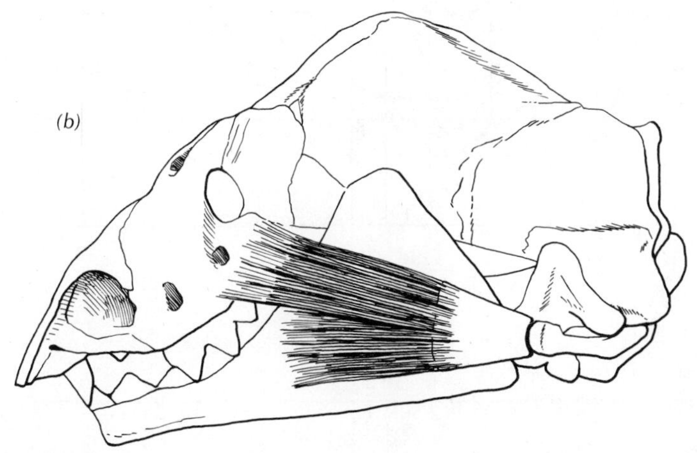

Figure 4-33 Diagram to show the position of the cartilaginous extracolumella in *Amphisbaena caeca* (*a*) and the bony extracolumella in *Diplometopon zarudnyi* (*b*). In both animals this homolog of the epihyal (Figure 4-32) provides the major linkage from the sound-receiving surface to the middle ear. (After Gans and Wever, 1972)

Sound cues hence may be used to locate sound sources within the soil. This supports behavioral observations (Gans, 1960) which suggest that amphisbaenians hear their prey, and may indeed hear it through the soil. It is interesting that among the species thus far examined, the two amphisbaenian genera with the least modified skulls (*Bipes* and *Trogonophis*) show the best sound reception, even though the number of inner-ear hair cells in their cochlea is relatively lower than in other amphisbaenians and the mechanical linkage to the skin absent or most simple. Although the absolute sensitivity is greatest in these species, it remains to be seen whether their ear is most effective in prey detection.

The anteriad shift of the prey-detecting "ears" from the back of the jaws to the side of the face reflects the parallel specialization of the amphisbaenian biting pattern. Observations with artificial tunnels show that amphisbaenians have two main methods of taking prey. The first

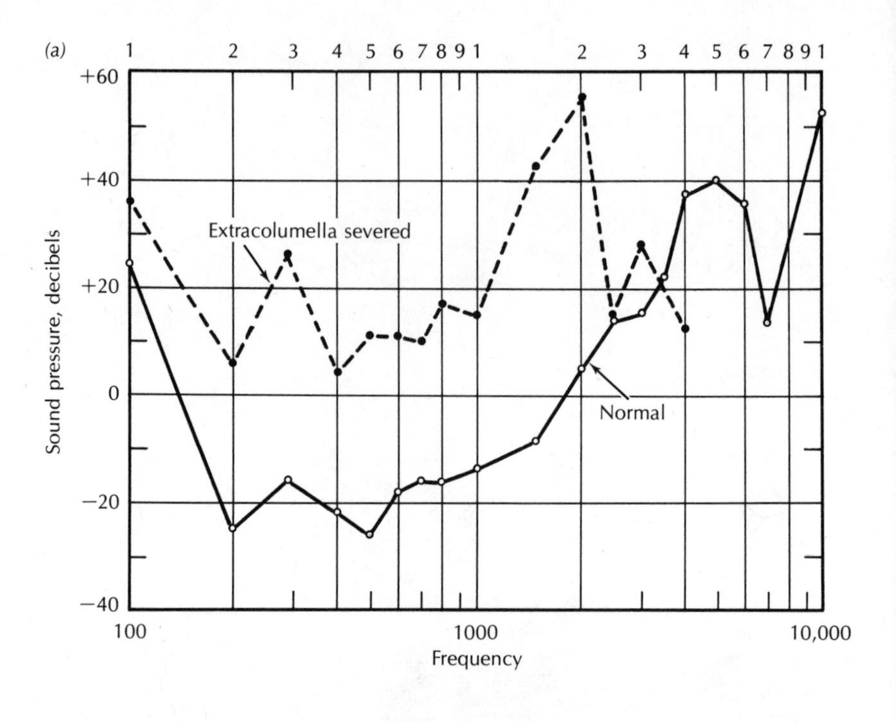

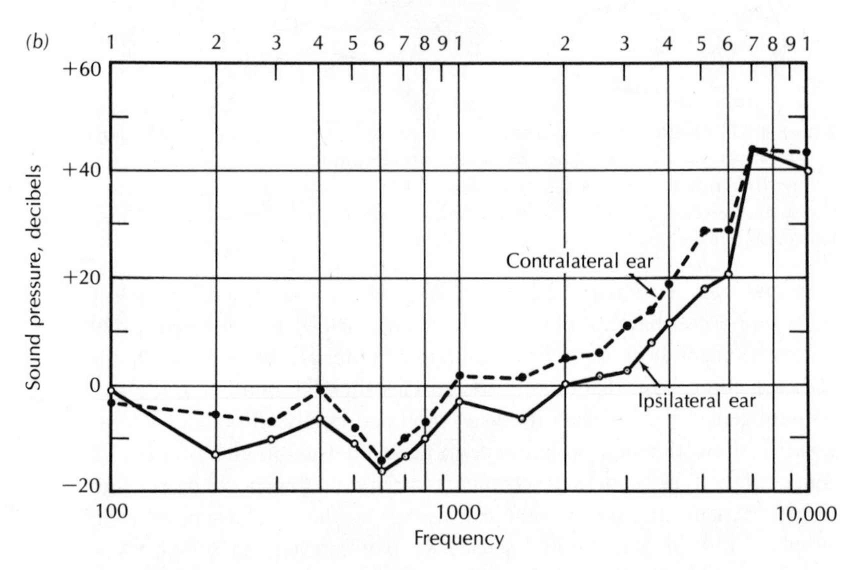

Figure 4-34 Sensitivity curves showing the amount of sound energy required to produce a cochlear microphonic of 0.1 microvolt: (*a*) the response of a single animal with the extracolumella intact (solid line) and after it has been severed (dashed line); (*b*) the response of the two ears to sound coming from one side.

is the attack bite seen also in snakes and lizards. In this, the amphisbaenian moves at the prey in a straight line with open mouth, as when the small Zululand species *Zygaspis violacea* emerges from the sand to pick termites off the surface, or when the giant South American *Amphisbaena alba* has driven mice into tunnel dead ends. Alternately, amphisbaenians move past prey or perhaps perceive (by its movement?) prey encountered alongside them. This prey is then forced against the wall, and the amphisbaenian backs up, keeping the prey immobile until a sideways bite permits a firm grasp. For both kinds of bite there is an obvious advantage to the shift that places the vibration detectors adjacent to the jaws. It is amusing to note that a similar shift of the sound-detecting site has occurred in whales, in which the sonar receptors again appear adjacent to the jaws.

Most amphisbaenians have conical teeth with those of upper and lower jaws interlocking. The staggered arrangement is clearest at the tip of the snout where the median premaxillary tooth is always enlarged and fits between the anterior cones of the left and right dentaries. The grip exerted is singularly effective in crushing prey. Many species twist after

Figure 4-35 South American spade-snouted amphisbaenians such as this *Leposternon microcephalum* will feed on insect larvae. They can also utilize larger prey as shown by their ability to bite chunks out of pieces of fish left on the surface of the soil.

achieving a bite, then tear loose the grasped portion. Consequently, they are much less limited as to prey size than are snakes, which have specialized to swallow their prey whole. An *Amphisbaena alba*, for instance, will stay with a mouse it has killed and bite pieces out of it again and again until it all has been ingested.

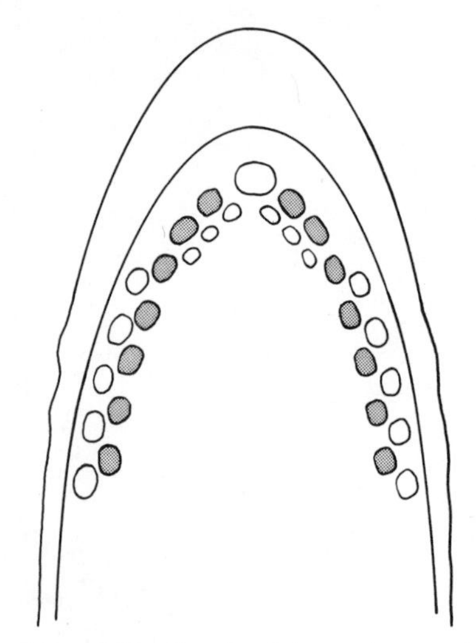

Figure 4-36 Sketch of the occlusion of the amphisbaenian bite. The teeth of the upper jaw are shown hollow and those of the lower jaw solid. The arcades can be seen to interlock so that a twisting of the head will cause the prey to shear along the line of the bite. This accounts for the formidable bite of these predators.

The dentigerous (tooth-bearing) bones of the upper jaw have some motility against the braincase, particularly in juveniles. There is a limited tendency toward cranial kinesis,[10] providing a slight increase of the gape. Adductors of the lower jaw originate on the braincase or quadrate. The latter is effectively locked into the side of the braincase on which it may rock but slightly. The pterygoid muscles are relatively small, suggesting that they stabilize the pterygoid arch rather than cause major protrusion or retrusion. All but three amphisbaenians have the lower jaw recessed into the upper so that soil-penetrating forces will not impinge on the mandibular symphysis. The depressed position of the lower jaw suggests that either the depressor of the mandible or some of the anterior slips of the hypaxial musculature may open the mouth. In some spade-snouted species the depressor inserts on a prominent retroarticular process that provides effective leverage for opening the mouth even when the snout is depressed.

Some of the motility of the upper jaw is supposed to reflect a shock-absorbing mechanism that reduces potential damage to the brain. Such mechanism need only be accessory, perhaps for contending with formidably large, struggling prey. The most effective and obvious control of the jaw should regulate the strength of contraction of the closing muscles; at least in mammals, this is modulated by the input of various proprioceptors (Kallen and Gans, 1972).

A major selective factor influencing the ways in which the jaws are supported would seem to be transmission of the forces imposed by the bases of the teeth when grasping or twisting prey. The bite does tend to exert ascending moments onto the snout (tending to bend it dorsally); these are generally in direct opposition to the bending moments induced in the tunnel-enlarging movements of spade-snouted species (tending to bend the head ventrally). Here is another instance in which it is difficult, if not impossible, to deduce all of the imposed forces from the structure's architecture. When several mechanical demands overlap in identical or

[10]Some lizards, indeed many lower vertebrates, retain a certain flexibility of their skull even as adults. This is known as cranial kinesis and generally involves movements of the snout, the tooth-bearing bones (such as maxillae, palatines, and pterygoids) and of various bracing elements (such as the ectopterygoids, epipterygoids, and quadrates) against the braincase. The actual site of kinesis differs between species and groups. The function of kinesis has been the topic of much argument (cf. Bock, 1964; Frazzetta, 1961; Iordanski, 1971). Some of the conflict results from the fact that kinesis may serve different functions in the adults of different species.

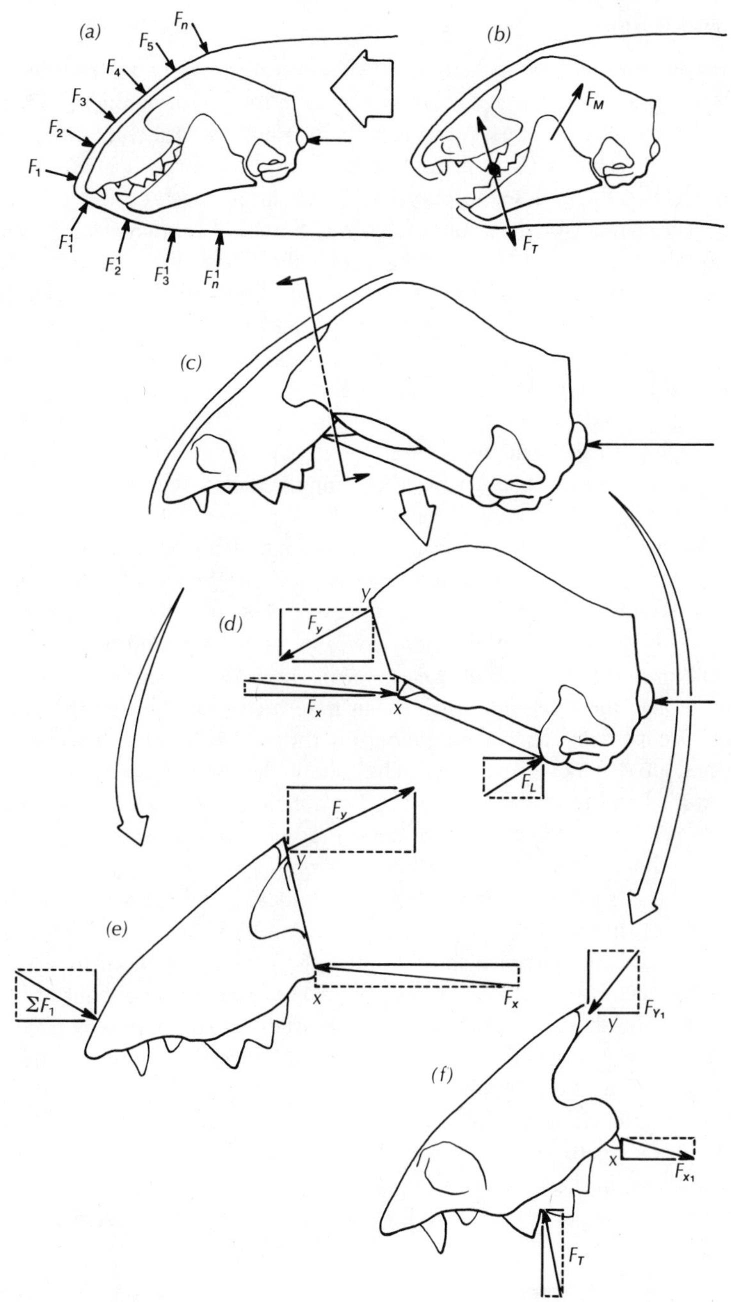

Figure 4-37 These diagrams of the forces exerted on the skull of *Diplometopon* when digging (*a*) or biting into prey (*b*) show that the loadings in the cranial roof (*y*) and on the maxillopterygoid arch (*x*) occur in opposite directions. Sketch (*c*) shows the skull with the line along which it is separated (into two "free bodies," cf. Figure 3-3) for analysis. Sketch (*d*) shows the posterior portion including the force imposed during burrowing, and (*e*) and (*f*) show the anterior portion loaded during burrowing and biting, respectively. Note that both direction and magnitude of vectors at points *x* and *y* have changed with the change in function. Force analysis has been simplified by considering joints pinned and neglecting moments. (After Gans, 1960)

in conflicting directions, it is difficult to know which demand imposes the critical load and whether the tensile or compressive strength is limiting the system.

Compromises

Interpretation is easier when one of the several functions achieves overwhelming importance. An example occurs in the Trogonophidae. Their oscillatory movement is induced both by muscular attachment to the back of the skull and by transmission of torsional forces via shortened cervical vertebrae. The morphologically advanced species show a diagonal arrangement of the cephalic sutures. Sutures between frontals and parietals are elongate and interdigitating; their directions diverge from a point on the midline. Such a connection of roofing bones obviously braces the skull against torsional stresses, and the sutures on each side may be seen to follow the tension lines imposed by torsion. (The nature of these sutures might have helped me to unravel the trogonophid burrowing methods; only after seeing the living animals burrow did it become clear to me that the sutural angulation represented a significant departure from the pattern in other amphisbaenian skulls.)

Each trogonophid genus retains a different stage of specialization for oscillatory burrowing. Those species showing advanced oscillatory patterns have short skulls that are apparently strengthened to reduce torsional strains and hence increase their stiffness in several dimensions. Figure 4-39 visualizes these differences by superimposing a grid on a lateral view of the skull of *Trogonophis*, the least specialized genus in the family, and drawing a similar but deformed grid through the position occupied by equivalent points on the skulls of more specialized trogonophid genera.[11] The sketches show that the skull has become shortened and the facial portion has been inclined ventrad, whereas the cranial component became tilted dorsad rather than compressed. In contrast to these relatively simple migrations of mechanical components, the resultant angulation between braincase and face forced severe modification of the biting apparatus. The shortening of the overall length of the skull and proportional shortening of the jaws required an increased angular opening to achieve an equivalent linear gape at the front of the mouth. The shortening of the tooth row reduces the number of teeth that may be positioned along its length and hence the potential area of biting surface. The trogonophids have compensated for this by fusion of the individual teeth. For instance, each set of two maxillary and five mandibular teeth of *Agamodon anguliceps* forms a single compound tooth acting in shear against its opposite. The jaws have become deeper, increased their dorsoventral thickness, to match different force patterns.

[11]This method of deformed coordinates is ordinarily ascribed to D'Arcy W. Thompson (1951), but actually dates back to Dürer's time (Richards, 1955). There is some question whether it should be used for composite animals (rather than individual elements) or for two- (rather than three-) dimensional reconstructions. In spite of its limitations, the method gives a facile visualization of proportional changes in ontogeny or phylogeny.

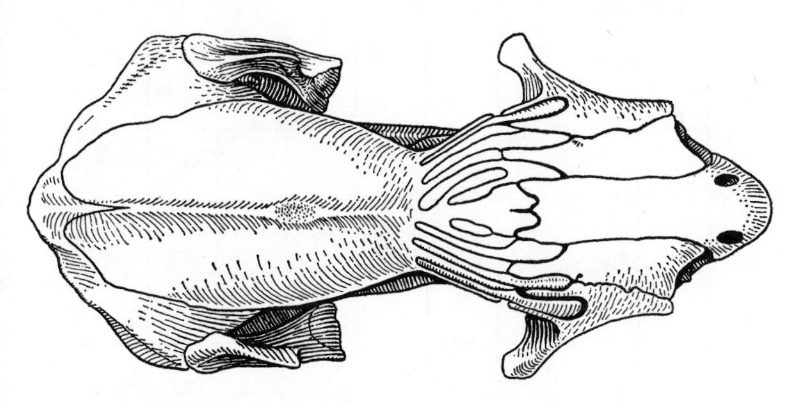

Figure 4-38 In dorsal view the skull of *Trogonophis* shows the elongated interdigitations among nasals, frontals, and parietals. Their radiate, rather than parallel, arrangement apparently reflects the skull's capacity to resist the torsional forces induced during oscillatory digging. (From Gans, 1960)

An increase in the relative height of the coronoid process increases the mechanical advantage of the temporal muscles.

The most important compensation for a shortened skull seems to be the depression of the mandibular articulation relative to the level of the tooth row. This induces a backward as well as the upward component to mandibular closing, which locks the prey between the sharp-edged tooth cusps so that the bite will shear rather than merely puncture the prey. All of these modifications for biting, as well as the parallel restructuring of the extracolumella into a wide bony plate, apparently represent compensations for a shift to a different burrowing method involving excavation by torsion. Perhaps the shortening of the trogonophid skull was possible only because the excavation method, which produces a tunnel by a single set of oscillating motions, demanded or permitted a simultaneous shortening of the trunk while maintaining a relatively constant body volume. It has been suggested above that the respective specializations of ears and jaws in the three advanced trogonophid genera are responses to the functional influence of different preferred burrowing methods. Each trogonophid genus apparently encountered different conditions and remained at a different equilibrium state between the advantage accruing from further perfection of the burrowing mechanism and the inherent limiting necessity of compensating for the associated inefficiency of prey perception and food handling. Is it possible, for example, that the geographically peripheral position of the Trogonophidae (Figure 4-14) results from their inability to spread from a limited environment to which they are highly adapted? Could the restriction reflect the innate disadvantage of a particular burrowing method, in this

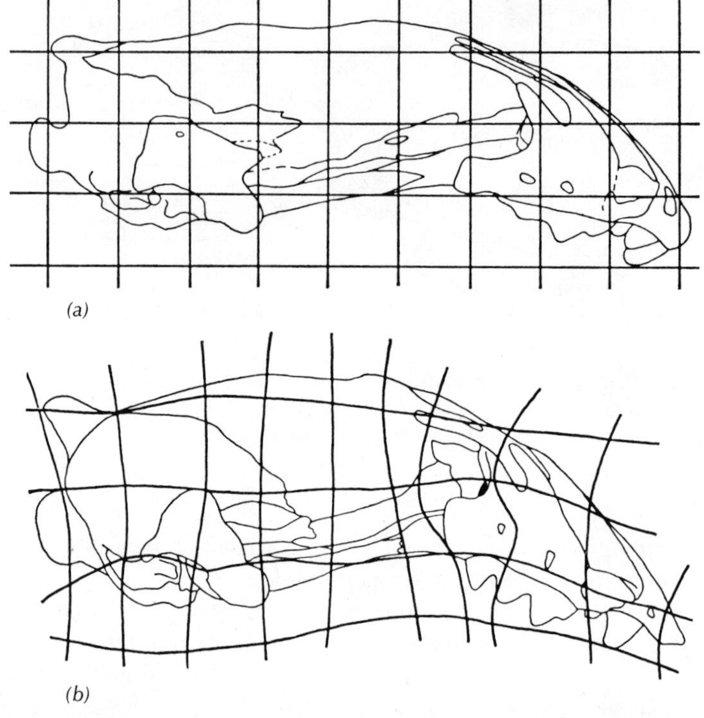

Figure 4-39 These diagrams of deformed coordinates suggest the way in which selection has reformed the skull of the advanced trogonophids *Pachycalamus* (*b*), *Diplometopon* (*c*), and *Agamodon anguliceps* (*d*) from an original condition that is likely to have been similar to that seen in *Trogonophis* (*a*). (From Gans, 1960)

case oscillation, that required them to increase rather than decrease their diameter per unit volume?

Similar questions may now be raised concerning the distribution of the remaining amphisbaenians (Gans, 1968). It becomes clear that all amphisbaenians seem to be very successful predators in the subterranean environment; their general pattern of adaptation allows them to compete successfully with other fossorial squamates. Their competitive success need not be considered exclusively in terms of the effectiveness of tunnel generation or prey capture; at least the middle-sized and larger amphisbaenians can deliver a formidable bite and are consequently capable of preying directly on practically any subterranean squamate. Amphisbaenians may well prefer particular soil types, such as tropical humus or sandy soils, as well as moist microclimates. Such amenable sites may

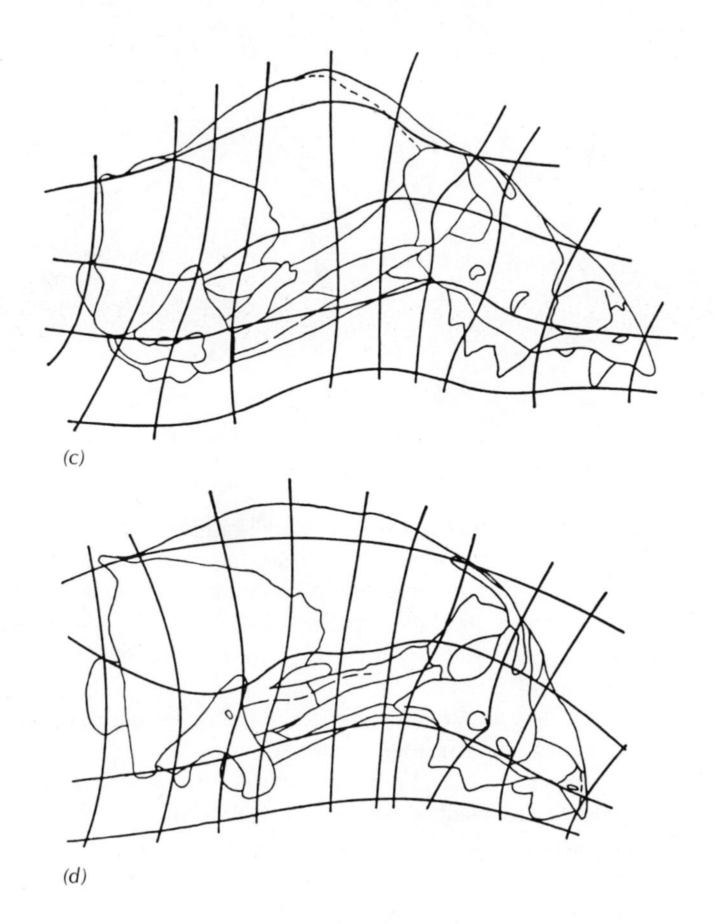

be occupied by any amphisbaenian. This suggests an explanation for the observation that the overall range occupied by amphisbaenians is well defined by the range of generalized and round-headed burrowers. Only the spade-snouted Floridian *Rhineura* and the members of the trogonophid radiation occupy areas outside of this zone (Figure 4-14). The generalized species apparently can occupy the subterranean biotope of the overall range without further specialization for burrowing.

Adequate resources in some regions may be partitioned and support several species; such sympatric amphisbaenians almost always differ in size (Gans, 1971) and may utilize distinct sorts of prey and portions of the ground. We need studies of such sympatric situations, but it is already clear that the probability of encountering more than one morphologically generalized species in a given spot increases as one moves closer to the geographical center of either the African or the American

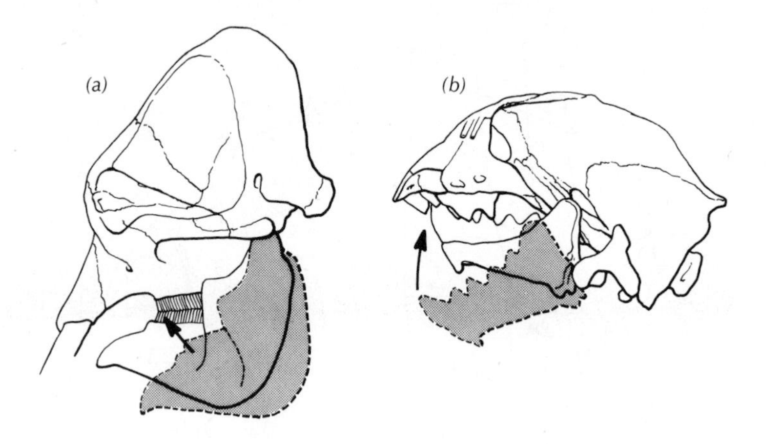

Figure 4-40 The position of the jaw joint affects the way the teeth impact on food objects when the jaws close. The teeth approach each other at right angles to the contact surface when the jaw joint lies on its posterior extension. Food objects are then exposed to compression (or shear) between adjacent cusps. In the giant panda and the elephant (*a*), the jaw joint lies far dorsal to the occlusion line. During closing, the dentary teeth then have a forward vector (arrow) which imposes shearing forces on food objects. In the elephant, the shear peels the bark off branches; in the panda, the shear allows the animal to crush bamboo stalks.

In amphisbaenians, such as *Diplometopon* (*b*), the jaw joint lies below the contact line. This causes the dentary teeth to approach the maxillary ones with a backward as well as upward motion (arrow). The action traps the prey and provides further shearing forces, thus cutting the chitinous armor of small arthropods.

composite range.[12] As stated above, the overall range of the generalized burrowers overlaps entirely that of the spade-snouted and keel-headed forms (Table 4-1). Furthermore, there are more generalized round-headed amphisbaenians than the combined numbers of spade-snouted plus keel-headed ones. Indeed, there are more round-headed forms than those belonging to the two latter groups plus all structurally intermediate forms. There are slight differences in the ratios, but the general pattern is equivalent in Africa and South America.

The spade-snouted and keel-headed amphisbaenians burrow by the two-cycle (penetration-compression) mechanism. This is mechanically more elegant, certainly it is more complex than the single-cycle drive used by the "generalized" round-headed forms. Why, then, do the generalized species occupy a greater geographical range and, within this, maintain a greater number of species? Do such standards of area and number relate to evolutionary success? Or should the definition of evolutionary success be restricted to the number of offspring contributed to the next generation in a particular species? One wonders whether the kinds of parameters here cited may not have long-range advantages to the survival of particular specializations. Or does the occupation of a larger geographic range (or of more niches) imply only that there is a geographically larger zone for which this specialization is suitable? In any case, the generalized species are more common and one encounters a much wider region that is occupied by them. How could such a state become established and how is it maintained?

[12]As the taxonomy of amphisbaenians gradually became clarified, it became obvious that sympatry of generalized burrowers is much less common in Africa than in the Americas. The Recent round-headed amphisbaenians of Africa, furthermore, are smaller than their American counterparts. While the smallest round-headed forms occur in Africa, the American fauna includes many round-headed species larger than any found in Africa. This nonparallel situation still requires explanation.

Table 4-1 Distribution of Amphisbaenids

	Distribution	
Amphisbaenid	Ethiopian Africa (excluding Mediterranean region)	South America (excluding Mexico and Caribbean islands)
Round-heads	26 species 4 genera	35 species 1 genus
Keel-heads	6 species 3 genera	4 species 3 genera
Spade-snouts	17 species 2 genera	8 species 2 genera

In answering this last question we must differentiate generalized organisms from those specialized for burrowing. It has already been noted (p. 134) that in a broader context all amphisbaenians are certainly specialized.[13] The analysis of the skull now suggests that the further modification of the animal's anterior end for digging proceeds at the expense of reduced efficiency of other cephalic functions. The flattening of the anterior shield encroaches on the space available for the olfactory system and, more importantly, forces a shift of the tooth rows to a position less advantageous for biting and crushing. The ear has been similarly restructured, perhaps at the expense of auditory acuity, and the various modifications of the amphisbaenian extracolumella are then explicable as compensatory devices for this consequence of cranial deformation.

Such different degrees of feeding, sensory, and burrowing efficiency are likely to occur when inhabitable subterranean zones are large enough and contain enough food to permit specialization. The shallower and perhaps loose subsurface zones would then be occupied by a generalized amphisbaenid and the deeper ones by a specialized one. If the environment provides resources for more than two species, each zone might be further partitioned. This sympatry might be expected to occur in those parts of the entire group's range where the environmental resources would be richest. The keel-headed and spade-snouted species could be expected to be better suited for burrowing in and occupying harder or

[13]This statement emphasizes the dangers inherent in such subjective terms as generalized and specialized, primitive and advanced. If members of a group become sufficiently specialized to establish themselves in a new habitat, there is a good chance that they could give rise to an adaptive radiation exploiting all of its aspects. Are the members of the invading group advanced (or better derived) or primitive? The example makes it obvious that the terms only have meaning as long as we define the group within which they are to apply. More than that, these terms will apply only to the states of one or another specialization, in this case those for burrowing.

coarser soils than the round-headed forms. They could presumably dig deeper tunnels and prey preferentially on the inhabitants of deeper zones. The round-headed forms might be more effective in surface zones, perhaps in terms of feeding specialization (though such factors as predator response might also be important).

Both of these expectations are confirmed. For example, the spade-snouted species of the African genus *Monopeltis* occur in deeper sands in zones where they are sympatric with the shallower-dwelling, round-headed *Zygaspis quadrifrons* (D. G. Broadley, personal communication). At Kalabo, Zambia, two sympatric species of *Zygaspis* show character displacement in size and scale patterns (Saiff, 1970). Both are sympatric with *Monopeltis ellenbergeri*, a slender, spade-snouted form that deeply burrows. In those patches of the overall range of *Monopeltis* in which *Zygaspis* is absent, the thickest, most pigmented, and least modified of the local species of *Monopeltis* seems much more likely to range immediately adjacent to the surface. This suggests that it is capable of withstanding surface conditions. Does it invade this zone only where more effective competitors are lacking? Would the "less specialized" surface forms prey on it at some stage of its life cycle?

The patchiness of these African sympatry patterns is still being studied. One does not observe simple allopatric replacement of one round-headed by another or by a spade-snouted form; there seems to be a series of up to five species involved, with differing but generally gradable capacities to deal with such environmental demands as tunnel formation. The number of species actually occurring at a particular site seems to be determined by the depth and diversity of the substratum layers. Whenever round-headed forms are absent from the surface soils, this zone is inhabited by the least modified of all the local spade-snouted forms; more highly specialized spade-snouted species may also occur, but in the deeper layers. In South America the situation is similar. Multiple species here may be sympatric in the central parts of the amphisbaenian range, but one mostly observes sympatry between several round-headed and only one specialized species. Also, there seem to be fewer sites at which the specialized species occupy the immediate subsurface zones. Does this reflect, or does it induce, the greatly increased range of sizes shown by the generalized forms on that continent? In any case, it does suggest reasons for the observation (based on morphology) that the several kinds of specialized forms derived from local generalized ones. Spade-snouted and keel-headed species each appear to have evolved at least once in the Old and once in the New World.

The sequential replacement of morphologically generalized by specialized species in the deeper, subterranean layers of some regions does not explain the absence of sympatry between spade-snouted and keel-headed amphisbaenians. The latter are clearly specialized relative to the round-headed species, and such limited ecological information as is available (Gans and Rhodes, 1964; D. G. Broadley, personal communication) suggests that they do occupy the deeper and denser soil types as might be expected. The keel-headed amphisbaenians are significantly smaller than the spade-snouted species, both in Africa and South America; indeed, there is almost no overlap in range of body diameters. Furthermore, in both Africa and South America there are about three times as many spade-snouted as keel-headed species, and the range occupied by the former is much larger.

For driving a wedge into packed soils, there is no apparent difference in mechanical effectiveness in having the edge positioned horizontally rather than vertically. The possible mechanical advantage must be sought, then, in the tunnel-widening mechanism. Here again there is a certain basic similarity between keel-headed and spade-snouted forms. The distance from the site of the muscle attachment to the animal's centerline—in other words, the moment arm of the force applied at that attachment—does not appear to differ significantly between the two forms. What does differ is the maximum mass of muscle (or number of muscle fibers) that can be employed in a single power stroke.

In the keel-headed forms there are two such muscle masses, one on each side; either can move the head toward its own side only. In the spade-snouted forms, the major fraction of the epaxial musculature of both sides can act as a single unit, driving the upstroke; the downward movement of the head could occur simply by gravity. Alternately, the head may be pulled ventrad against minimal resistance by the weak hypaxial muscles. The vertically keel-headed forms thus employ a two-phase compression acting on the left and the right walls of the tunnel alternately; each compression phase can be driven by no more than 50 percent (the right or the left half) of the total mass of muscle. The horizontally spade-snouted forms, in contrast, have only a single-phase compression stroke. They compress the roof and the floor of the tunnel in a single movement; this stroke may be powered by up to 100 percent of the anterodorsal (close to 80 percent of the total axial) muscle mass. The mechanical limitation of the keel-headed forms might then be seen in the kinds of forces that can be exerted for tunnel widening, and perhaps also in the energy levels that may be continuously maintained by their muscle system (Gans and Bock, 1965).

If the keel-headed and spade-snouted forms have indeed been specialized for penetrating denser, harder, more cohesive, and less compressible soils than round-headed forms, such a mechanical limitation may well be important. There could be a compensating advantage to the keel-headed pattern; we note that both kinds of animals are found. Unless it involves aspects beyond the present comparison, such an advantage may well be related to size, and perhaps to the fact that the

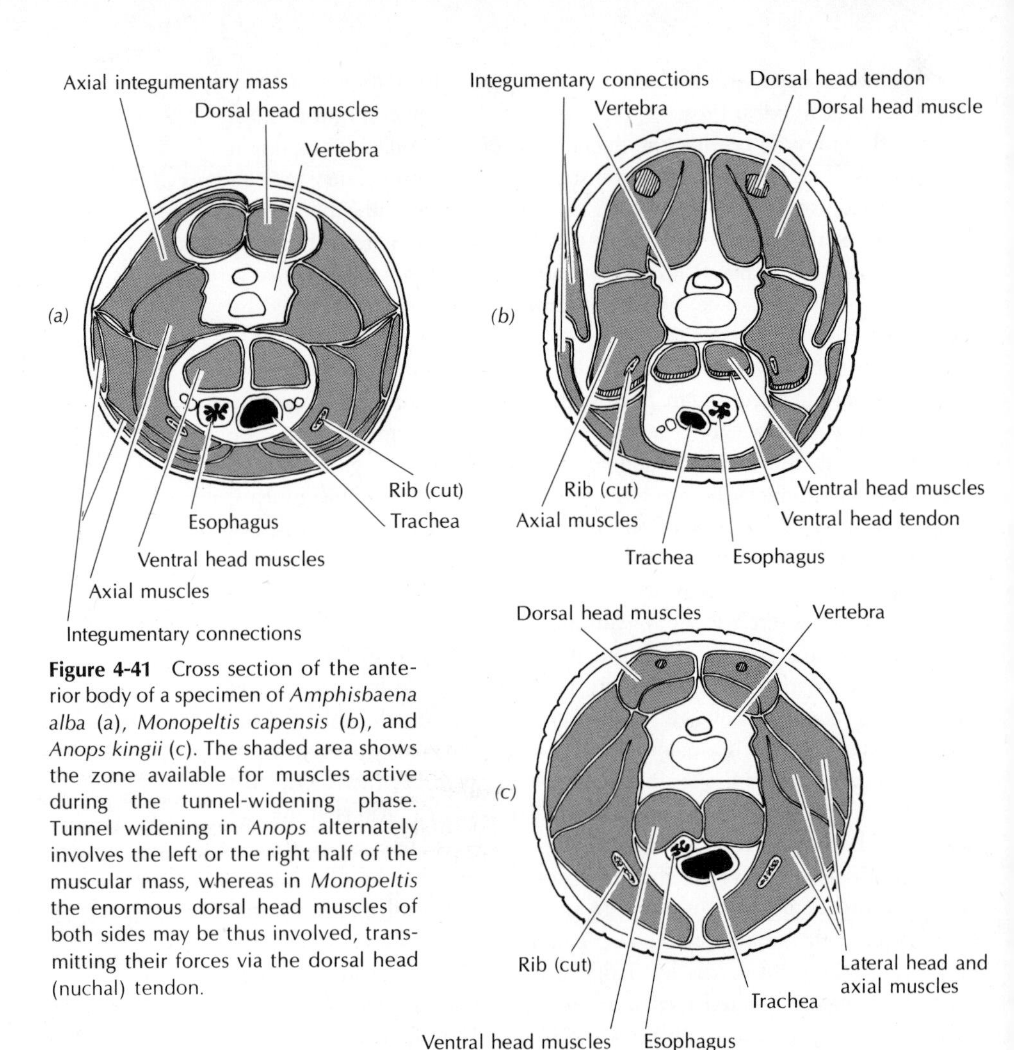

Figure 4-41 Cross section of the anterior body of a specimen of *Amphisbaena alba* (a), *Monopeltis capensis* (*b*), and *Anops kingii* (c). The shaded area shows the zone available for muscles active during the tunnel-widening phase. Tunnel widening in *Anops* alternately involves the left or the right half of the muscular mass, whereas in *Monopeltis* the enormous dorsal head muscles of both sides may be thus involved, transmitting their forces via the dorsal head (nuchal) tendon.

compaction stroke need only widen one half of the tunnel's absolute diameter. Indeed, all the keel-headed species are of smaller body diameter than almost any of the spade-snouted forms. Still other factors, such as opportunity for invasion, predators, and other nonmechanical aspects of the organisms, probably complicate the real situation; however, the present analysis has thus far been based mainly on burrowing mechanics.

This analysis has led us through an explanation for the success of the generalized amphisbaenian forms, as well as for differential success among several species showing various degrees of specialization. Purely mechanical efficiency of a burrowing apparatus hardly confers an absolute selective advantage; rather, the changes necessary to specialize in this direction have had adverse effects on other aspects of the animal.

It is impossible to be a most efficient burrower and simultaneously a most efficient prey biter and a most efficient prey perceiver. Indeed, some of the sensory amplifiers, initially assumed to represent high degrees of specialization for auditory acuity, seem in this context to be little more than compensating devices for conditions accompanying burrowing modifications. Such external compensation probably keeps the sensory input to the brain close to the level it would have been if the animals had not evolved radically restructured anterior ends.

In all cases selection favors the fitness of the entirety of the organism. When further improvement of two systems requires opposite morphological changes, the structures observed will represent a compromise. Each set of environmental circumstances establishes maximum fitness for a particular set of compromise solutions, namely a locally optimized genotype, perhaps combining one level of hearing effectiveness with another level of feeding modification and a third level of burrowing efficiency. This explains why one often sees "a series of intermediate conditions" in a series of Recent species.

Structure and History

The results of the preceding analysis, derived from a comparative look at the specializations of the major subgroups among the approximately 130 species of amphisbaenians, permit certain speculations about the nature and magnitude of various historical factors in the evolution and present distribution of these animals. The kinds of comparative evidence presented here permit one to formulate hypotheses regarding the presumptive path of past evolution. Fossil specimens then provide a test of such hypotheses. Before dealing with these speculations, it must be indicated that the results of this kind of comparative analysis and their reliability depend heavily upon a preliminary taxonomic review. Until species have been characterized and specimens correctly identified, a comparative functional-morphological analysis cannot proceed. Properly assigned specimens are obviously as critical in these studies as are properly mixed and purified solutions in physiology.

The analysis suggests that there is no intrinsic reason to assume that the earliest stage of either the keel-headed or the spade-snouted burrowing was more advantageous than the other. Presumably a shift in the direction of either method of tunnel widening would have conferred equivalent additional fitness, limited only by the possibility of developmental rather than immediate functional modification. However, the two evolutionary pathways would have differed in the secondary adaptations that became advantageous after the method of tunneling was fixed.

In the absence of evolutionary prescience in ancestral forms, the selection between the two possible options was probably random. Alternatively, it might have been based on the genetic plasticity of the particular genotypes—that is, on their current ability to be further affected by selective influences. It is unlikely that selection was directly related to the ultimate functions permitted by it. Yet the burrowing methods leading to the keel-headed and the spade-snouted modifications were not equally capable of further development. Specialization almost inevitably led to a condition where one or another line was restricted from further modification, and consequently restricted in the environments it could occupy. It is unlikely that the specialized population could revert to the original generalized conditions, and in this sense (rather than the condition at a particular genetic locus) one may well talk of the "irreversibility" of evolution.

The concept of compromise leads to a distinctive view of the way in which further modifications may then occur. Once a group of organisms is being selected toward one or another primary solution to an environmental demand, selection will obviously enhance any additional modification that will increase the overall fitness of the organism. Such secondary modifications need not merely supplement a particular primary option; they may indeed compensate for its inherent disadvantages.

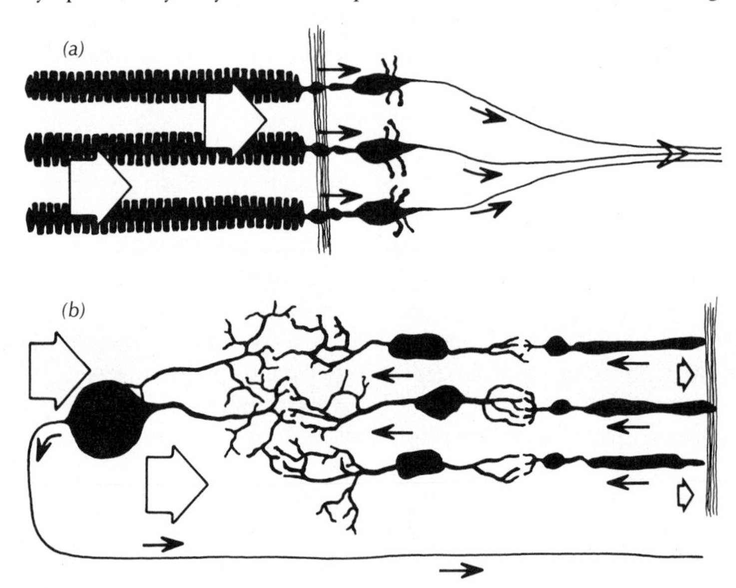

Figure 4-42 Light (open arrows) reaching the retina of the octopus (*a*) immediately impinges upon the photosensitive portions of the retinal cells. The stimuli (solid arrows) continue to pass down the axons in the same direction. In contrast, light impinging on a vertebrate retina (*b*) must penetrate multiple layers of connecting neurons before reaching the photosensitive portions. The stimuli (solid arrows) pass this zone in the opposite direction to that of the impinging light (open arrows). This clearly involves significant scatter for which signal-discrimination circuits must provide some compensation.

The vertebrate eye may be used as an example. From a purely engineering viewpoint, the eye of cephalopods seems infinitely more efficient. Light passing the lens impacts immediately onto the light-sensitive portion of the cell. The cellular connections to the brain are arranged at the side opposite to the light input; hence they do not disturb or refract the incident beam. In contrast, the cells are reversed in the vertebrate eye so that the light impinging on the vast bulk of the retina (peripheral to the fovea centralis) must filter through several layers of cells and processes before triggering the receptor cell. This passage clearly involves significant scatter, and one wonders how many of the neural connection patterns seen in the various kinds of vertebrate eyes and neural mechanisms, such as peripheral inhibition, are significantly more than compensations for an initially "wrong" decision.

The concept of compromise, furthermore, has some importance in the interpretation of isolated fossils that seem to show a condition intermediate between those from earlier and those from later epochs or that seem to show an incomplete approach to a particular specialization seen in Recent animals. Such specimens are generally referred to, explicitly or implicitly, as intermediate or possibly ancestral stages in the evolutionary development of the conditions seen in the Recent, particularly when the specimen or the analysis emphasizes one or two structural or

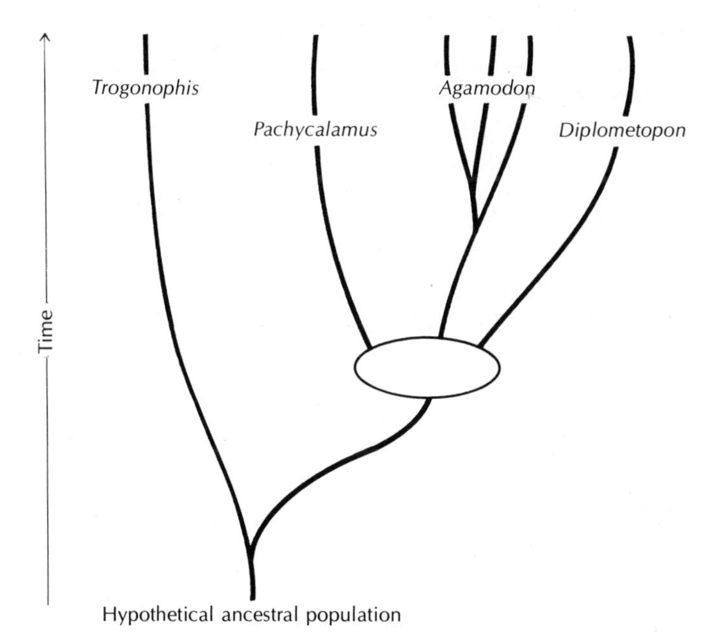

Figure 4-43 This phylogenetic sketch of the trogonophid amphisbaenians suggests that the three advanced genera clearly represent parallel developments rather than arrested stages of a presumed phylogenetic sequence.

functional systems of fairly limited extent. The results of the preceding section on compromises suggest the dangers of this assumption. Let us suppose that the trogonophids we know today had survived without modification across several geological periods. If we had only two fossil specimens to document their existence, with the earlier one deriving from *Trogonophis* and the later from *Agamodon*, it would be most plausible to proceed to the conclusion (correct on the basis of other evidence) that *Trogonophis* represents a more "primitive" stage of the trogonophid adaptive pattern than does *Agamodon*. Unfortunately, the same conclusion would be reached if the earlier level were represented by the Socotra island genus *Pachycalamus*. Yet analysis suggests that only for burrowing is *Pachycalamus* less specialized. Actually, some of its other organ systems have shifted in a different direction so that it is really the result of a distinct adaptive line that differs from *Agamodon* in major characteristics of almost all systems examined. Most of these differences do not parallel the superficial gradation suggested in skull shape or burrowing mechanism.

It is even more upsetting, of course, to consider the results of interpretation of the fossil record should a specimen of *Agamodon* accidentally be preserved in a layer earlier than that of the fossil *Trogonophis*. Better than anything else, such considerations document why modern paleontologists tend to be cautious in interpreting their evidence. The greater the number of specimens discovered and included in the analysis, the less the likelihood that blatant deceptions due to inadequate material will confuse interpretation.

The detailed analysis of multiple species has also pointed to a second kind of risk. The need to interpret the functional pattern of a now-extinct, lizardlike "predator" discovered perhaps in the Permian, Triassic, or Eocene has sometimes led paleontologists to ask about the feeding, locomotor, or digestive pattern in modern varanids (the largest of the now-living predacious lizards). When the detailed analyses required to answer such a question were lacking, the common solution has been to look at "*Varanus*," namely that species among the 57 forms of monitor lizards most easily obtainable from dealers or, perhaps, least likely to be endangered. This might not be equivalent to selecting one amphisbaenian at random, but it will only by accident lead to a true understanding of the adaptive state of the group, particularly if it is diverse. Only when a group's full range of structural modification and the parallel range of functional correlates begin to be understood is it safe to extrapolate at length to the fossil record.[14] A perfect example

[14]It should be obvious by now that these remarks apply with even more force to any comparison (ecological or physiological, behavioral or biochemical) which might attempt to contrast the conditions in fishes, amphibians, reptiles, and mammals, by comparing a goldfish to a bullfrog to a turtle to a rat.

of arbitrary usage would be the reference to amphisbaenian remains as indicators of a mesic (humid) environment, based almost entirely on the report that *Rhineura floridana* steadily lost weight in dry soils (Bogert and Cowles, 1945). As we have seen, this case is hardly typical of the Recent Amphisbaenia as a group. The occurrence of a seasonal and perhaps subterranean water course on the edge of the Desierto de la Sechura of Peru or the Kalahari Desert of west Africa hardly qualify such regions as moist!

A more detailed understanding of the importance of historical accidents to explanations of the present distribution of animals may well be an important corollary of such comparative analysis. For instance, amphisbaenians range into western Iran but are absent from eastern Iran and Pakistan. In parts of India, we find the subterranean biotope occupied by the efficiently burrowing small snakes of the family Uropeltidae, a curious assemblage of animals with pointed heads (Gans, 1973a). It is superficially plausible to explain such distributions by arguing that either the amphisbaenians or the uropeltids are superior for the particular environment, especially since the two groups differ in their burrowing mechanics. (The uropeltids burrow by a vibratory ramming of the head and widen their tunnel by a variant of concertina curvature—refer back to Figure 3-35). Could there be an intrinsic advantage to either method?

More recent studies, however, suggest that the distribution need not reflect burrowing effectiveness at all. The uropeltids have become adapted to environments of lower temperature; they live in subterranean niches in the rich and moist humus of the central Indian hills. They obviously were derived from groups of snakes now found farther to the east and may well have originated in southern Asia. Their current restriction to the Indian peninsula appears clearly due to moisture and temperature tolerances rather than to aspects of burrowing mechanics. The easternmost amphisbaenians, on the other hand, are trogonophids highly specialized for sandy environments, as long as they may find a minimum of moisture. Although *Diplometopon zarudnyi*, the easternmost trogonophid species, ranges across the Arabian peninsula, it has apparently not been able to deal with the harsher conditions of coastal Iran and Pakistan. Since the easternmost extreme of the amphisbaenian range was occupied by a trogonophid rather than an amphisbaenid, we are, furthermore, ignorant of whether amphisbaenids could have been successful in India had they managed to reach that continent before the climatic barrier had become established. Each group in question is hence limited by factors related to its own past evolution, and their distribution cannot be explained by present or past competition even though we are dealing with two groups of burrowers, some members of which occupy superficially rather similar environments.

The preceding discussion has dealt with an analysis based on the comparison of conditions in numerous members of a major adaptive radiation. Many aspects of their distribution may well reflect aspects of their burrowing mechanics; in contrast, other patterns observed prove to be due to factors only indirectly associated with the subterranean environment. Consideration of the trogonophid adaptive pattern documented the difficulty of deriving function directly from structure. The analysis has also suggested the existence of multiple analogous modifications of the burrowing system; in other words, many of the functional patterns—for instance, spade-snouted burrowing—developed more than once.

The understanding of amphisbaenian specialization clearly required more than a simple biomechanical analysis. As ecological and behavioral factors were included in the analysis, the resolving power of the functional analysis was markedly increased; an associated morphometric analysis (statistically separating the effects of minor differences of shape and size) might take this kind of study to its logical conclusion. Even simple movement analysis for burrowing patterns markedly increased the plausibility of the ultimate conclusions. Yet the present chapter, as well as the two preceding ones, retains the limitation that (except for studies on the ear) the behaviors of most of the muscles and bones were only observed or deduced, not measured or recorded directly. The error inherent in such extrapolation represents a clear area of uncertainty, reducing the resolving power of all the aforementioned analytical approaches.

REFERENCES

Alexander, A. A. (1966). Problems involving vertebral segmentation in amphisbaenids and primitive snakes. Ph.D. diss., State U. N.Y. at Buffalo, 147pp.

Allee, W. C., A. E. Emerson, O. Park, T. Park, and K. P. Schmidt (1949). Principles of animal ecology. W. B. Saunders and Co., Philadelphia and London, xii + 837pp.

Bellairs, A. d'A. (1972). Comments on the evolution and affinities of snakes. In Studies in vertebrate evolution (K. A. Joysey and T. S. Kemp, eds.). Oliver and Boyd, Edinburgh, pp. 157–172.

Bentley, P. J., and W. F. C. Blumer (1962). Uptake of water by the lizard, *Moloch horridus*. Nature, 194:699.

Bock, W. J. (1964). Kinetics of the avian skull. J. Morph., 114(1):1–41.

Bogert, C. M., and R. B. Cowles (1947). Results of the Archbold Expeditions. No. 58. Moisture loss in relation to habitat selection in some Floridian reptiles. Amer. Mus. Novitates (1358):1–34.

Carroll, R. L. (1970). The ancestry of reptiles. Phil. Trans. Roy. Soc. London, 257B(814):267–308.

Cowles, R. B., and C. M. Bogert (1944). A preliminary study of the thermal requirements of desert reptiles. Bull. Amer. Mus. Nat. Hist., 83(5):261–296.

Dallavalle, J. M. (1943). Micromeritics. The technology of fine particles. Pitman Publishing Corp., New York and London, xxvii + 555pp.

Frazzetta, T. H. (1961). A functional consideration of cranial kinesis in lizards. J. Morph., 111(3):287–319.

*Frost, H. M. (1967). An introduction to biomechanics. C. C. Thomas, Springfield, Ill. 167pp.

Gans, C. (1960). Studies on amphisbaenids (Amphisbaenia, Reptilia). I. A taxonomic revision of the Trogonophinae and a functional interpretation of the amphisbaenid adaptive pattern. Bull. Amer. Mus. Nat. Hist., 119(3):129–204.

———— (1965). Notes on a herpetological collection from the Somali Republic. I. Introduction and itinerary. Mus. Roy. Afrique Centrale, Ann. 8° (1934):1–14.

———— (1966). The functional basis of the retroarticular process in some fossil reptiles. J. Zool. (London), 150:273–277.

———— (1967). *Rhineura*, p. 42. In Catalogue of American amphibians and reptiles. Amer. Soc. Ichthyol. Herpetol.

———— (1968). Relative success of divergent pathways in amphisbaenian specialization. Amer. Nat., 102(926):345–362.

————(1969). Amphisbaenians—reptiles specialized for a burrowing existence. Endeavour, 28(105):146–151.

———— (1971). Studies on amphisbaenians (Amphisbaenia, Reptilia). 4. A review of the amphisbaenid genus *Leposternon*. Bull. Amer. Mus. Nat. Hist., 144(6):379–464.

———— (1973a). Locomotion and burrowing in limbless vertebrates. Nature (London), 242(5397):414–415.

———— (1973b). Uropeltid snakes—survivors in a changing world. Endeavour, 32(116):60–65.

————, and W. J. Bock (1965). The functional significance of muscle architecture—a theoretical analysis. Ergeb. Anat. Entwicklgesch., 38:115–142.

————, and J. J. Bonin (1963). Acoustic activity recorder for burrowing animals. Science, 140(3565):398.

————, and H. Pandit (1965). Notes on a herpetological collection from the Somali Republic. V. The amphisbaenian genus *Agamodon* Peters. Mus. Roy. Afrique Cent., Ann. 8° (134):71–86.

————, and C. Rhodes (1964). Notes on amphisbaenids (Amphisbaenia, Reptilia). 13. A systematic review of *Anops* Bell, 1833. Amer. Mus. Novitates (2186):1–25.

————, and E. G. Wever (1972). The ear and hearing in Amphisbaenia (Reptilia). J. Exper. Zool., 179(1):17–34.

Gaymer, R. (1971). New method of locomotion in limbless terrestrial vertebrates. Nature, 234:150–151.

Gehlbach, F. R., J. F. Watkins II, and H. W. Reno (1968). Blind snake defensive behavior elicited by ant attacks. BioScience, 18(8):784–785.

Gordon, A. M., A. F. Huxley, and F. J. Julian (1966). The variation in isometric tension with sarcomere length in vertebrate muscle fibers. J. Physiol., 184:170–192.

Greene, H. W. (1973). Defensive tail display by snakes and amphisbaenians. J. Herpetol., 7(3):143–162.

Iordanski, N. N. (1971). A contribution to the functional analysis of the skull in lizards (Lacertidae). The skull characteristics related to cranial kinesis. (Russian.) Zool. Zh., 50(5):724–733.

Jones, G. M., and K. E. Spells (1963). A theoretical and comparative study of the functional dependence of the semicircular canal upon its physical dimensions. Proc. Roy. Soc., B., 157:403–419.

Kallen, F. C., and C. Gans (1972). Mastication in the little brown bat, *Myotis lucifugus*. J. Morph., 136(4):385–420.

Krakauer, T., C. Gans, and C. V. Paganelli (1968). Ecological correlation of water loss in burrowing reptiles. Nature, 218(5142):659–660.

Kritzinger, C. C. (1946). The cranial anatomy and kinesis of the South African amphisbaenid *Monopeltis capensis* Smith. South African J. Sci., 42:175–204.

Kummer, B. (1966). Photoelastic studies on the functional structure of bone. Folia Biotheoretica, 6:31–40.

Norris, K. S., and J. L. Kavanaugh (1966). The burrowing of the western shovel-nosed snake *Chionactis occipitalis* Hallowell, and the undersand environment. Copeia, 1966(4):650–664.

Northcutt, R. G., and A. B. Butler (1974). Retinal projections in the northern watersnake, *Natrix sipedon sipedon* (L.). J. Morph., 142(2):117–135.

Pough, F. H. (1969). Physiological aspects of the burrowing of sand lizards (*Uma*, Iguanidae) and other lizards. Comp. Biochem. Physiol., 31:869–884.

Ramsey, R. W., and S. F. Street (1940). The isometric length-tension diagram of isolated skeletal muscle fibers of the frog. J. Cell. Comp. Physiol., 15(1):11–34.

Richards, O. W. (1955). D'Arcy W. Thompson's mathematical transformation and the analysis of growth. Ann. N.Y. Acad. Sci., 63(4):433–636.

Rosenberg, H. I. (1973). Functional anatomy of pulmonary ventilation in the garter snake, *Thamnophis elegans*. J. Morph., 140(2):171–184.

Saiff, E. I. (1970). Geographical variation in the genus *Zygaspis* (Amphisbaenia: Reptilia). Herpetologica, 26(1):86–119.

Stebbins, R. C. (1943). Adaptations in the nasal passages for sand burrowing in the saurian genus *Uma*. Amer. Nat., 77(768):38–52.

Tannes, K. (1971). Notizen zur Pflege und zum Verhalten einiger Blindwülen (Amphibia: Gymnophiona). Salamandra, 7(3/4):91–100.

Thompson, D. W. (1951). On growth and form, 2nd edition. Cambridge Univ. Press, Cambridge, 2 vols., 1–464 + 465–1116pp.

Trendelenburg, W., and A. Kuhn (1908). Vergleichende Untersuchungen zur Physiologie des Ohrlabyrinthes der Reptilien. Arch. Anat. Physiol., Physiol. Abt., 1908:160–188.

Vanzolini, P. E. (1951). Evolution, adaptations and distribution of the amphisbaenid lizards (Sauria, Amphisbaenidae). PhD. diss., Harvard U., Cambridge, Mass., 148pp.

Wake, D. B., and I. G. Dresner (1967). Functional morphology and evolution of tail autotomy in salamanders. J. Morph., 122(4):265–306.

Werner, Y. L. (1968). Regeneration frequencies in geckos of two ecological types (Reptilia, Gekkonidae). Vie et Milieu, C., 19:199–222.

Wever, E. G., and C. Gans (1972). The ear and hearing in *Bipes biporus* (Amphisbaenia, Reptilia). Proc. Natl. Acad. Sci. (USA), 69(9):2714–2716.

———— (1973). The ears in Amphisbaenia (Reptilia): further anatomical observations. J. Zool., 171:189–206.

Wickler, W. (1968). Mimicry in plants and animals. McGraw-Hill, New York and Toronto, 255pp.

Zangerl, R. (1944). Contributions to the osteology of the skull of the Amphisbaenidae. Amer. Midland Natur., 31(2):417–454.

Zug, G. R. (1970). Intergradation of the two *Rhineura* (Reptilia) populations in central Florida and comments on its scale reduction. J. Herpetol., 4(3-4):123–129.

5 ANALYSIS BY QUANTIFICATION: AIR BREATHING AND VOCALIZATION IN FROGS

Problem

Frogs are descendants of the group of vertebrates that first made the transition from water to air, that changed from fishes occupying an aquatic environment to tetrapods inhabiting terrestrial habitats. Frogs are very different from the truly transitional amphibians. Yet they have been so widely used in the determination of basic physiological parameters that the data thus provided inevitably influence our interpretations of conditions in ancestral forms now extinct (Gans, 1970). Therefore, one must ask which of their modifications reflect past history and which reflect present adaptation?

The "external gas exchangers," consisting of diverse lungs, gills, pouches, and integumentary vascularizations, represent one system that had to change in the transition to terrestrial life. Most tetrapods obtain their respiratory oxygen from the air. Breathing movements force fresh air across the surfaces of the external gas exchangers, which most often derive from diverticula of the anterior alimentary canal. In the frog, air movements also serve for sniffing (olfaction), coughing (cleaning), and vocalization. The ventilation of the system is produced by various kinds of buccal (mouth cavity) flutters, pulses, and oscillations. These activities involve muscles that also participate in feeding, swallowing, and vocalization; the relative utilization of these muscles for each of these diverse functions then poses a problem of obvious interest to functional morphologists.

Individual muscles, and even their component motor units, should be considered potentially independent entities. The sequences, time courses, and magnitudes of each muscle's activities are parameters to be used in analysis. Such a detailed approach to the muscles of the buccal apparatus is practicable because it involves only some half-dozen major muscles, and most of these have a relatively limited number of slips and divisions (see Figure 5-15). Beyond this, the individual muscles are easily accessible, and their electrical activities can be recorded as a sequence of electromyograms. Finally, there is the heritage of one and one-half centuries of frog physiology, including numerous procedures for anesthesia and tranquilization of these animals. Quantification and analysis of these muscular activities then seems soluble, but how should one proceed?

Unfortunately, these logical considerations do not represent the sequence along which our approach really developed. The projects here described started differently, as a combined study of the mechanics of the feeding mechanism and tongue muscles of the bullfrog, *Rana catesbeiana*. We wanted to use electromyography to test a hypothesis of tongue movement developed on the basis of dissection and observation alone (Gans, 1962). Yet the animals' breathing involved some of the same muscles we were beginning to study, and the breathing movements represented an annoying side effect in our records. Hence it quickly appeared useful to characterize these "more regular" muscle activities. But the animals obstinately showed an activity pattern that differed from that described by most previous students of frog breathing, and which furthermore posed some major and interesting problems. This caused us to shift the focus of our project. Indeed, the feeding pattern has yet to be treated!

Approach

The literature on frog breathing is vast (de Jongh and Gans, 1969; Gans, 1971). However, there is an equally vast area of disagreement regarding the cycles that occur and regarding the induction of the cycles; this in spite of the fact that many past studies involved experimental approaches. The basic difficulties are fourfold: (1) Many past records were taken on anesthetized animals; (2) the animal's normal behavior was generally ignored by recording from animals that were immobilized or restrained in unnatural postures (e.g., spread out upside down); (3) many of the operative procedures severely traumatized the animals (upper or lower jaw removed in its entirety); and (4) many workers proceeded to measure the wrong components (e.g., external movements rather than internal pressures—see p. 197).

It seems instructive to consider these difficulties in order to understand alternate approaches. The first three form a unit. The objection, in this case, is that an organism is capable of performing under a variety of circumstances. For instance, man can breathe with a tidal volume of 500 ml or less, but this may be increased almost sevenfold both by reducing the air residual in the lungs at the end of exhalation and by increasing the depth of inhalation. From an evolutionary viewpoint, the basic adaptations of an animal can be understood only by determining which natural activities are compatible with the physiological scope of the organism. At what levels within this range do various activities occur, and what is the energy demand on the system?

Such questions must be answered by intact animals; even anesthetics affect physiological responses and thus introduce errors into the interpretation of the adaptation.[1] Traumatized animals tend to compensate for their discomfort; hence neither their behavior nor their physiology is typical. We have more than enough documented cases of such effects to question measurements taken during any departure from natural conditions.

The behavior of an unrestrained animal presents the baseline to be determined, but the behavior of a semirestrained animal may represent the closest practicable approach to this. Anyone aware of the technology involved in the various space and nuclear projects knows that it is potentially possible to isolate an experimental animal, to monitor its responses on television, and to perform all manipulations by electromechanical servomechanisms. Anyone aware of the cost of such approaches will also know that it will rapidly exceed even the most generous level of resources available to the average experimenter. One obviously has to assess the practical possibilities and be prepared to compromise.

How does one decide on the cost effectiveness of a potential experiment? In other words, how does one decide how complicated an experiment must be to obtain the desired result? Since some animals respond naturally if they do not see the experimenter, it may be that a simple blind will be sufficient screening. Other animals respond to olfactory and auditory cues; to study these organisms one needs to remove all disturbing signals. When should one provide such massive doses of extraneous white noise[2] that signals emitted by the observer become insignificant? Is it possible to habituate a particular animal so that the investigator's presence is no longer associated with disturbance? Perhaps one should ask: What is the maximum level of disturbance or other insult that can be tolerated by an animal without causing a major distortion of the responses to be studied? What are the relative merits of the possible approaches?

Preliminary value judgments about such questions can be achieved in several ways. One of the best is the temporary use of some type of remote observation (e.g., an inexpensive closed-circuit TV system). The organism may then be observed by one person while another experimenter approaches the back of a blind or while the sounds of the experimental apparatus are masked by white noise. Does the arrival or departure of the investigator change the animal's behavior? Do sounds or changes in the level of the white noise elicit obvious reactions? What parameters seem to induce disturbance?

Indeed, how is disturbance to be recognized by the experimenter?

[1]Anesthetics are essential for any painful procedure. Yet each has numerous secondary effects, and it is necessary to take these into account in evaluating recovery time from anesthesia.

[2]Sound containing all audible frequencies at random intervals over a period of time.

One obvious way would be to record surface signs, which in many animals are very inconspicuous. Snakes, for instance, were long assumed to be deaf because they failed to respond when observers entered or left a room. Snakes have probably undergone selection for remaining immobile when faced with strange stimuli. It remained for Cowles and Phelan (1958) to show that they do respond to such events by marked changes in their heart rate. Often, then, heart rate, or similar parameters not under immediate conscious control, will provide the experimenter with a good indicator of the animal's perception level when it is faced with diverse experimental situations.

A record of heart beat, endocrine level, or brain waves may thus provide critical information; almost as useful will be the experimenter's experience with the behavioral responses of the animal. This may seem to be a curiously unscientific view, but it reflects an important issue. Vertebrates, indeed many other animals, signal their internal physiological and behavioral state by postural and similar gross signs. These signs are often species- or group-specific, and prior study of the animal under natural conditions may help one to recognize them. We soon learn to recognize when a pet dog is indisposed, when it is behaving normally, and when it seems to have been disturbed. The capacity to recognize such behavior may mark the difference between a good and a bad animal keeper, between a successful and an indifferent experimenter.

As one studies the attitudinal pattern of a particular organism, one must also remember that the animal's perceptual world is not only different from that of man, but that it also may change depending on the animal's psychological as well as physiological state. A change in internal program may then induce a different response to the same cue.[3] For instance, a "hungry" egg-eating snake, *Dasypeltis*, will respond to a fresh egg by tonguing and swallowing; "frightened" or "disturbed" *Dasypeltis* respond to the same object by attempting to hide beneath it!

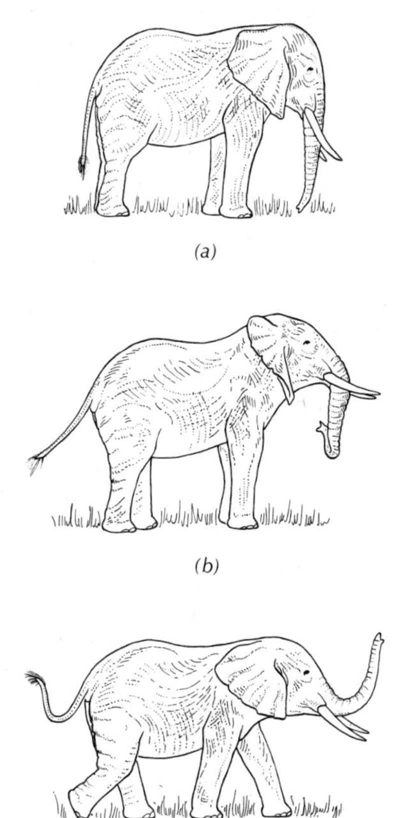

Figure 5-1 Recognition of postural signs in an elephant may be important to hunters and, in these days, to tourists as well. The bull at the top is resting, whereas that in the middle figure is excited and may charge, as indicated by the lifted head with spread ears, the curled trunk, and the outward-pointed tail. The bull in the bottom figure is running in fright as indicated by the lifted trunk and tail. Similar indicators could be sketched for lizards, dogs, or humans.

The foregoing considerations suggest that even when physiological baselines are being monitored, it is still desirable to construct environments in which the disturbed animal will most quickly calm down. The greater one's understanding of the animal's signs, the easier it becomes to induce the activities that are to be studied. Construction of a semi-natural environment will ordinarily increase the chances that the appropriate cues are perceived in the proper fashion. Shelterers will feed best if they need not expose all of their body when they approach their food. Thigmotropic species, those that react positively to touch, may rest when their body contacts a perforated or transparent surface that leaves

[3]Many books refer to the Harvard Law of Animal Behavior (under carefully controlled conditions animals behave as they please). Presumably this reflects the frustration of experimenters untrained in observing and motivating the experimental animal.

them otherwise exposed to observation. This principle can be simply applied to tortoises that tend to wedge their shell against a wall when disturbed. If they are made to contact the wall at an angle, they will wedge only one side of the anterior shell; one may then observe and record head and limb movements during unrestrained breathing, rather than having them obscured by the wall. Nocturnal animals will start their activity cycle when the ambient light level drops; a shift from white light to red or infrared illumination may serve as well as a shift to total darkness. It is well to consider that an environment totally "neutral" or "empty" in human terms may contain multiple disturbing and frightening cues to another species; some understanding of the organism's ecology is therefore necessary to assure that the first cue will be read as "display," "move," or "feed," rather than as "hide."

Situations that are particularly difficult to manage may also be dealt with by temporary tranquilizations of the animal. We have found, for instance, that the banded krait (*Bungarus fasciatus*) would feed during recovery from anesthesia, induced for venom extraction, even though the same individuals regularly rejected food when fully conscious. When disturbance is an inevitable concomitant of an experimental procedure, it may be masked by habituating the animal in order to avoid the recording of a startle reaction as the procedure is started. The flashes of a stroboscopically coupled movie camera (Gans, 1966), for instance, will affect the feeding responses of some lizards. This may be overcome by use of a small commutator that causes the light to flash even when the camera is inactive. As soon as the animal starts to ignore the flashes, one may offer food objects and record feeding with less risk of artifact.

Assuming that one can induce nearly normal behavior, one still has to decide how and what to measure. Procedure and interpretations of electromyography are discussed in the next section. Under any but the most unusual circumstances there are always more muscles or parts of muscles than can be monitored simultaneously. One might initially ask which of these components should be recorded at one time? What other information is necessary on the record to make the results most meaningful? In many cases one has to select two, four, six, or eight muscles for simultaneous recording out of a total of perhaps twelve to thirty. Are there desirable combinations?

In answering these questions we must consider why one uses electromyography. Its basic task is to determine the sequence and magnitude of the muscular activities that elicit or inhibit movements (by inducing acceleration on some part of the animal). Solution of the latter question involves difficulties. One needs to demonstrate that the electromyogram from a particular muscle is causally rather than accidentally associated with a particular movement. This, in turn, requires that closely associated events be studied. For instance, when a frog suddenly contracts any major group of its trunk muscles, all of its soft tissues are likely to wobble; therefore a record of deflections on the frog's skin tells only as much about movements created by the activity of a specific muscle as a seismogram tells about slippage along a distant fault line somewhere in the earth's crust. It thus becomes important to decide which mechanical event is to be recorded. The closer one comes to measuring the mechanical event immediately produced by muscular activity, the more direct and conclusive will be the interpretation of cause and effect. When the muscles reduce the volume of a closed chamber, one must measure the pressure change; when the muscles turn bones about a joint, one must measure the angular change; when the muscles effect displacements, one must measure the distance moved.

A study of isolated events may become the study of history. Only when an association may be observed and recorded repeatedly does one have the prerogative of calling it science. This applies particularly to the association of electromyogram and movement. When motion patterns are regularly cyclic, as are the motor sequences of some invertebrates, one may test the association repeatedly during diverse cycles. Once one electromyogram has proved to be invariably associated with an event, one may even use the measure of this muscle's activity as a substitute for a record of the mechanical event. Analysis becomes far more difficult if the activity studied is irregular or only randomly repetitive. This is particularly true of combined motions in which distinct displacements are induced by several different muscles. A masticating bat not only opens and closes its jaws, but protrudes and retrudes them, as well as shifting the jaws from side to side. The magnitude of each of the several displacements depends upon the nature of the food and the side on which the animal chews. No one jaw muscle is involved in only uniaxial movements; consequently, the activity of no one muscle may then serve as a substitute for all masticatory activity (Kallen and Gans, 1972).

The solution to more complex patterns of combined motions lies in an increase in the recording time and repeated scanning of the record to check for simple or more complex associations. First one obtains a record of displacement, velocity, and acceleration; then one checks for muscular activities that always synchronize with the particular mechanical events. Ultimately, one may define groupings of muscles that seem to induce particular mechanical events. As the information becomes refined so may the questions. Do the muscles of the left and right temporalis of the animal fire simultaneously? Do anterior and posterior digastrics act in concert? Why and how is the mandible protruded in closing? Such comparisons are best made on simultaneous records; it is useful to keep this in mind during interim analysis.

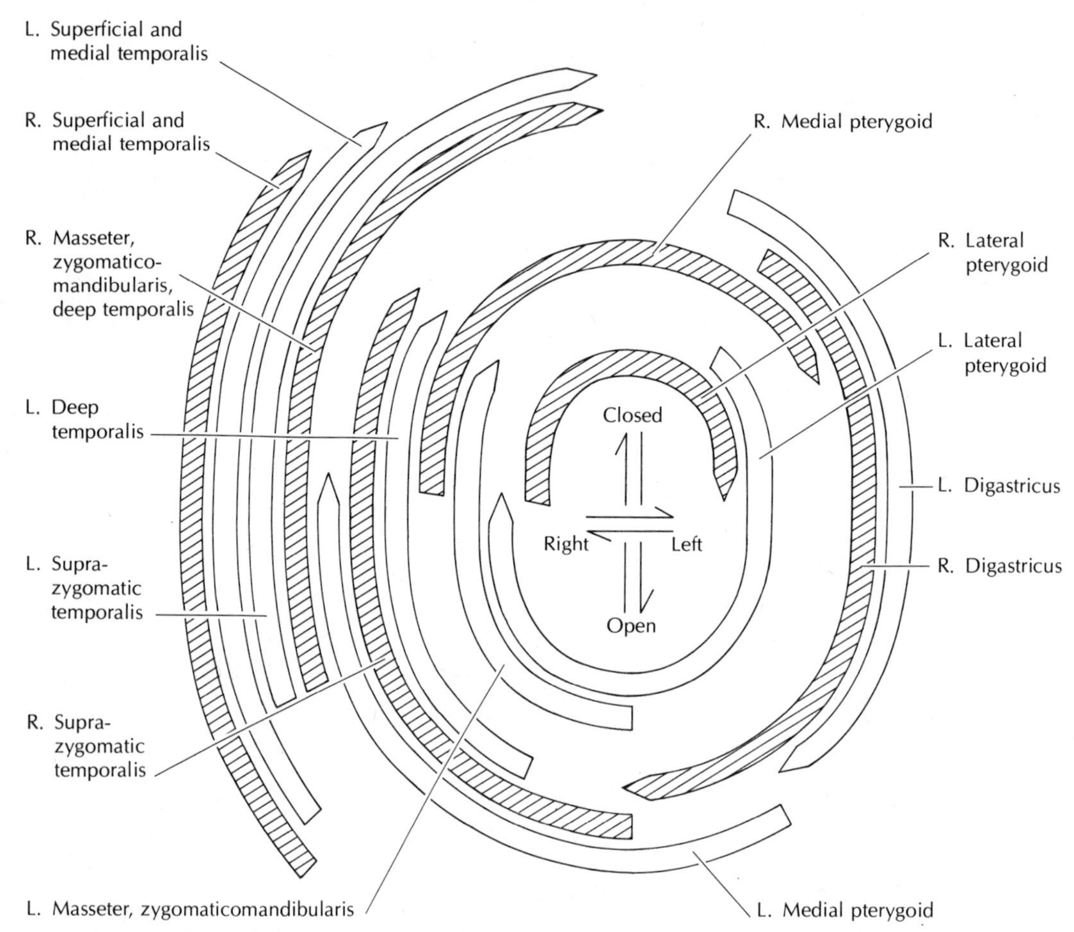

Figure 5-2 Sketch to show which masticatory muscles fire during the orbit shown in Figure 5-3. Note that no muscle or pair of muscles serves only for opening or closing. All engage in complex starts and stops, and several muscles fire at each portion of the cycle. Different food types and reversal of orbit leading to mastication on the opposite side change the firing sequence. Consequently, one cannot, before analysis, utilize the behavior of any one muscle as an indicator of the mechanical event. (After Kallen and Gans, 1972)

Myographical Concepts

Most skeletal muscle fibers (single, multinucleated muscle cells) appear to be associated with only a single, large motor neuron, the body of which lies in the ventral horn of the spinal cord, and here receives the input from hundreds of other cells. (See Wilkie, 1968, for a brief and lucid treatment of muscle.) The axon of such a motor neuron splits into many terminal twigs. Ultimately each twig forms a motor end plate on the cell membrane of a muscle fiber, so that the motor neuron potentially controls between fifty and five hundred fibers. Only when a motor neuron conducts an impulse (fires) will its set of muscle fibers be triggered; hence the combination of a single motor neuron and of its associated fibers is called a motor unit. All instructions, whether resulting from proprioceptive (originating inside the body) or exteroceptive (originating outside the body) input, must pass to the muscle fiber via the cell body of its motor neuron. Since this nerve cell collects, processes, and acts upon input data from many systems and transmits the resultant impulse directly to the muscle, Sherrington (1904) referred to the motor neuron as the "final common path."

Stimulation of the motor neuron causes a wave of depolarization to travel the length of its axon and induce the release of a chemical transmitter at the motor end plates. As this substance is received at the surface of the cylindrical motor fiber, it induces a secondary electrical event that sweeps across the muscle cell membrane, and apparently inward along the endoplasmic reticulum, to trigger the shortening response of bundles of actin-myosin filaments. The response temporarily shortens the fiber by generating a force which tends to pull the fiber's ends toward each other, an action which displaces the fiber's sides and increases its diameter (Box 4-3). The overall summated electrical event in the muscle fibers is of sufficient magnitude to produce localized voltage changes in the muscle. Electromyography is a relatively simple[4] technique for extracellular recording of these electrical events. A "spike" on an electromyogram ideally represents the discharge or "firing" of a single motor unit; it commemorates a single occurrence on Sherrington's common path.

Figure 5-3 The little brown bat (*Myotis lucifugus*) swings its lower jaw in a complex curve. The cycle shown indicates that it is chewing on the right side (see Figure 5-2).

Box 5-1 Motor Units

The cell body of the motor neuron lies in the ventral horn of the spinal cord. Its axon passes out to a particular muscle, splitting as it goes. Individual branches terminate as motor endplates on individual fibers of a striated muscle. The motor neuron thus provides the main efferent (outflowing) pathway controlling the contraction of an assemblage of muscle fibers; the neuron and muscle fibers are a *motor unit.*

The dendrites of the neuron's cell body receive the synapses of other cells. Similarly many cell endings synapse directly to minor hillocks on the cell's membrane, the so-called *buttons.* In this fashion the motor neuron receives a complex input, partly from the higher levels of the brain and partly from pathways that involve only simple reflex arcs employing afferent (incoming) signals that enter the spinal cord,

[4]The simplicity is due in large part to the development of modern electronics which now make available numerous "off the shelf" units for recording and amplifying electromyograms. Earlier workers who studied muscle action potentials in the 1920's started building their own myographs by modifying telephone receivers.

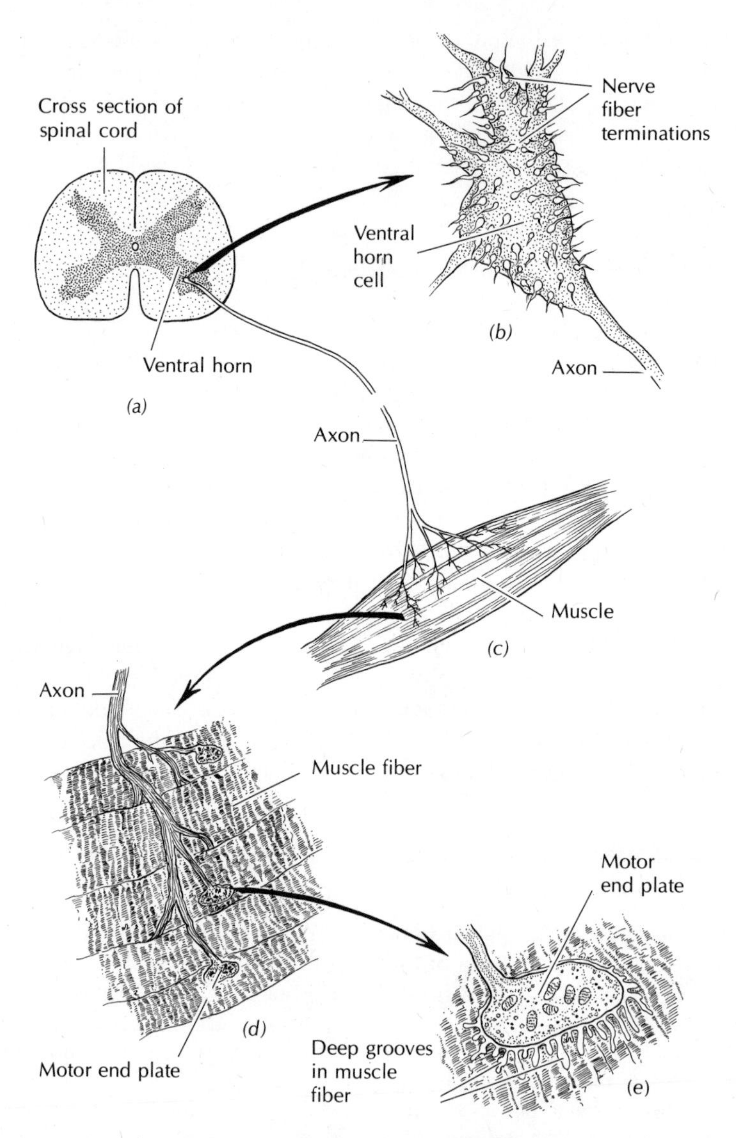

close to the cell's position. This complex input reflects instructions from higher control levels that may regulate whether an animal trots or gallops, but also regulates whether it opens its mouth to bite or to snarl. The motor neuron also receives "proprioceptive" direct input from motor spindles. These are sense organs located within the muscle itself; they indicate how extensively a muscle is contracting. Other input comes from tendon organs that signal the contraction force that the muscle is generating. This dual aspect is important since the force produced by a muscle depends on the extent to which the muscle is allowed to shorten (Box 4-3). The balance of these inputs provides a feedback, steering positional shifts as well as recruitment of the motor units required to generate particular levels of shortening or force. Thus motor neurons serve as the ultimate integrators and transmitters of multiple levels of input.

Each motor neuron may send axon terminals to hundreds of muscle fibers. The motor endplates contain vesicles that release a chemical transmitter when an excitation wave reaches them. This transmitter traverses the intersynaptic cleft and activates the folded cell membrane of the muscle fiber. Here it starts another series of electrical events that culminates in the shift of the actin and myosin rods relative to each other and, with this, the shortening and tension development between the ends of the fiber.

The number of motor fibers innervated in this fashion ranges from 50 to over 500. In general, small muscles and those that require more rigorous control have fewer muscle cells per motor unit, whereas the larger muscles and those in which control is not critical have a higher number. Examples are 50 cells per motor unit in the finely tuned human eye muscle and several hundred in the less precise muscles of the human back. Still, these figures represent means, since they are determined by counting the number of axons somewhere along the nerve and dividing this by the number of muscle fibers. Even when such studies carefully eliminate errors due to the passage of afferent axons in the same nerves, they suffer from the uncertainty that a single muscle may contain several types of fibers—for instance, those contracting rapidly and those contracting slowly. Consequently we still need studies showing how and why the endplates of individual motor neurons are distributed throughout particular muscles.

Most myograms represent a mixed record of multiple events. The magnitude (voltage) of a single firing sequence will differ between species; among other factors, it involves the number of muscle fibers of a motor unit in the immediate vicinity as well as the number of motor units. We have observed it as low as 30 μV in a teleost and as high as 3 mV in a snake. The magnitude of the actual signal will increase with the size of the electrodes, decrease with increasing distance between the event and electrode, and increase with the number of events occurring per unit time. It will also reflect the position of the electrodes relative to the volume occupied by the motor unit, and the material making up the electrode used (cf. Osse, 1969, for a discussion of equipment limitations).

The record of electrical events on tape, paper, or a film of the face of an oscilloscope hence will reflect the size, conformation, and position of the electrodes and the size and length of the leads transmitting the signal to the preamplifier. This means that neither two separate pairs of electrodes nor two placements of the same electrodes will ever produce exactly comparable records of voltage. Only the rate of firing—i.e., the number of discrete events per unit time—may be compared; integration of records (collecting the number of events per unit time) ideally should be for the rate rather than for the magnitudes of individual events. Comparison of absolute magnitudes generally must be restricted to events recorded by a single electrode pair over time.[5]

There is a finite time between the arrival of the stimulus at the motor end plate and the beginning of mechanical shortening in the fiber. For most descriptions of movements this time is insignificant, particularly since different firing rates will produce different rates of tension development (Buller, 1965). Myography is inadequate for distinguishing between different types of fibers, such as fast and slow or red and white muscles, which differ in the rate at which tension is developed or in the number of times that a muscle may shorten without firing; these have to be determined by different physiological and ultrastructural methods (cf. Goslow et al., 1972). Electromyography, then, tells only whether a muscle fires, and how long and how strongly the excitation proceeds.

Box 5-2 Instrumentation

Electromyographical analysis requires: (1) equipment that will detect and filter from extraneous noise the electrical event coincident with the firing of a motor unit; (2) equipment suitable for transmitting this signal to a recording device; and (3) equipment that will amplify the signal to a level that can be visualized or recorded.

The first aim requires both a pickup and a preamplifier that may measure the difference in electrical potential between two closely adjacent sites. The pickup and preamplifiers must also reject the signals from more distant fibers and avoid the production of local currents or metal deposition that may adversely affect the behavior of the tissue. The second aim is met by various loose wire loops and connectors or by wireless telemetry in which the raw signal is transmitted from animal to preamplifier. The amplifier should be stable, yielding an equal amplification of a weak or a strong signal. Its amplification should be constant over time and not affected by the number of recording devices connected to the output.

Since one ordinarily wants to correlate the muscular activity with one or more movements, these electrical and mechanical events are best recorded simultaneously. Consequently, one needs a variety of transducers to change mechanical or light sig-

[5]Even so, care is necessary to assure that the electrodes do not gradually deteriorate due to death of injured cells, metallic poisoning of the tissues, or surface polarization.

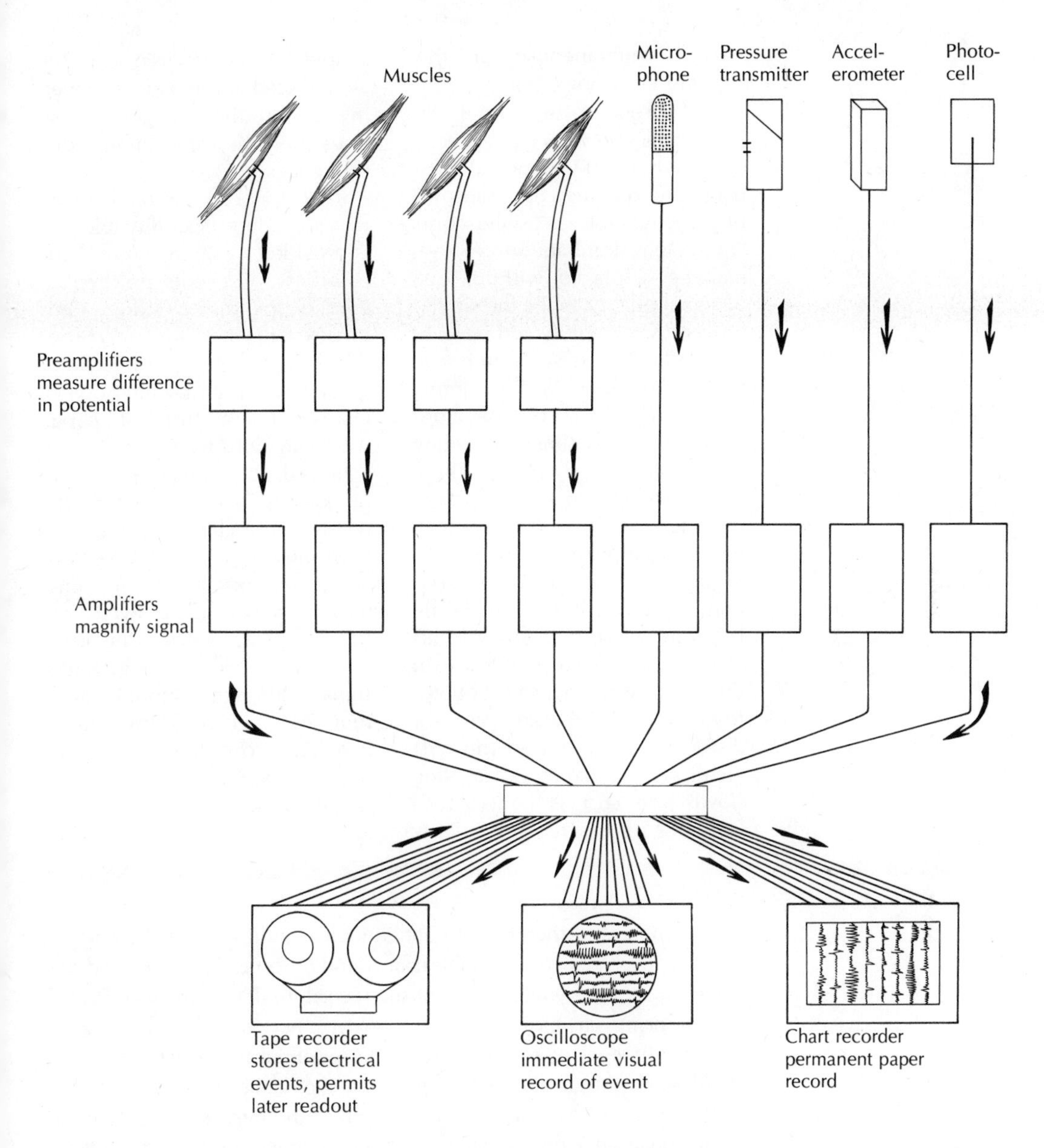

nals to electrical ones, and appropriate amplifiers are necessary to raise them to a level equivalent to that deriving from the muscles.

The block diagram shows four such transducers dealing respectively with sound, pressure, acceleration, and light. If multi-dimensional movements are to be recorded by motion pictures, television, or x-ray movies, these records must be coordinated with the myograms. The records of the muscles are either included on the same film or tape as is the image, or an irregular signal is marked simultaneously on the film and electromyogram.

The output of the several amplifiers may be displayed directly or stored on tape. Immediate displays are obtained by transmitting the signals to an oscilloscope. Up to eight simultaneous tracings may be displayed without loss of signal quality. Since the signals proceed very rapidly and need to be studied at leisure, one can photograph them with a simple camera. However, the development of film is time-consuming and expensive; therefore, there are advantages to utilizing chart recorders writing with ink, or with light or temperature. Such techniques provide a record that is immediately visible. One should keep in mind that some chart recorders do have limitations in their response time, so that high-frequency signals are lost or distorted. There is also the cost factor. If the paper is traveling slowly the cost is limited and scanning of the record is fairly simple, but the details discernible are reduced. If the paper is traveling rapidly, the cost goes up, as does the time for scanning long sequences.

It is obviously desirable to retain maximum flexibility and this is provided by multichannel tape recorders. All events are then recorded for future study. Only those events that ultimately prove to be of interest need be photographed from an oscilloscope or charted on a strip of paper. An even further refinement is achieved, of course, by simultaneous or later analysis of the records on a computer. One can then more quickly ask questions about the correlation between a particular mechanical event and the onset, as well as the firing rate, of an electromyographical signal. Although computerization requires a high initial investment, it may be cost-effective in terms of the time required to solve specific problems.

It is important, therefore, to obtain a clear indication of the onset of the series of stimuli, or "stimulus train," and of its cessation. Beyond this, it may be useful, by integration of the record, to determine whether the rate of firing changes during the activity period. This will suggest whether different motor units are being recruited to the activity during the muscle's stimulation; remarkably few such data are yet available for systems that are correlated with a specific functional pattern. It is also useful to consider the distribution of motor units; most specifically this may be determined by recording simultaneously from two or more places within the muscle (there is also the question of whether or not there is bilateral symmetry in the firing pattern within the animal).

Any comparison between electromyograms and mechanical events must, furthermore, take into account the differential delays of the several types of transducers (gadgets that produce an electrical signal in response to a mechanical event). Although some microphones may have response times on the same order as electromyographic preamplifiers, strain gauges used to measure pressure or acceleration vary in their compliance, and gas-filled tubing connecting animal to pressure transducer may involve both damping effects and time delay of the signal.

When a mechanical event is difficult to measure directly, it may be desirable to substitute an associated measurement—for instance, an electromyogram for a pressure change or the pressure in the visceral cavity for that in the lung. Such substitution becomes less satisfactory as the time interval between the signals being compared becomes smaller. Comparison of multiple parts of a single muscle or of the same muscle from the two sides of an animal should always be on the basis of simultaneous recordings from the several electrodes.

The literature contains dozens of methods for structuring electrodes to pick up myographical signals (cf. Jonsson and Reichmann, 1968). In general terms, a "firing" muscle will induce an electrical event that spreads through the adjacent tissues and will induce voltage change between electrodes placed in or upon such tissues. Consequently, it is possible that two electrodes on the surface of, or even some distance from the periphery of, a large muscle mass will record electromyograms, but the magnitude of the signals will decrease with the distance from the actual site of the electrical event and the number and kind of internal discontinuities between the site and the electrodes. Surface electrodes placed upon the skin without otherwise interfering with the animal give only very undifferentiated records of all muscles adjacent to the recording zone. It is similarly impractical to implant one electrode within the muscle and compare its voltage to ground; such recording methods tend to pick up a scattering of electrical activity from all the intervening muscles, including movement artifacts (as the electrodes slip within the tissue) and the often powerful electrical field of the heart.

As a consequence of these problems, it is necessary to place both electrodes within the muscle to be studied. A steady record requires

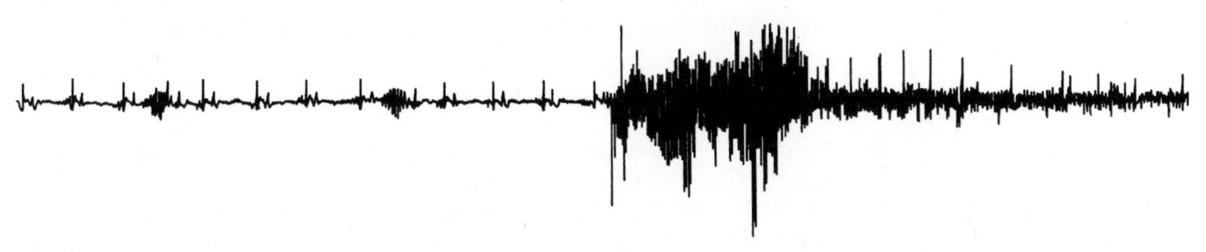

Figure 5-4 When the electric potential is recorded across a turtle, one obtains a simultaneous electromyogram and electrocardiogram. The heart beat may be read quite clearly from the low signal on the left side of the record. About the middle of the sequence the animal was suddenly disturbed, and the electrical output of the multiple muscles between the two electrodes completely masked the electrocardiogram. Such unfiltered, coarse recordings tell us about the onset of major events. They do not permit us to dissect the function of individual muscles as would records taken with bipolar needle or wire electrodes. (After Ireland and Gans, 1973)

that the exposed surface area of metallic anode and cathode be constant and that the distance between these exposed areas remain reasonably constant as well. This may be achieved by using parallel or bipolar pin (tube) electrodes with different portions of the surface freed of insulation; the insulated connectors normally pass centrally within the core of the bipolar tube. Such pin electrodes are easily implanted and retain relative area and position, but their rigid shape may interfere with the contraction effect of the muscle and thus be traumatic to the animal. Only where a muscle may be sampled close to an attachment site at which movement is minimal or where the pin may travel freely with the muscle's movement are pin electrodes useful in unanesthetized animals.

In other cases it is preferable to use a pair of very fine insulated metal wires that have been twisted or glued together along their length and have had the tips bent into one or another shape of single or double hook. A specified area on each wire may then be freed of insulation. Such wires may be implanted from the outside by means of a hypodermic needle (Basmajian and Stecko, 1962), or may be threaded into the muscle under the operating microscope. The latter technique obviously requires anesthesia as well as suitably aseptic conditions, whereas placement of electrodes by hypodermic needle involves no more trauma than any other kind of injection. It may proceed very rapidly and painlessly if the hypodermic needle is kept small.

The benefits of wire electrodes are increased if attention is paid to the movement of the insertion site relative to its surroundings. Not only must the bare terminals be firmly fastened within the muscle's belly, but the leads to the outside should be loose or coiled, so that the terminals may move and the leads will not stress the recording site mechanically when the muscle contracts. Convenient sites for entry of electrode wires are often suggested by the entry of blood vessels and nerves of the muscle; these conductors also have to move with the contracting muscle. The leads best pass to the skin within the loose superficial fascia and may well perforate the skin through a pinhole, even when the actual implantation requires surgery; the natural elasticity of the skin tends to hold the leads in place at the point of exit and reinforces the effects of sutures and surgical adhesives used for attachments.

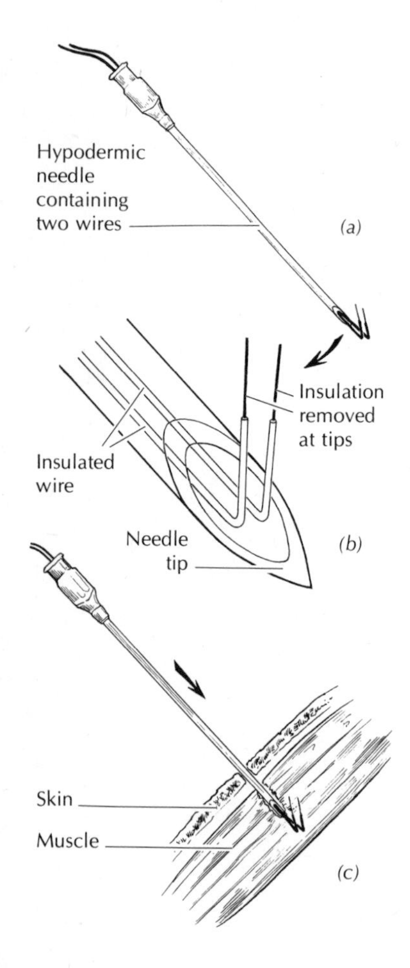

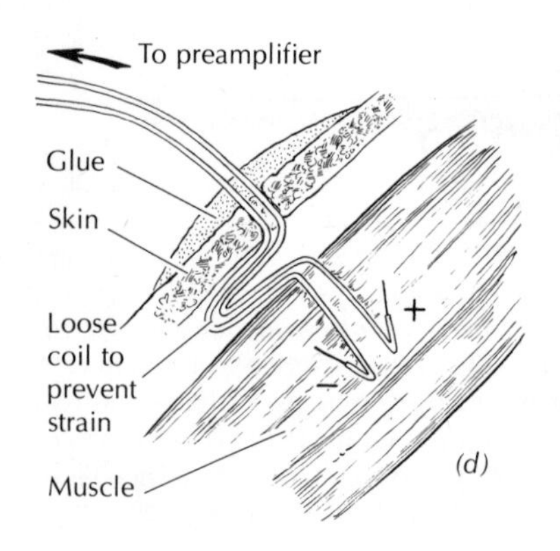

Figure 5-5 Since an electromyogram is intended to record events occurring within a muscle, the uninsulated tips of the electrodes should be placed there. One of the simplest ways of achieving this is by inserting two very fine wire hooks by means of a hypodermic needle (*a*). The insulation is removed from the tips of the hooks only (*b*), and the needle and the containing wires are inserted into the belly of a muscle (*c*). The hypodermic needle is then withdrawn, leaving the hooks anchored and a loose coil placed between muscle and skin. Clips or glue will anchor the wires at the skin; alternately, they may be soldered to a connector or a telemetry unit within the animal. The free ends of the electrode (or the telemetry signal) pass to the preamplifier (Box 5-2).

Various groups of animals have muscles of different textures, and the texture will be an important factor in determining the stiffness of wire actually selected for electrodes. The terminals must remain permanently within the muscle during repeated contractions; they should not be allowed to deflect enough to contact one another and short out. Neither terminals nor leads should induce undue strain; any pain will affect the normality of the animal's behavior. However, insulated stainless wire sized at 0.0009 inch (0.00035 cm) is commercially available. This is so fine that a human patient does not notice retention after insertion. Careful observations have shown no alteration in the feeding pattern of little brown bats (*Myotis lucifugus*, average body weight 7 g) when these wires were placed into eight masticatory muscles at a time (Kallen and Gans, 1972). On the other hand, this wire does not implant well in frogs, and we used a much heavier but weaker and less elastic silver wire for the experiments that will now be discussed.

Movement and Pressure

Undisturbed frogs and toads[6] show pulsating movements of the floor of their mouth when resting; these oscillations are often the only obvious movements seen during long periods. The oscillations proceed with the mouth closed and the nostrils open. After attaching a bit of tissue paper or foil near the nostrils, one can observe a slight inward current when the buccal floor drops and an exhalation when it rises; the nostrils do not change configuration. There is no concurrent movement of the frog's flanks; thus the glottis appears to remain closed, and the lungs are not directly involved. Numerous records of this and other cycles involved in respiration were analyzed from films taken of frogs either sitting on the bottom of small glass aquaria or half-floating in water.

The nature of the pressure changes during such cycles could be determined in unrestrained bullfrogs (*Rana catesbeiana*) by recording the pressures in the body cavity adjacent to the lung and within the buccal cavity. The first measurement was taken indirectly, using a small balloon inserted into the body cavity and connected via a polyethylene tube to an air-pressure gauge; the buccal pressure was recorded directly by inserting an open polyethylene tube through the soft tissues of the mouth. As the techniques became refined, we learned to insert a third (open) catheter directly into the bronchial portion of the lung; this direct measurement confirmed both that the pulmonary pressure is always higher than atmospheric and that neither pressure nor volume of gas in the lung shows significant change during oscillating movements of the buccal apparatus. The pressure within the buccal cavity fluctuates above

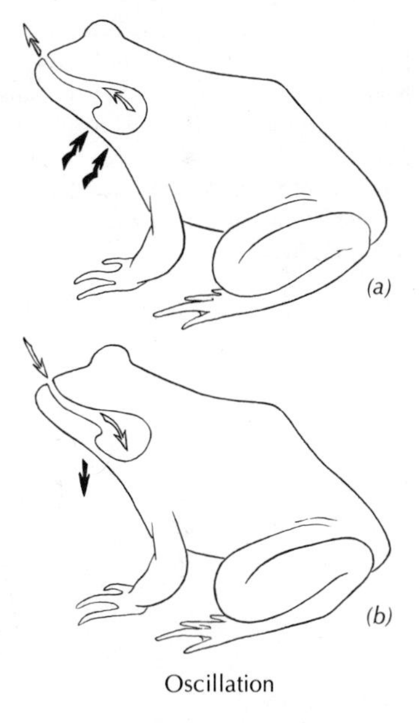

Figure 5-6 The oscillation cycle consists of a (muscle-activated) lifting of the floor of the mouth (a), which reduces the volume of the buccal cavity and forces air out of the nostrils. This is followed by a dropping of the floor of the mouth (by gravity), which increases the volume of the buccal cavity and allows air to flow back in (*b*). The larynx remains closed and the nostrils open. Frogs show oscillations of varying amplitude, whether active or resting.

[6]A few groups of frogs do not seem to follow the scheme here described; since their breathing mechanism is still under study, it seems premature to discuss it.

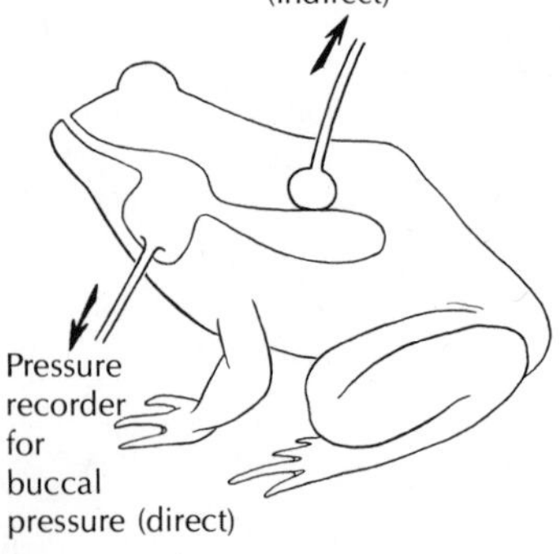

Figure 5-7 Visceral pressure may be monitored by placing a small balloon adjacent to the lung. Since the pressure in the balloon need not be the same as that in the visceral cavity, the measurement is *indirect*; however, the pressure changes in the balloon will reflect the changes in the lung. The open tube inserted into the buccal cavity allows a *direct* monitoring of buccal pressure by a transducer.

and below atmospheric pressure; it rises slightly above this while the buccal floor rises and drops below atmospheric as the buccal floor falls. The oscillations of the buccal floor apparently serve to aspirate and flush air from the buccal chamber. Does this gas movement serve olfaction, buccal gas exchange, or some other function?

The nares close briefly at irregular intervals; this signals a true *ventilatory* cycle interspersed among the purely oscillatory ones. Such ventilatory cycles occur at intervals ranging from a few seconds to several minutes. Nasal closure always starts just before the buccal floor rises. Closure is preceded, furthermore, by an inward dimpling of the animal's flanks; this reverses after the nostrils have blinked closed, simultaneous with the rise of the buccal floor. Pressure recordings (Figure 5-8) showed that the pulmonary (lung) pressure starts to drop just before the inward dimpling of the flanks but that it never descends to atmospheric. Shortly after the nostrils close and the floor of the mouth rises sharply, the pulmonary pressure again rises and may very briefly overshoot the previous base level.

The pressure changes in the buccal cavity are more complex than the pulmonary changes. The first indication of the start of a ventilatory cycle is an unusually large drop of the buccal pressure as the buccal floor is lowered (phase 1). This is followed by a sharp rise in buccal pressure, coincident with the fall in pulmonary pressure, but not associated with any significant movements of the buccal floor (phase 2). The buccal pressure steadies about the level of the lowest pulmonary pressure during a variable interval of perhaps one-tenth second (phase 3). This is also the period during which nasal closure starts. Within the next tenth of a second the buccal pressure rises sharply coincident with the closure of the nostrils, a sharp rise of the buccal floor, and a rise in the pulmonary pressure (phase 4). The buccal pressure, in this phase, is always greater than the pulmonary and, at its end, significantly exceeds the latter's resting level. The end of the fourth phase is marked by the opening of the nostrils, the descent of the buccal floor, and a sharp drop in buccal pressure to or below atmospheric (phase 5). The oscillations of the buccal floor then continue, interspersed with further ventilations.

This description of movements and pressures during ventilation leaves open three questions: (1) What is the state of the glottal valve connecting buccal and pulmonary spaces? (2) What is the direction, sequence, and magnitude of air flow at the nasal and glottal valves? (3) What muscles drive the system?

The last question clearly requires electromyographical analysis and is dealt with in the next section. Before our studies, there had been some information about the first question from an interesting experiment by Schmidt (1965). In this, an eye of the frog was removed, so that the

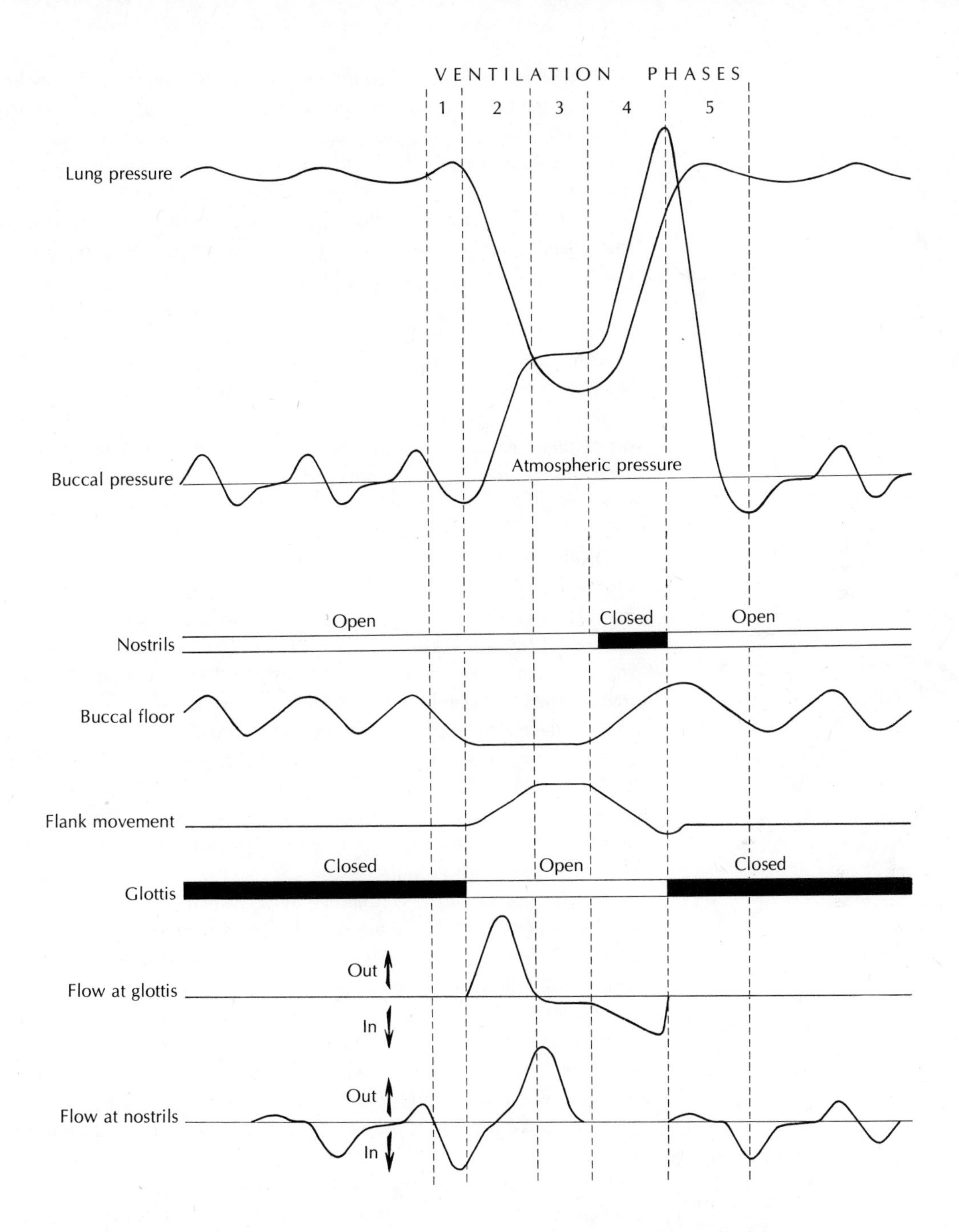

Figure 5-8 Combined diagram of changes in pulmonary and buccal pressure during oscillation and ventilation cycles. Below this are shown the movements of the nostrils, the buccal floor, the flanks, and the glottis, as well as the magnitude and direction of flow between the chambers and the outside of the animal.

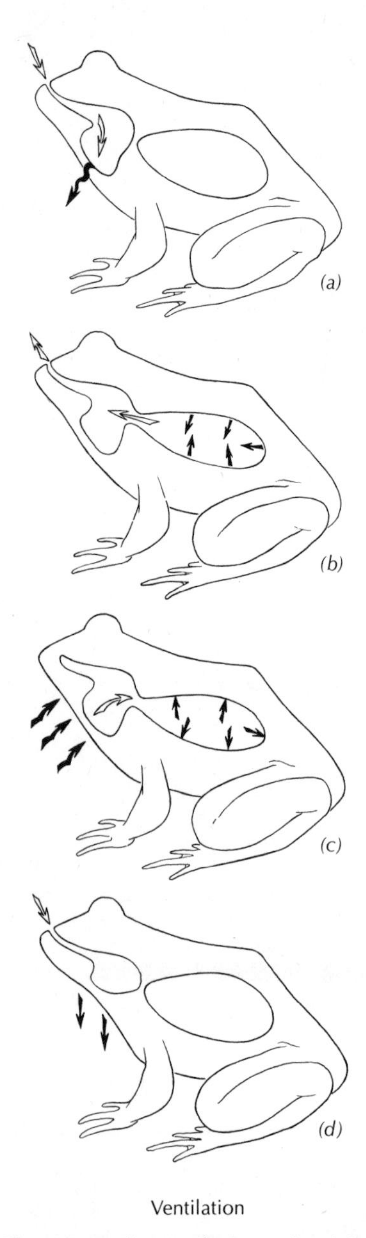

Figure 5-9 The ventilation cycle of the bullfrog takes less than one second and is powered by the muscles of the floor of the mouth (c). Since these distend the elastic tissues of the lung, some of the muscular energy from each cycle powers the pulmonary efflux during the second phase (*b*) of the next cycle.

position of the glottis could be directly observed. (Later—1972—Schmidt repeated the observations through a plexiglass plug placed in the orbit.) These observations suggested that the glottis opened at the end of phase 1 (when pulmonary pressure first sinks) and remained open until the start of phase 5 (when pulmonary pressure returns to and exceeds base level). The sequence determined by direct observations of the glottis has since been confirmed as correct by electromyographical records from the laryngeal muscles (see p. 222).

The direction of gas flow mainly had to be extrapolated from the expectations that the narial and laryngeal valves only permit flow when they are open and that gas flows from regions of high to regions of lower pressure. One direct observation may be made during the start of phase 2, when the expiratory current of air from the nostrils will deflect a bit of tissue paper or foil. Another direct confirmation results from an x-ray movie film (kindly prepared by Dr. P. Dullemeijer) which indicated that the inward dimpling of the sides of the frog occurred in synchrony with an emptying of the lungs.

Dealing with the pulmonary gas first, we note that it must leave the lung when the larynx opens at the start of phase 2. It first flows to the buccal cavity (causing its contents to rise in pressure) and then to the outside via the nares. During phase 4 the buccal pressure rises faster than the pulmonary pressure and presumably reinflates the lungs. As the contents of the buccal cavity are above atmospheric pressure from the instant the larynx opens until the nostrils close, and are also at or above the buccal pressure, the air then entering the lungs must have been aspirated into the buccal cavity during phase 1, before the larynx opened. This curious pattern then admits the possibility that inhaled and exhaled air mix in the buccal cavity and that part of this mixture is pulse-pumped into the lung by the mouth.

The fifth phase of the ventilatory cycle probably involves some slight outflow from the nares if these open before the buccal floor has dropped enough to restore the buccal contents to near atmospheric pressure; inflow will follow as the buccal pressure again passes below atmospheric. It is clear that these shifts of gas to and from the pulmonary cavity could have respiratory function. Yet why the curious mixing of gases within the buccal cavity? Is frog respiration really as inefficient as the records would make it appear? Direct observation did not seem to provide an answer, so we had to set the question aside until we could design some new experiments.

When recording visceral and pulmonary pressures during sequences of ventilatory cycles, we were annoyed with "drift" in the records. Rather than maintaining a respectable and repeatably constant level during the interventilatory apnea (temporary suspension of ventilation), both these pressures would increase or decrease slightly from cycle to cycle. We checked the strain gauges for stability and considered the possibility of slow leaks in the connecting lines; we blamed equipment malfunction (always a useful excuse when things do not go as expected) and turned off the air conditioner to avoid slow pressure cycles in the gas-filled tubing. However, whenever we seemed to have solved the problem, the "drift" reappeared. This was particularly annoying when we had all amplifiers near maximum setting, as the trace would then quickly drift off scale over a period of a few ventilations. Happily, we decided to run some long-range, low-amplification records of a cooperative resting frog. These records indicated that the phenomenon was real and led to rediscovery of the inflation cycles described below. Our experience also documented the merit of studying and reconsidering seemingly aberrant phenomena. Animals have almost infinite potential for curious solutions to problems of survival; only observation allows us to understand what is happening.

The pressure of the pulmonary contents and consequently the volume of the lung do undergo regular, medium- to long-term cyclic changes in undisturbed animals, superimposing an *inflation* cycle upon a series of perhaps fifty or more ventilatory cycles. During most of an inflation cycle, the pressures (and volumes) remain relatively steady at a min-

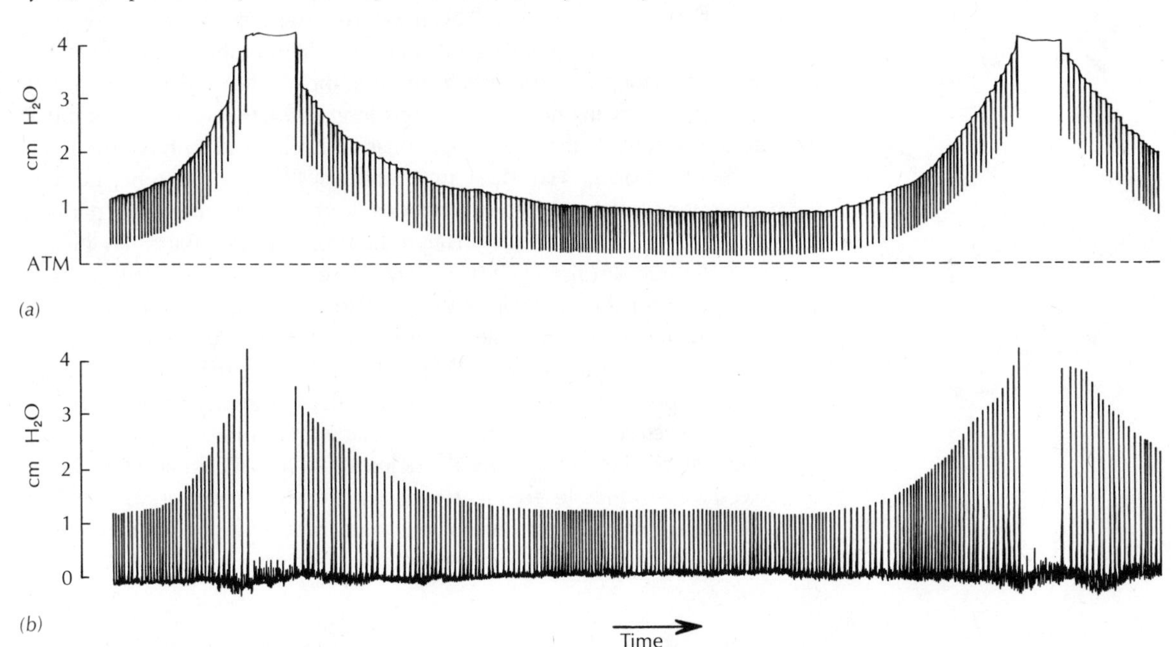

Figure 5-10 Continuous recording of pulmonary (*a*) and buccal (*b*) pressures during an inflation cycle. Time passes from left to right, and the time bar denotes one minute. Note that the pulmonary pressure reaches a peak followed by an apneic period and that the baseline (*a*) pulmonary pressure closely reflects the maximum pressure during ventilation (end of phase 4).

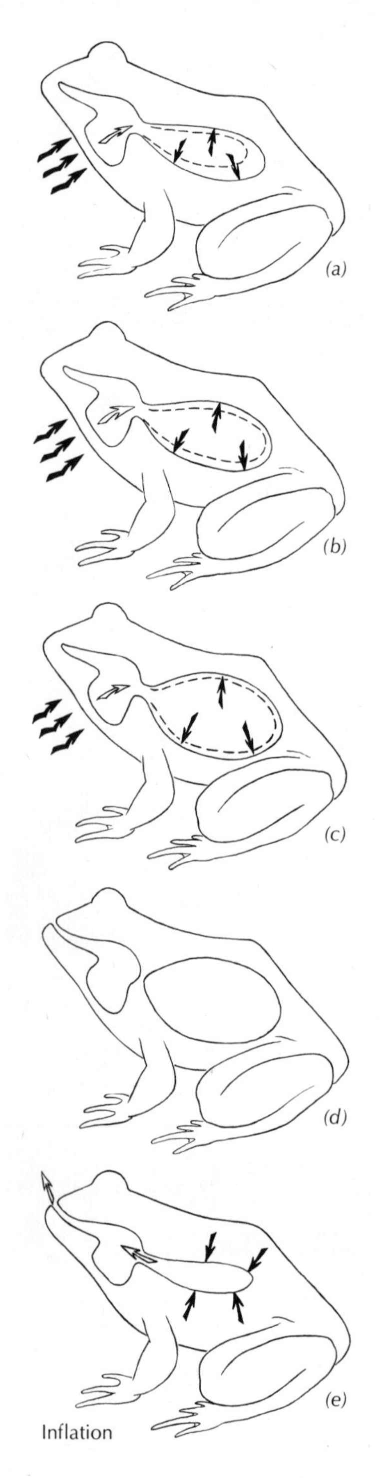

imum level of perhaps 2 cm of H_2O (in adult *Rana catesbeiana*). During this period the volume of gas pumped by the buccal pulses is just equal to the amount needed to inflate the lung. Toward the end of the inflation cycle, however, this pressure level starts to rise. Each successive buccal ventilatory pressure reaches a greater peak (at the end of phase 4), and the volume of the lung consequently increases. This is nicely seen by the flotation level of the animal if it is resting in water. The pressures rapidly and exponentially increase to a maximum in successive ventilatory cycles, and ventilations then stop for an apnea of variable length (Figure 5-10). The first postapneic ventilation may be of a magnitude only slightly smaller than the ventilation immediately preceding the suspension, with the following ones gradually dropping to a midcycle base level. Alternatively, the postapneic ventilations may immediately proceed near the resting level, not reflecting the preceding high pressure at all. Rising pressures obviously involve reduced pulmonary outflow and more effort of the buccal muscles to pump against increasing resistance. The ventilatory cycles then continue at a fairly constant level until the next inflatory period.

These inflation cycles obviously affect the mixing and composition of pulmonary gases. The bullfrog's lungs are fairly smooth-walled chambers, but they are traversed by muscular crossbars, or trabeculae. It is unclear whether the regular, low-amplitude filling movements of resting-level ventilations leave a significant amount of unmixed (i.e., high O_2, low CO_2) gas in the center of the lungs. Do the greater postapneic deflations toward the end of the inflation cycle tend to flush the intertrabecular spaces? May these inflation cycles have a different and nonrespiratory function? What variations do they show over long periods?

Thus the air movements through the frog's buccal cavity may involve one of three distinct cyclical respiratory events, here defined as oscillatory, ventilatory, and long-term inflation cycles. This accounts for some of the confusion in the literature on breathing and also reflects the complexity of the system. These observations, furthermore, indicate that the pressure of the buccal cavity reflects a greater number of different mechanical events than does that of the lung. Buccal pressure, then, offers a better mechanical baseline to which to compare the various electromyograms. Its frequent cyclic nature helps association; one soon has enough cases to ascertain whether a particular myographic sequence is always synchronous with a particular pressure change.

Figure 5-11 The pressure increases during the long-time inflation cycles are produced by peaks of muscular activity during the fourth phase of successive ventilation cycles. With each peak, the frog swells noticeably as its sides distend (*a, b, c*). During apnea the animal rests with glottis closed (*d*). Deflation thereafter may be sudden or stepwise (*e*).

Solids and Spaces

Dissection of the bullfrog discloses that the respiratory events involve two sets of chambers and two sets of valves (Figure 5-9). The first valves encountered by entering gases are those in the nares visible from the outside, and the first chamber is the buccal one lying between them and the larynx. The second valve is the larynx which separates the buccal cavity from the anteriormost portion of the paired, ovoid lungs. It is easy to see that the respective chambers (buccal and pulmonary) are delimited and that their volume is controlled primarily by extensive muscular areas, but the muscles and connective tissues clearly exert their action by transmission of forces via the slender skeleton.

The bullfrog's skeleton represents a rather delicate suspension for the bulging visceral mass. The flat skull is circumscribed by a U-shaped, tooth-bearing bony arch. Anteriorly this is loosely supported by the nasals and central elements of the vomeropalatine series; posteromedially the quadratojugal extensions of the U's posterior tips are braced by pterygoid and squamosal connections to the delicate T-shaped braincase. The enormous orbit lacks bony separation from the buccal cavity, and there are large fenestrations lateral to the braincase.[7] The mandibles are

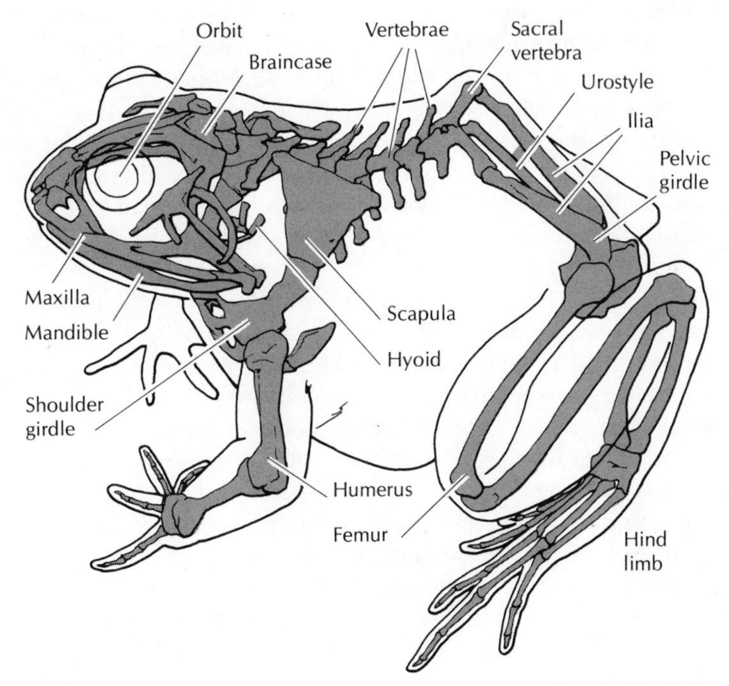

Figure 5-12 Sketch of the skeleton more or less in place within the outline of a frog. Note the absence of ribs and the unsupported bulge of the viscera.

[7]The extent of cranial ossification varies markedly among frogs and toads. Toads (*Bufo*) generally have much more robust skulls than do ranid frogs, and terrestrial species show more solid framing than aquatic ones.

slender, tubular elements; they define and support only the lateral edges of the wide lower jaw. The symphysial tips of the mandibles are enlarged into cartilaginous bosses; in some species of the genus *Rana* these regions are represented by separate mentomeckelian ossifications that are mobile upon the mandibular tips.

The skull is supported by a fairly rigid articulation with the anterior vertebrae. The short, ribless vertebral series terminates in a urostyle flanked by slender ilia. As the latter articulate with the lateralmost processes of the posterior vertebrae, only the anterior half of the trunk permits either bending or torsion. The femoral attachment is to the posterior axial apparatus and lies very close to the midline. Most of the propulsive force is transmitted via the attached and buttressed pelvic girdle, and the vertebral column consequently is subject to compression during locomotion.

The pectoral girdle, in contrast, is independent of the vertebral column. The large superficial cartilages of the sternal system are braced by ventral connections between the laterally placed fossae with which the humeri articulate. More dorsally the girdle extends into wide scapular cartilages that flank the vertebral series, to which they show muscular and loose connective-tissue attachments. The girdle also has muscular attachments to the large and relatively rigid skull and, with this, some restriction of spatial position. The forelimbs move much more freely toward than away from the frog's midline. In *Rana* they act as shock absorbers when landing but lack a significant role in propulsion. They do support the anterior end of a sitting animal, keeping skull and visceral mass off the ground. Beyond this, they assist feeding, as they can swing forward and inward to stuff food objects into the wide mouth. Finally, they have a critical function in reproduction, as the males use them to clasp and stimulate the females.

The posterior portion of the buccal floor incorporates the hyoid elements, remnants of the ancestral branchial skeleton. Its cartilaginous plates and processes support a larynx which lies at the very rear of the buccal cavity, just ventral to the esophageal entry. The posteriormost portions of the hyoid, which brace the posterior aspect of the buccal cavity, lie dorsal to the sternal system, and the posterior hyoid horns swing around to connect to the sides of the skull.

The nostrils are oval tubes of soft tissues, bounded dorsomedially by the immobile oblique cartilage and ventrolaterally by the movable alary cartilages (Figure 5-13). The latter facilitate closure of the nostrils by a most indirect series of events. They are suspended by short cartilaginous rods from the nasal capsule. The medial tip of each alary cartilage also has a dense connective-tissue connection to the facial process of the premaxilla, which rises vertically from the dental and palatine portions of this bone. Each palatine portion of the premaxilla is tightly connected

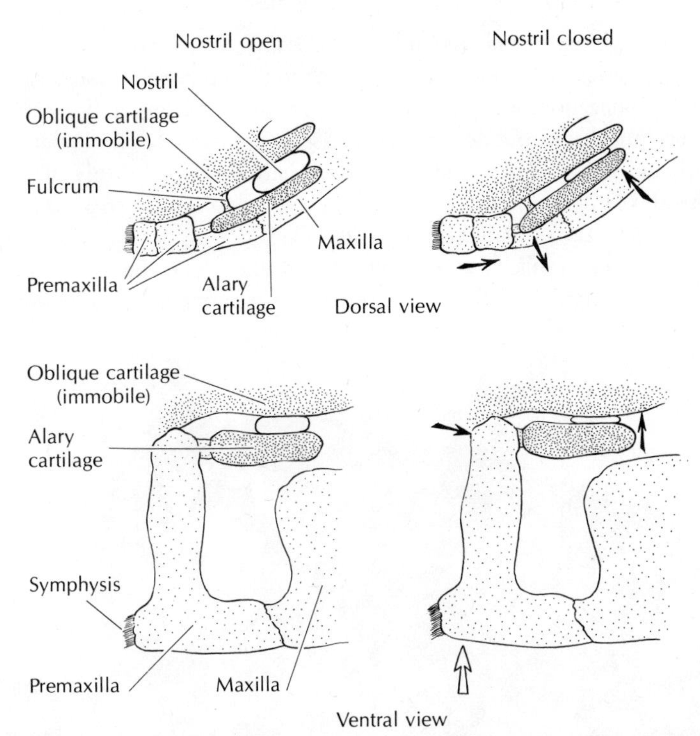

Figure 5-13 The frog's nostril is a slit-shaped tube, positioned between the fixed oblique and movable alary cartilages. These cartilages are connected by an elastic cartilaginous rod that serves as a fulcrum when the premaxilla is rocked about its attachment to the maxilla by an upward movement of the tip of the mandible. The alary is then forced laterally and simultaneously rotated about the fulcrum so that the nostrils are squeezed shut. When the mandibular pressure is removed, the system returns to rest by the elastic return of the cartilaginous fulcrum.

with the tip of the maxilla on one side and with its lateral counterpart on the opposite. The maxillary connection does not permit much lateral movement. Any pressure directed dorsally upon the palatine and dental portions of the premaxilla (in the subrostral fossa into which the lower jaw fits) hence results in a dorsoposterior movement of the central region of the premaxilla about a center of rotation at the maxillary-premaxillary joint. The dorsal tip of the facial process then rotates laterally, and the attached superior prenasal cartilage starts to slide on the nasal capsule, rotating the medial portion of the alary cartilage about the alary's attachment to the nasal capsule.

Since the tubular nostril lies on the opposite side of the center of rotation of the alary cartilage, the movement of this cartilage occludes the nostril against the oblique cartilage. Rotation of the premaxilla increases the torsion of the inferior prenasal cartilage, and the elastic return of this cartilage can rotate the premaxilla and other tissues back to the resting position. The nostril thus returns to the open condition without further application of force, suggesting an active closing and a passive opening mechanism. The bullfrog lacks any muscles of the upper jaw capable of inducing the kinds of motions here described. Only upward pressure of the mentomeckelian region will close the nostrils. This is easy to observe by manipulating a live frog.

The orbits are filled by the bulging eyeballs which themselves form part of the buccal roof. Retractor muscles may pull these into the buccal

(*a*)

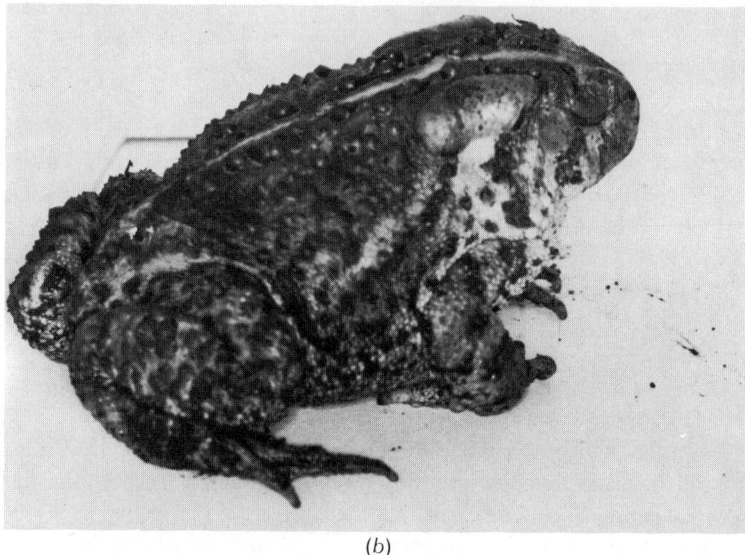

(*b*)

Figure 5-14 Swallowing in frogs and toads involves an inward push of the eyeballs. This American toad is about to pick up the meal worm (*a*); he then swallows it with what appears, to us, as a pained expression (*b*).

cavity (thus protecting them during a jump or reducing buccal volume during feeding). However, no such retractions were observed concomitant with the oscillatory, ventilatory, or inflatory cycles of breathing. The other posterior apertures in the skeletal roof are blocked by the mandibular adductor muscles, which show no activity during breathing.

The extensive intermandibular space of the lower jaw is filled by two layers of muscles. A deep layer originating on the medial edge of the symphysial regions stretches posteriorly as some six parallel strips of geniohyoid muscles that insert on the ventral surface of the hyoid elements. These should potentially lift the hyoid apparatus when they contract. The posterior aspect of the lateral geniohyoids surrounds the insertions of two additional sets of muscles. The first consists of the omohyoid and petrohyoid groupings and swings ventrally around the lateral aspects of the hyoid from origins on various posterolateral aspects of the skull. In shortening, these muscles should lift the middle of the hyoid toward the skull. A pair of larger muscles, the sternohyoids, inserts adjacent to these, but swings ventrally from a fleshy origin upon the xiphoid cartilage and the sternum; these muscles should depress and pull the hyoid posteriorly when they shorten.

The more superficial intermandibular muscles are all transverse. The posterior intermandibular is the largest; it forms a thin sheet extending from the mediodorsal border of the mandible to a thin and irregular medial aponeurosis. The simple statement that the fibers of the intermandibular muscle are transverse represents an oversimplification. These fibers make a sharp turn adjacent to the mandibles, first dropping ventrally and then curving much more sharply anteriorly as they pass bluntly posterior to the medial region. Along its posterior aspect the posterior intermandibular muscle is continuous with and parallel to the interhyoid muscle. Contraction of the posterior intermandibular and interhyoid would tend to impart maximum vertical displacement of the buccal floor. Anteriorly the posterior intermandibular is crossed transversely by a thick, spindle-shaped anterior intermandibular muscle that attaches through short aponeuroses to the Meckelian cartilages where these are exposed along the ventral circumference of the jaw (between the mentomeckelian and angular bones). Contraction of the anterior intermandibular would serve to draw the mandibular tips together and secondarily shift the position of the mentomeckelian bones.

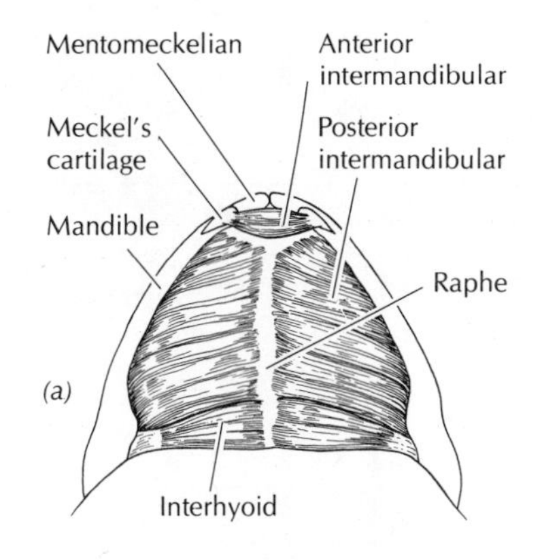

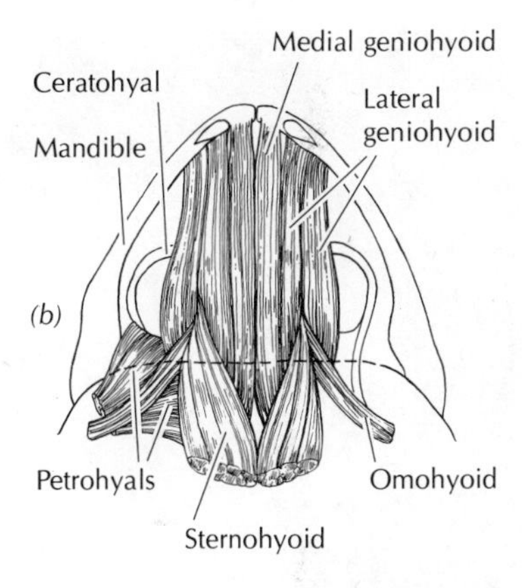

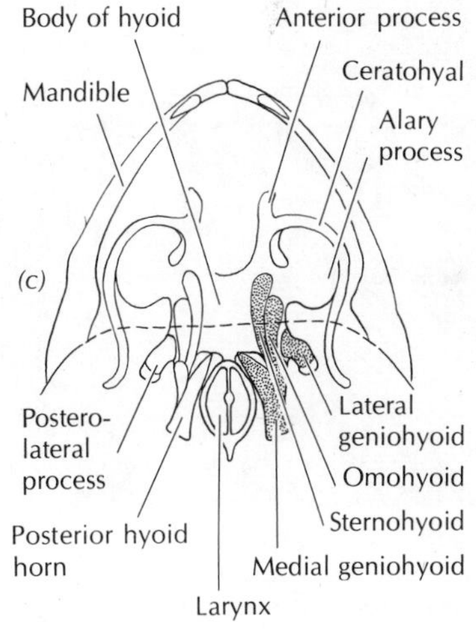

Figure 5-15 These sketches show a ventral view of the throat of a bullfrog that was dissected deeper for each sketch. The top view (*a*) shows the major transverse intermandibular muscles. Their removal (*b*) reveals the longitudinal muscles which lie just beneath the tongue. The removal of these muscles (*c*) exposes the hyoid cartilage and, at its rear, the larynx nestled between its posteromedial processes. The tongue muscles have been omitted as they do not participate in ventilation.

Summarizing these aspects of anatomy, we can see that the dorsal roofing surface of the buccal cavity must be immobile in ventilation. The ventral surface may be lifted against the buccal contents by contraction of the geniohyoid, intermandibular, interhyoid, petrohyoid, and omohyoid muscles; only the sternohyoid muscle is in a position to depress the posterior aspect of the buccal floor by shifting the hyoid cartilage ventrad. In doing this the sternohyoid acts antagonistically, primarily against the elements of the petrohyoid and omohyoid complex.

The larynx consists of the cricoid, a cartilaginous ring which surrounds the laryngeal chamber, and a pair of triangular shell-shaped arytenoid cartilages. The arytenoids articulate with the anterior edge of the cricoid and form the basal support for the margins of the glottis. Just posterior to the laryngeal aperture lie the vocal cords, which are discussed subsequently. The soft tissues framing the arytenoids provide for elastic closure of the larynx; opening occurs by an outward rotation

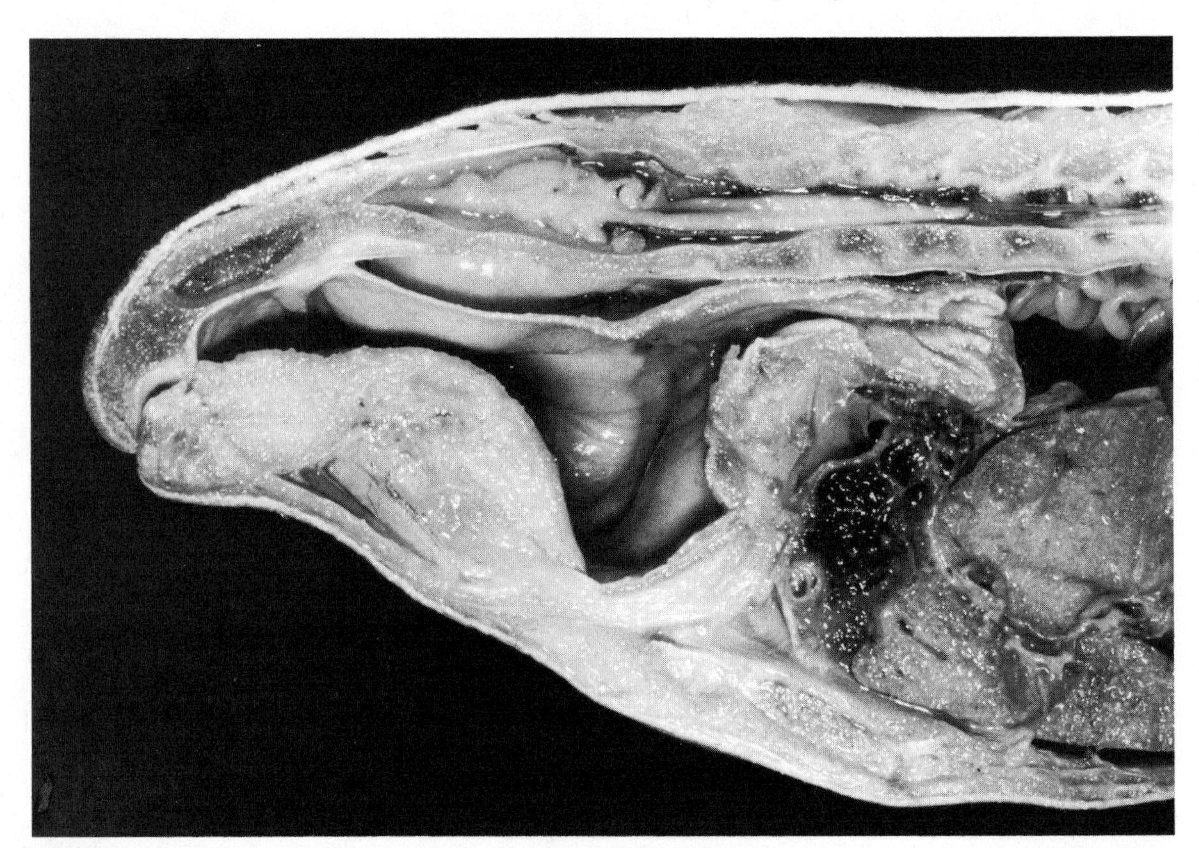

Figure 5-16 A sagittal section of the head of a bullfrog shows how the roof of the buccal cavity is formed entirely of bone. Only the muscular sheets of its floor may change the volume of the buccal cavity. This picture also shows the anterior portion of the buccal cavity, which is relatively shallow compared to the deep, posterior one. (From De Jongh and Gans, 1969)

of each of the two arytenoid shells. Three sets of constrictor muscles, originating from the hyoid and cricoid elements, pass across the surface of these shells. They insert onto a median raphe of soft tissue between the arytenoids, and their contraction would tend to push the two arytenoids together into medial closure and to pull the entire laryngeal apparatus back into an anterior aperture in the hyoid plate. The dilatators of the larynx are a pair of fairly thick muscles that run from the hyoid processes to insert on the lateral aspect of each arytenoid near its apex. Contraction of these muscles should pull the arytenoid cartilages laterally, opening the glottal slit widely.

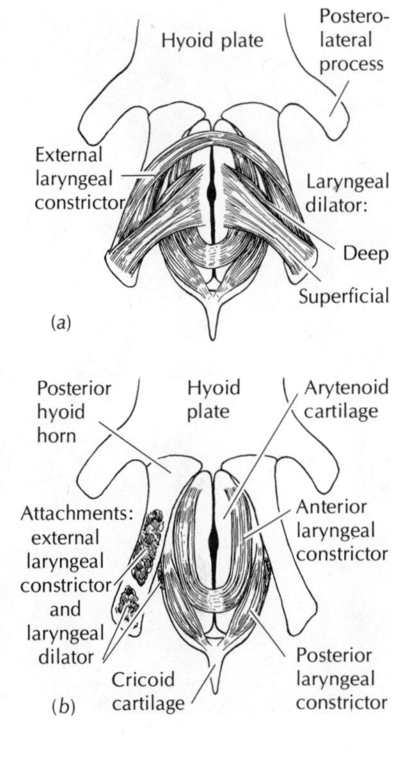

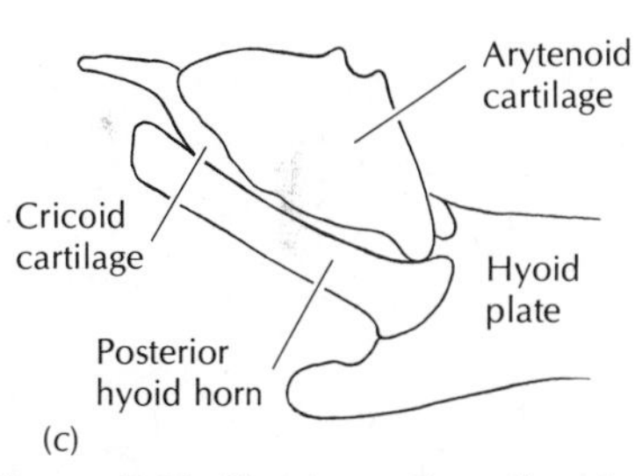

Figure 5-17 Sketches attempting to show the position of the arytenoid and cricoid cartilages between the posterior hyoid horns in dorsal (*a* and *b*) and lateral (*c*) views. The more important activators of the larynx are also shown in superficial (*a*) and deep (*b*) view.

The ovoid lungs connect directly to the laryngeal chamber, as frogs lack a real trachea. The lungs are internally crossed by muscular trabeculae, forming a honeycomb of tissue on their walls (Figure 5-18). During most of the inflation cycle, the lungs extend to a point well beyond the sacroiliac joint and, when maximally inflated, they account for 25–30 percent of the body's volume. (This figure can be very much exceeded in certain tropical frogs such as species of *Physalaemus*, in which the volume of the animal can more than double during inflation.) The visceral mass as such is suspended from the lateral processes of the dorsal vertebrae and the iliac bars. Much of the body wall is reinforced by the thin sheets of the external oblique (outer) and transverse (inner) muscles (Figure 5-31). Ventrally there is the wide sheet of rectus abdominus muscle which may be subdivided by several transverse aponeuroses. The body wall, then, forms a soft sac with but minimal skeletal support. Rather than attaching to discrete skeletal elements, the component muscles attach on each other, forming a tube. Gutmann (1969) notes that this phenomenon is far more common in lower than in higher vertebrates; he suggests that lower forms have not yet achieved independent control over separate divisions of their visceral mass.

Myograms and Gas Flow

The *oscillatory* cycles involve synchronous contractions of the anterior muscles of the buccal floor (de Jongh and Gans, 1969; Figure 5-21). The geniohyoid, posterior intermandibular, and interhyoid muscles fire at a low level with some slight activity in the anterior intermandibular as well. The electrical activity is synchronous with (or may slightly precede) the lifting of the buccal floor, the rise in buccal pressure, and the outflow of gas from the open nares. Muscular contraction raises the buccal floor, thus decreasing the buccal volume and inducing the pressure rise and gas outflow. It is interesting that the omohyoid, petrohyoid, and sternohyoid muscles are inactive during oscillatory cycles.

The dropping of the buccal floor occurs independently of any muscular activity. It occurs by gravity, sometimes enhanced by some elastic

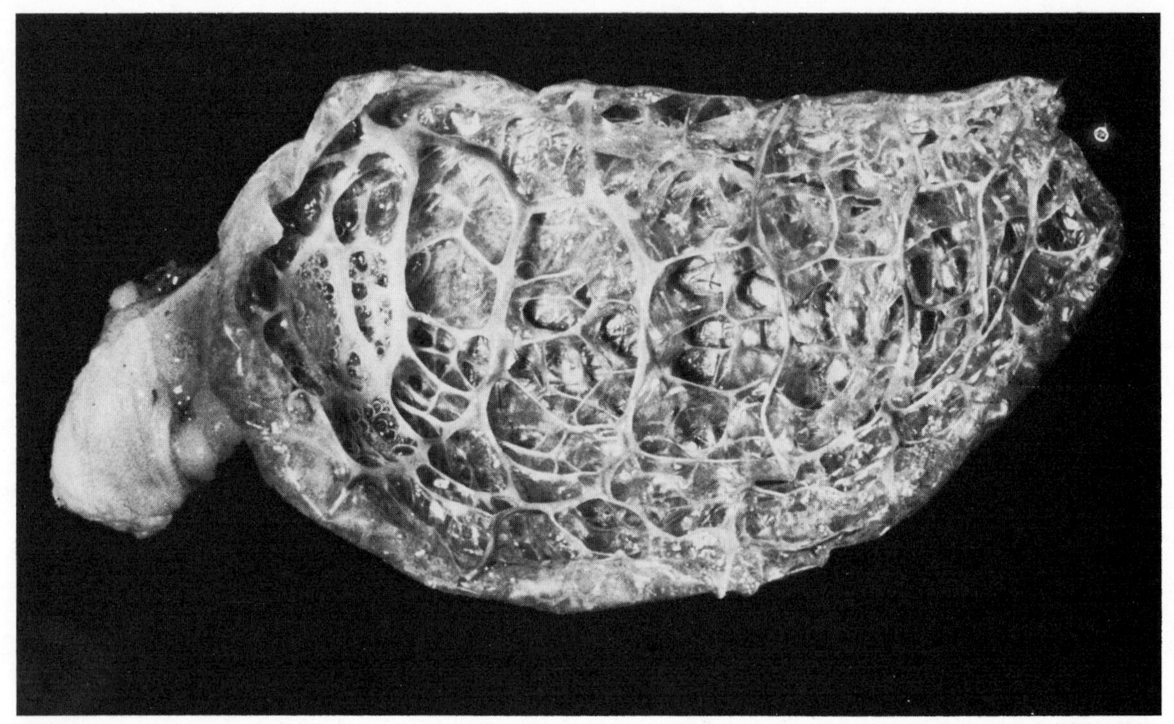

Figure 5-18 This lung of a bullfrog, *Rana catesbeiana*, was specially preserved and then opened. One can thus see the thickened trabeculae supporting the honeycomb that arises from the pulmonary wall. Both the trabeculae and wall are reinforced by smooth muscles that provide for the innate elasticity of the lungs. (After De Jongh and Gans, 1969)

return of tissues (depending on the positioning of the hyoid plate). This dropping of the floor again increases the volume of the buccal cavity and with this decreases the pressure below atmospheric. Air then flows in via the open nostrils.

The start of a *ventilatory* cycle is always signaled by activity in the sternohyoid muscle synchronized with a period of (exaggerated) pressure drop. Its activity changes the period of pressure drop of an oscillatory cycle into the first phase of the ventilatory cycle. The beginning of the second phase (lung emptying) is marked by activity in the laryngeal openers as well as by some slight activity in the anterior intermandibular. The sternohyoid continues to fire throughout the initial portion of this phase, but none of the muscles of the body wall show any activity. The variable third phase (equalization of buccal and pulmonary pressures) shows almost no muscular activity except for low-level responses from the dilatators of the larynx (keeping this distended against the elastic closing forces). Just before and at the beginning of the fourth phase (lung filling), one observes strong muscular activity not only in

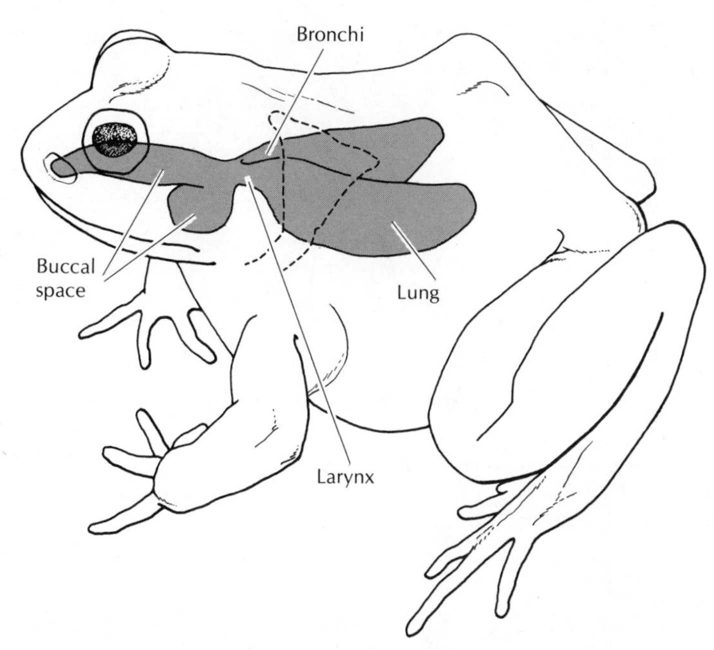

Figure 5-19 Sketch showing the buccal cavity, larynx, bronchi, and lungs in their approximate position within the body of a frog. Note that the lungs lie in a space that is restrained laterally only by soft tissue (compare Figure 5-12).

both intermandibulars, interhyoids, and geniohyoids, but in the omohyoids and petrohyoids as well. The activity is greater by several orders of magnitude than that seen during the oscillatory cycles. Most of these muscles continue to fire into the beginning of the fifth phase (nostrils reopen); the phase transition from fourth to fifth is marked by very strong activity in the several constrictor muscles of the larynx. This activity continues beyond the end of ventilation and into the beginning of the first postventilatory oscillation. Ordinarily the muscles of the body wall do not fire during ventilatory cycles.

Comparison of the electromyographical record with pressure events during ventilation confirms the curious flow sequence (Figure 5-8). During phase 1 the depression of the hyoid plate by the contracting sternohyoid muscle dilates the posterior portion of the buccal cavity. This reduces the buccal pressure and promotes an inflow of air. At the start of phase 2 the dilatator of the larynx causes the arytenoids to spread. The pulmonary air, which has been at superatmospheric pressure, flows into the buccal cavity and raises its pressure. Since the nostrils are still open, the rising pressures in the buccal cavity cause an immediate outflow of air through the nares. The outflow continues into the third phase, during which flow is reduced as the pulmonary pressure approaches that of the buccal cavity. It is interesting that the pulmonary outflow is

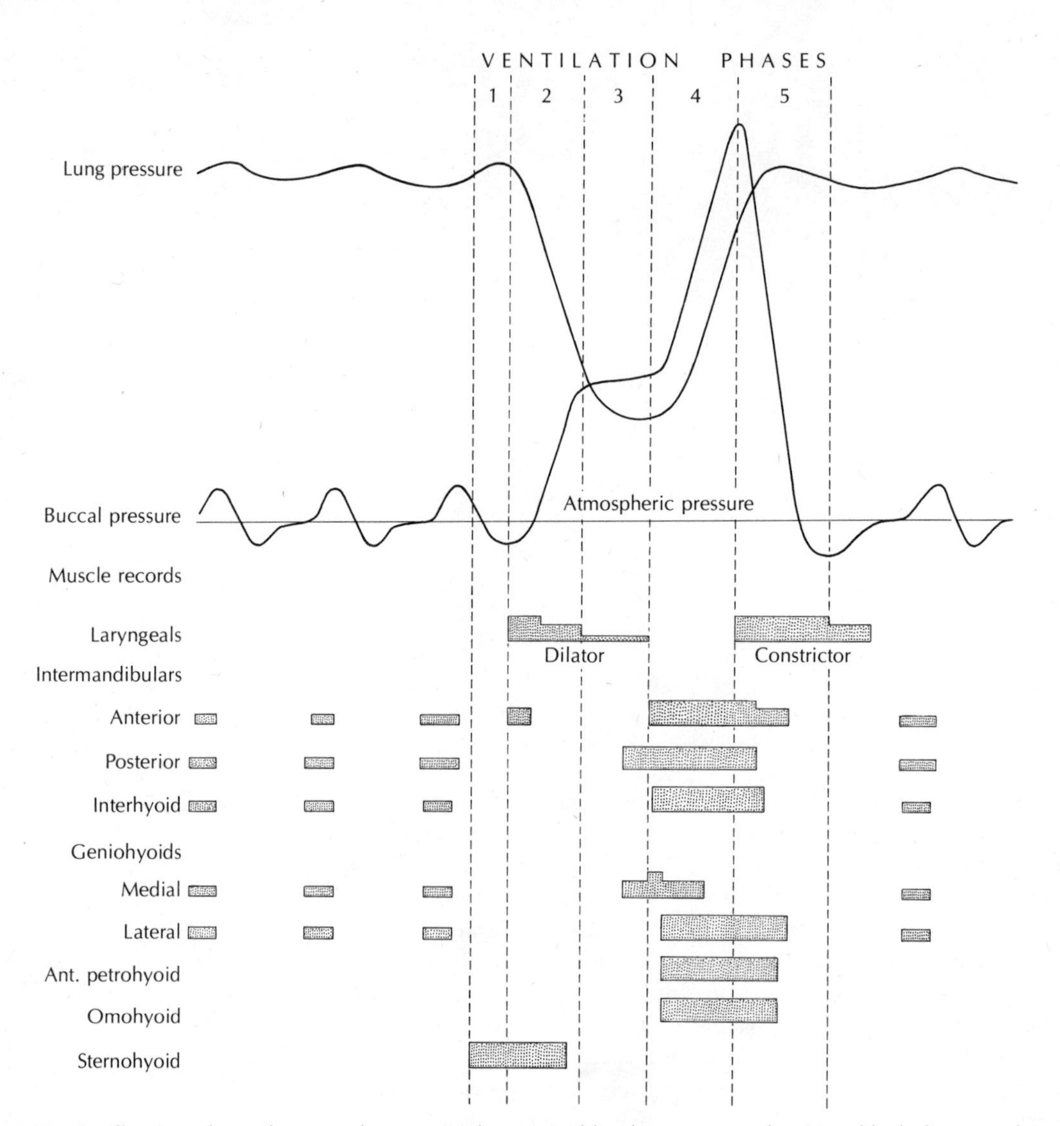

Figure 5-20 Oscillatory and ventilatory cycles are mainly powered by the same muscles. Here block diagrams show the firing times of various muscular groupings simultaneous with the pressures in the buccal cavity and in the lung (compare Figure 5-8). Note that mainly the activity of the sternohyoid, omohyoid, and petrohyoid differentiates oscillation and ventilation.

powered entirely by the innate elasticity of the lungs (and surrounding tissues) as well as by the presumably continuous contraction of the pulmonary smooth muscles (Gans, 1971). The muscles of the body wall are not involved in pulmonary emptying.

The anterior intermandibular muscle shows a sharp burst of activity at or just before the beginning of the fourth phase, this phase being defined by the closing of the nostrils; intermandibular activity appears associated with closed nostrils. Contraction of the anterior intermandibular apparently forces the mentomeckelian tips of the mandible dorsad, which deflects the premaxillae so that their facial processes spread, which in turn deflects the alary cartilages to close the nares (Figure 5-13). The simultaneous firing of the muscles associated with the buccal floor greatly reduces the volume of the buccal cavity. The strong activity in the omohyoid and petrohyoid muscles specifically tends to reduce the volume of the buccal cavity's posterior portion; this is the space that was previously enlarged by contraction of the sternohyoid muscles. The presence of activity in these two sets of essentially mutually antagonistic muscles marks a fundamental difference between ventilatory and oscillatory cycles.

The extreme activity in the laryngeal constrictor muscles, marking the end of phase 4 (just before the nostrils reopen), apparently closes the laryngeal valve. The larynx may be retracted as well since both constrictors (and perhaps also the dilatator) tend to pull the arytenoid and cricoid cartilages ventrally between the wings of the hyoid plate. This occurs just at the peak of buccal pressure and results in maximal buildup of pulmonary pressure. The lung-inflating phase (4) takes about 0.1 sec. The simultaneous contraction of all the buccal muscles and the resultant sharp pressure rise is certainly a pulse; hence the name pulse pumping for this method of breathing.

Almost all muscles become inactive after phase 4. Only the laryngeal constrictors fire sharply, closing the larynx near the peak of buccal pressure. As the anterior intermandibular relaxes, the elastic energy stored by twisting the inferior prenasal cartilage returns it to the resting position. This forces the premaxilla ventrad and the alary cartilage outward so that the nostril again becomes patent (open). If the nostril opens before the end of phase 5, there will be some slight outflow of gas; generally, by the time this occurs, the continuing relaxation of the buccal muscles will have returned the buccal pressure to or below atmospheric. As the nares open, air may enter, signaling the inflow phase of the following oscillatory cycle. Only if the frog is extremely excited, and ventilation follows upon ventilation, will the downward movement of the buccal floor again be associated with contraction of the sternohyoid muscle.

With the exception of the additional increase in volume of the buccal cavity produced by the sternohyoid muscle, all of the preliminary aspiration of gases into the buccal cavity is induced by the elastic and gravity-induced dropping of the buccal floor. The effect of gravity is most clearly seen, of course, when the bullfrog sits in a normal position. Significantly, no normal frog will voluntarily lie on its back. Yet the vast majority of past experiments dealing with the mechanics of the frog's respiratory system involved the crucifixion of these animals upside down on a "frog board." This, of course, reversed the effect of gravity and got the experimental artifacts off to a flying start.

As lung volumes increase during the terminal ventilations of an inflation cycle, electrical activity in the buccal musculature rises directly with the pressure being generated. This supports the concept that the magnitude of the electromyogram reflects the magnitude of the exerted force. Beyond this, these terminal phases of the inflation cycle apparently involve slight shortenings of phases 2 and 3 of the ventilatory cycle (shorter time for pressure equilibration between lungs and mouth). Thus efflux is limited while more gas is pumped in with each stroke. The reverse occurs during periods of dropping pulmonary volume; the efflux is then more than the inflow. The electromyograms of the respective ventilatory cycles making up an inflation cycle are otherwise identical.

The Jet-Stream Bypass

Once we had proceeded to these levels in our analysis, we began to wonder about the reality of the results (Gans et al., 1969). The experiments showed conclusively that the frog aspirated a slug of air into the buccal cavity, where it was retained. Simultaneously the pulmonary contents flushed through the same space as they passed the buccal cavity from lungs to nostrils. No secondary buccal inhalation was intercalated between the pulmonary outflow (during phase 2) and the pulse-pumping event (of phase 4). How then did the frog avoid reinhaling the spent gases that had just left the lung?

Reinhalation would return to the lungs a volume of gas that contains a considerable admixture of spent gases. Such mixing had already been mentioned in the literature and its occurrence troubled us. Everything we know about ventilation suggests that the energy expended upon it is a significant item in an animal's budget. One might hence expect selection for ventilatory efficiency. The greater the mixing, the greater the pumping work needed to provide the frog with a unit of oxygen. A pulse-pumping system "should" not operate just as described. Had we overlooked something?

The first observation in pursuit of this problem was that the laryngeal slit in a relaxed frog appears to be positioned fairly high at the back of the buccal cavity and that it faces directly forward. A jet of pulmonary air leaving the glottal slit might then have some significant chance of passing along the roof of the mouth. It might pass above the tongue to impinge directly upon the nares and exit with relatively little mixing. In particular, there should be little mixing with the gases contained in the posteroventral part of the buccal cavity, the space back of the tongue. The idea that this posterior space might indeed hold most of the inhaled air and be bypassed by the pulmonary outflow became more plausible when we remembered that electromyograms of ventilatory cycles differed from oscillatory ones mainly by showing activity of the omohyoid, petrohyoid, and sternohyoid muscle groupings. These are the very muscles in a position selectively to affect the volume of the posterior part of the buccal chamber.

Yet such speculations, although plausible, do not represent proof. We attempted to test the hypothesis by sampling gas from the frog's lung and parts of its buccal cavity. However, the relatively low oxygen depletion during gas exchange in the lung (the efflux is still 19 percent oxygen) and the rapidity of the process left us with suggestive but inadequate data. The courtesy of some friends in a neighboring department led to the design of an improved experiment. In order to sample the effluent from the nostrils we fitted some frogs with a mask formed of a dental prosthetic plastic. This mask surrounded the anterior portion of the head but had a wide cutout opposite the nostrils and around the eyes. It was attached to the upper lip and did not seem to inhibit the animal's functions. The cutout over each nostril contained a short sleeve of plexiglass tubing through which the narial gas flow had to pass. The tubing served to center a very fine (#32) probe that continuously

Figure 5-21 Photograph of bullfrog with mask and sampling tube, readied for checking the narial efflux. See the thin gas probe connecting the tube to the mass spectrometer.

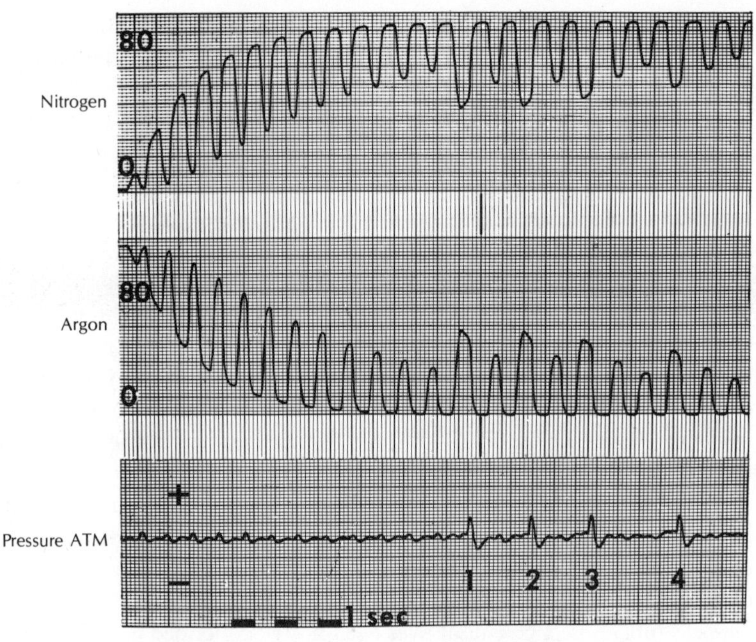

Figure 5-22 Record of gas concentration flowing from the nares of a frog sitting in an atmosphere that is changing rapidly from 80 percent argon (20 percent oxygen) to 80 percent nitrogen (20 percent oxygen). Some 15 seconds after initiation of change, the animal is seen to interrupt a series of oscillations and to ventilate for the first time (event 1), followed by three more ventilatory cycles (2, 3, 4), each set off by one or more oscillations of the buccal floor. The discharge from the lung contains a higher concentration of argon than do the contents in the buccal cavity. The degree to which this efflux is mixed in the buccal cavity may be seen from the slight rise in the argon concentration in the following oscillation. Four ventilations are shown. The pressure record has been shifted relative to that of gas concentration to compensate for the different delay times in the recording apparatus.

sampled the tube's contents. The sample passed to a respiratory mass spectrometer which gave us a continuous record of gas concentration, just before entering and just after leaving the nostrils.

In order to increase the differential in gas concentration between narial inflow and outflow, we prepared two separate atmospheres (Gans et al., 1969). The first consisted of 20 percent oxygen and 80 percent nitrogen and was then generally equivalent to air. The second consisted of 20 percent oxygen and 80 percent argon. Although these appeared identical from the viewpoint of respiratory behavior, the mass spectrometer could readily discriminate between nitrogen and argon. We next positioned the frog in a vacuum jar and pulsed gas through the jar fast enough to flush out 99 percent of the gaseous contents within less than 15 seconds. The hope was that we could replace one ambient gas mixture with another during a period when the frog was oscillating rather than ventilating. Fifteen seconds after the switch from argon to nitrogen we should then have a nitrogen atmosphere containing a frog with argon in the lungs and a decreased concentration of argon in the buccal cavity. Shortly thereafter, the gas concentration in the frog's buccal cavity should approximate that of the outside environment. The gas concentration in the first ventilatory cycle thereafter would then indicate how much mixing occurred (Figure 5-22).

Initial results indicated that between 25 and 75 percent of the gas remaining in the buccal cavity at the end of a ventilation cycle resulted from the pulmonary efflux which had taken place during that cycle. However, this did not directly address the issue of mixing because much of this gas presumably occupied the anterior space above the tongue, the contents of which we felt would have relatively low probability of being pumped into the lung. The shapes of the efflux curves for single exhalations provided the critical clue to the flow pattern. During an oscillatory cycle (with nitrogen ambient) the concentration of argon kept on rising during narial outflow. It reached a maximum just before the flow reversed. This was to be expected since, when gas is cleared from a simple oscillating piston, the farther down the cylinder the less diluted the contents. Even though the gas concentrations in the buccal cavity rapidly approached the outside levels, the shape of the curves remained characteristic.

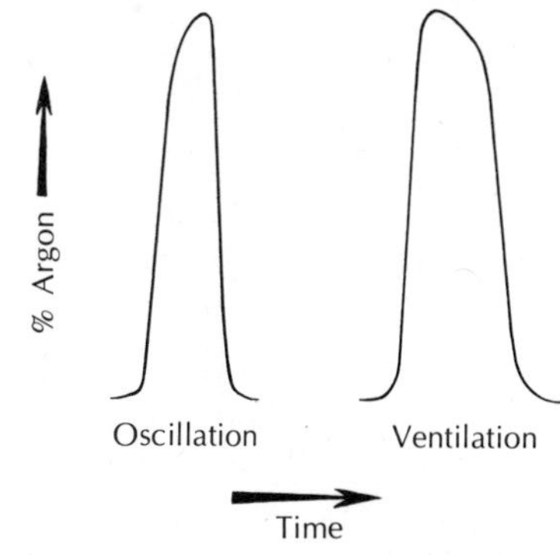

Figure 5-23 One of the most significant indicators of different flow patterns is seen in the shape of the gas concentration curves observed when the animal is oscillating (left) and ventilating (right). Note how, during oscillation, the argon concentration gradually increases until it reaches a peak and then drops sharply. In contrast, during ventilation the argon concentration rises sharply to a peak; it then shows effects of mixing as it drops gradually to baseline.

The concentration curve for the efflux from a ventilatory cycle had the opposite shape (Figure 5-23). Thus the argon concentration was highest at the instant flow began and then dropped gradually until flow ceased. This confirmed that a jet stream of relatively unmixed pulmonary air must pass through buccal cavity and nostrils. The decrease of argon concentration with time could only have been due to subsequent mixing with pockets of buccal gas. It indicated that such pockets of gas had survived the initial outflow from the lungs in a relatively unmixed state.

Jet flow was thus demonstrated as an effective bypass mechanism for discharge of the pulmonary contents; this despite the fact that the absolute degree of partitioning of the respective gases from one another remains to be determined. The analysis also suggested a function for the oscillatory cycles. These must serve to reduce the concentration of pulmonary effluent gas remaining in the buccal cavity at the end of a ventilatory cycle. Indeed, the amplitude of the oscillations does decrease with time (and residual concentration of effluent gas) after each ventilation. This is a somewhat better explanation than the previous one that oscillatory cycles power respiration through the buccal membranes. Such respiration seems most unlikely in view of the relatively poor vascularization of the bullfrog's buccal membranes. Gas analysis thus served as a nonmechanical tool to check a mechanical process.

It must be admitted that we almost failed to get any results from the mass spectrometer. The frogs quickly became accustomed to the mask and associated tubing. They also tended to ignore the bell jars within which the experiments were performed. But just about the time the gas change was complete, they would jump! They apparently perceived objects in their visual field and jumped when one of us moved (perhaps from spectrometer to chart recorder). They also jumped, even more often, when we surrounded the entire chamber with a paper curtain. Since bullfrogs are powerful animals and the mask and probe were relatively fragile, we soon got tired of replacing the probe. Accidentally, we discovered a simple solution. When a plastic berry basket was inverted over the animal, the frog would sit still without attempting to jump; indeed, it would squat down farther when it observed an extraneous movement. Apparently it interpreted the crossbars of the basket as a shelter rather than as a confining enclosure.

Mechanism and History

The experiments document that frog ventilation is effected primarily by the muscles of the buccal floor. Not only do these power pulmonary distension, but they also induce sufficient potential energy to empty the lungs later. Filling the lungs results in distension of the elastic fibers and of the tonically tensed smooth muscles of the pulmonary wall; later, the contraction of these elements powers the outflow.

The occurrence of a pulse-pumping device in which all the energy for inhalation and exhalation is imparted by a single muscular movement makes one wonder about its origin (Gans, 1970). How did such a complex system arise? Did the earliest (and now extinct) amphibians pulse-pump? If so, could pulse-pumping have evolved into the two different types of aspiration patterns seen respectively in mammals and birds? Or, as suggested by some authorities, might the earliest amphibians first have developed cutaneous respiration as they became terrestrial and only later an aspiration breathing mechanism?

Such questions are best approached by recourse to the comparative method and a look at the respiratory patterns in diverse groups of fishes. Some kind of air chamber or lung appears to have existed in the now-extinct placoderms, ancestors of the bony fishes (Denison, 1941). Similar diverticula of the anterior alimentary canal are seen in teleosts, chondrosteans, holosteans, dipnoans, and crossopterygians—in other words, in representatives of each group of bony fishes. This leads to the suggestion that all such structures are homologous derivatives of a primitive lung. In many teleosts, the dominant fishes since the Cretaceous, the air sac is used as a swim bladder and changed its function from gas exchange to buoyancy control. In other teleosts, however, it appears to have disappeared altogether (perhaps as an adaptation to ocean life, since a bigger fish has to expend more work in swimming—cf. discussion in Alexander, 1967). Here we see a spectrum of new (and nonhomologous) external gas exchangers, such as vascularizations of the tongue, diverticulae of the buccal cavity, modified branchial pouches, and even a gas-exchange surface in the intestinal wall. The ubiquity of such devices, some of which must have arisen anew after the lungs had once been lost, suggests that air utilization must long have been important to fishes. Its selective advantages can best be considered on ecological grounds.

An external gas exchanger should provide an organism with a ready means of discharging carbon dioxide as well as a ready source for the

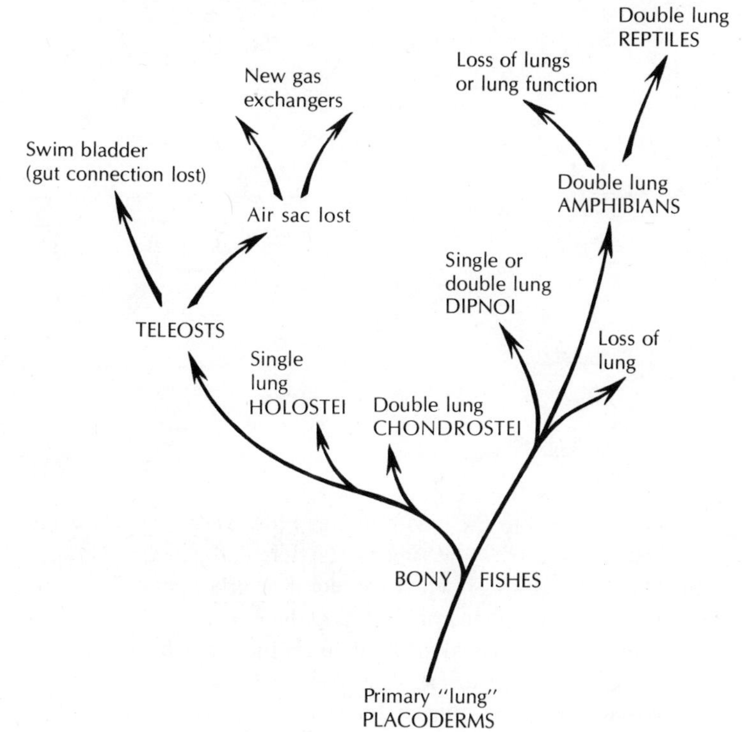

Figure 5-24 The phylogenetic diagram shows the kinds of lungs seen in lower vertebrates. Loss of lung or lung function has occurred a number of times in bony fishes and again in amphibians. A curious thing is that although some teleosts initially modified their lung into an air sac and others then lost it, a number of species have developed new gas exchangers in the mouth, in the gill chamber, and even in the intestines. Obviously, the ability to utilize atmospheric oxygen was often advantageous in phylogenetic history. As obvious, there are occasions when the spatial or buoyant properties of air sacs or lungs make these structures disadvantageous.

Figure 5-25 Aerial respiration cycle in a garfish, *Lepisosteus osseus*, traced from a film taken at 32 fps. The numbers refer to frames in an air-exchange sequence in which the fish approaches the surface at an angle (0). The gar depresses the floor of the mouth (3) and allows air to escape from the glottal aperture by hydrostatic pressure. This air then escapes back of the gill cover (6). After the snout "breaks" the surface (8), the mouth opens (10) and then closes rapidly (11), distending the buccal cavity with air. This air is then driven into the lung (13, 15) by the buccal pump, distending the body of the fish. The first post-breathing gill ventilation (29) drives some residual bubbles of air from the opercular cavity (29, 31). The breathing pattern of this fish differs from that observed in lungfish in which inhalation precedes exhalation. (After Rahn et al., 1971)

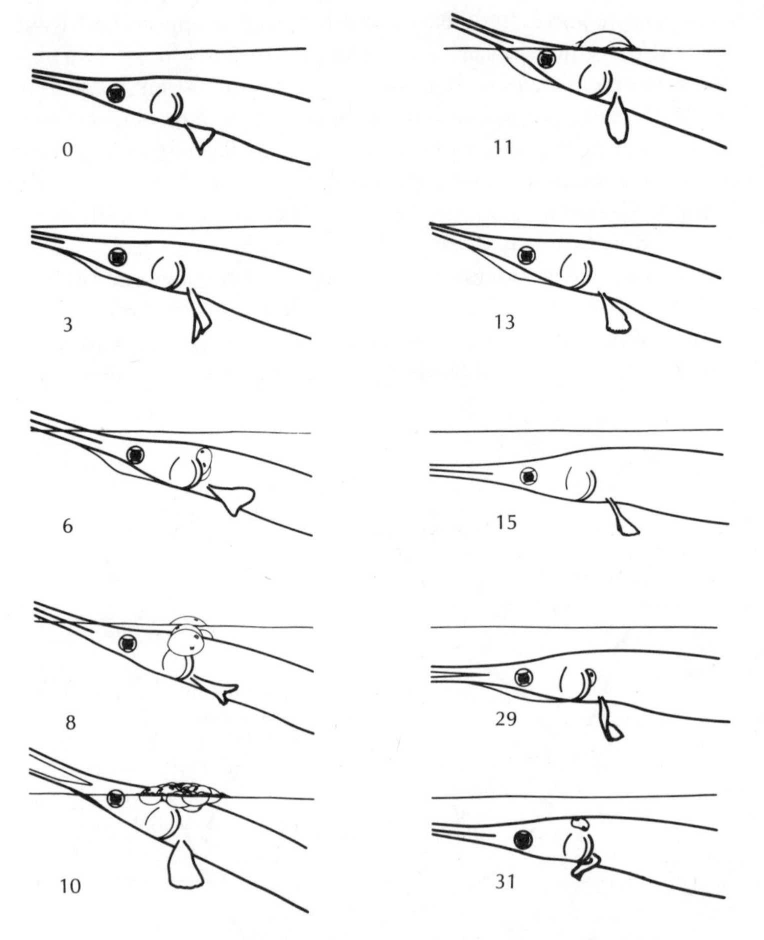

pickup of oxygen from the environment. However, both of these processes need not proceed at the same site. After all, the physical properties of the respective gases are quite different. Carbon dioxide is almost infinitely soluble in most natural waters; their buffering capacity takes it out of solution very quickly. Oxygen, on the other hand, must enter bodies of water from the surface or be released from aquatic plants during photosynthesis; it is utilized by innumerable organisms, and the processes of decay deplete the dissolved supply. Consequently, many bodies of water show drastic drops in oxygen concentration during the warm season or at other times when organic breakdown reaches temporary peaks.

The resources of a body of water that occasionally shows such a reduced oxygen level will then be unavailable to any absolutely aquatic species of fish that could not adequately oxygenate its blood at this minimum level. Even though the condition persists for only a few hours or for a day, and even though the condition occurs only once a year or once a generation, such fishes could not become permanently established. Any ability to utilize atmospheric oxygen during such a period of stress would consequently be of great advantage to the species involved. Only species capable of utilizing aerial oxygen could successfully invade, survive in, and exploit the energy available in such marginal environments.

It is fairly easy to see how bubble-pickup methods might have developed. Many Recent fishes can distinguish between oxygen-rich and oxygen-poor waters. As the level of dissolved oxygen drops, such species start to ventilate their gills with the more highly oxygenated water which they can suck up from just beneath the air-water interface. In such species bubble ingestion could obviously start by accident; indeed it would be difficult to avoid when feeding from the surface film. The bubbles could be intensively scrubbed within the buccal cavity, and the oxygen enrichment of the water taken in for gill ventilation would represent an advantage to the fish; a shift from accidental to facultative bubble biting would represent an obvious subsequent adaptation.

Electromyographical study of air respiration in appropriate fish supports such an origin of air breathing; the air pickup by fishes using aerial oxygen is generally powered by a muscle sequence identical to that used in aspirating water to perfuse the gills (cf. Ballintijn and Hughes, 1965; McMahon, 1969; Osse, 1969, 1972). In both water and air, inhalation proceeds with the jaws but slightly separated and powered by a depression of the buccal floor and a concomitant lateral spread of the branchial bars. The mouth then closes, either by activity pressing the jaws together or by the passive movement of a mandibular flap. The simultaneous raising of the buccal floor and narrowing of the branchial chamber force the water through the gill arches, or the air into the various kinds of gas-exchange pockets. (The movement of the air is pulsatile; the movement of the water obviously proceeds continuously but more slowly.)

The spent gas is later discharged from the buccal cavity by hydrostatic pressure. The fish approaches the surface of the water at an angle with open mouth and relaxed muscles and then opens the aperture to its gas chamber. This allows the spent air to escape as hydrostatic pressure

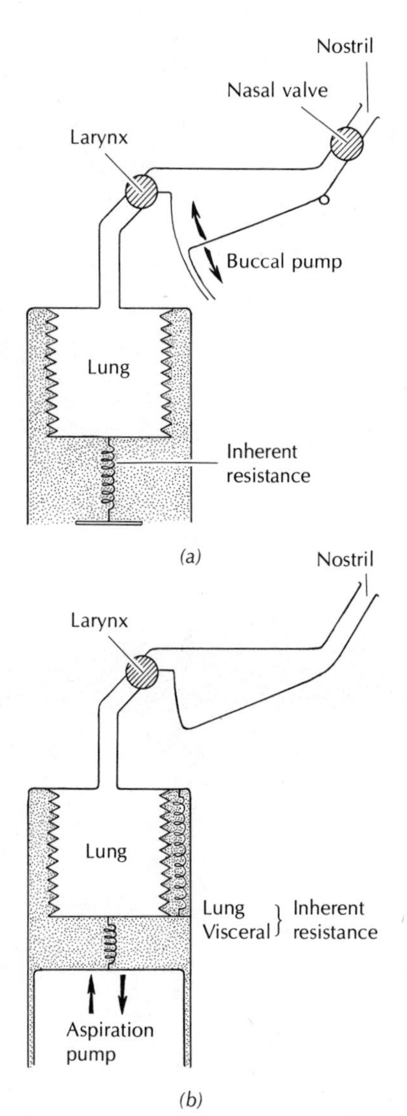

Figure 5-26 Simplified sketch to show the fundamental differences between a buccal (*a*) and an aspiration (*b*) pump. In a buccal pump, air is shifted from outside to lung by a two-valve, single-piston compressor. The energy for distending the lung and overcoming the inherent resistance of the viscera has to be transmitted via the compressed pulmonary contents, which ordinarily remain at supra-atmospheric pressure. In aspiration, the pump is coupled directly to the walls of the lung, thus overcoming the resistance of lungs and viscera. The laryngeal valve serves more to keep food from entering the pulmonary system than to constrain the pulmonary contents, which remain near atmospheric pressure.

squeezes the chamber's walls together. Lungfish, among the closest surviving representatives of the line leading to the early amphibians, apparently thrust their snout above the surface and fill the buccal cavity with air before opening the glottis and discharging "pulmonary" contents (McMahon, 1969). This sequence is similar to the "buccal inflow-total outflow-pulmonary inflow" breathing sequence seen in frogs, but differs in that inhalation proceeds via the mouth rather than the nostrils. It also differs, for instance, from the process in the gar pike, which empties its air sac under water before taking on a new bubble at the surface (Rahn et al., 1971). Perhaps lungfish fill the buccal cavity with air to avoid an increase in their relative density as they exhale; similarly they may obviate the risk of water entering the air sac (hydrostatic pressure would be inadequate, of course, for clearing liquid).

The pulse-pumping method can be shown to be relatively ineffective mechanically. Its high cost of energy per unit of gas moved makes it insufficient to deal with the continuous discharge of carbon dioxide (Gans, 1971). This inefficiency does not seem to have been critical for fishes that use only pulse pumping to supply emergency oxygen, nor perhaps for cool-temperature amphibians of small body size. Neither is it critical for species that could lose a significant fraction of their car-

Figure 5-27 The male Gulf Coast toad, *Bufo valliceps*, produces its mating call with the vocal sac fully distended. (Photo by J. P. Bogart)

bon dioxide by cutaneous or gill respiration. Larger species or the later life stages of the early tetrapods clearly must have developed aspiration (or suction) breathing as a solution to this problem.

It is easy to trace intermediate steps that could have led from pulse pumping to such aspiration breathing (Gans, 1971), but that takes us beyond the situation in frogs. More pertinent in the present context is the question of why frogs failed to develop aspiration breathing but instead perfected the curiously complex jet-stream bypass. The answer again seems to involve selective compromises between multiple and perhaps conflicting functions, in this case those for breathing and those for vocalization. The unusual or exceptional activity pattern here observed during an experiment becomes important in analysis. Thus, although it is true that the muscles of the body wall were generally inactive during bullfrog exhalation, they did in fact fire in isolated instances; in each such case the frog was highly excited, and in most instances such contraction of the body wall occurred when the frog was calling or had just vocalized!

Toad Vocalizations

Frog (and toad) vocalizations have great intrinsic interest because the literature suggests that they serve as the major factor in anuran social organization. More than half a dozen categories of calls have been described (cf. Gans, 1973). Phonograph records, such as that of Bogert (1958), document some of these diverse sounds.

The most important call is the mating call, produced generally by male frogs on their way to and at breeding sites. This call stimulates and attracts the female to approach the male. There is also some debate about a secondary aggregating function of this call, inasmuch as other males may home in on such choruses as well; this may allow them to orient their movements toward temporary bodies of water discovered by the first males[8] that reach suitable breeding sites. It should not be surprising that the calls differ, even those produced by closely related species, and that the respective calls of the more than a dozen species that may breed in a single area always differ sharply; the mating call represents an excellent example of a species-isolating mechanism.

Aroused males use their forelimbs to clasp the axial or inguinal (groin) region of any frog (or in exceptional instances they seize frog-sized objects) that approaches them when they are excited. When such clasp-

[8]This function of the call does not confer an advantage on the calling individual, as it introduces additional competition to the individual producing the sound. On the other hand, there might be an advantage to the responding male in species utilizing temporary and seasonally varying bodies of water, and perhaps in those in which the females move toward the greatest (or loudest) chorus.

Figure 5-28 In the southern toad, *Bufo terrestris*, the adult males are much smaller than the females. Amplexus is in the axillary region so that the male will be carried about as the female moves from one egg-laying site to another. (Photo by C. M. Bogert)

ing, or amplexus, occurs with conspecific females (those of the same species) that have ovulated, it leads to mating. However, excited males do not discriminate well. They may thus clasp males or females—of their own or different species—as well as other moving objects. There are reports of toads dying after their amplectant forelimbs had entered and locked into the orbits of a large carp! This response to moving objects caused Spallanzani (1784) to perform an extensive series of ablation experiments to study the motor-control levels of European toads.

As an apparent compensation for this nondiscriminating behavior pattern, male frogs, particularly at breeding time, emit a release signal when clasped by another male. The signal consists of body vibrations some of which may be associated with sounds. The sound pattern is basically similar to the tones of the mating call but involves a more restricted repertoire. Aronson (1944) long ago showed that conspecific male toads will release when a clasped individual emits the release sequence.[9] Male frogs will indiscriminately produce these vibrations when axial or inguinal pressure is applied from any source as, for instance, when a human experimenter picks up a frog by placing two fingers back of the forelimbs. The advantage to the emitter of the sequence is that in being released it is freed from distraction while searching for females with which to mate; the advantage to the receiver is similar, and the risk of wasting gametes reduced.

[9]Females, at least of some species, also produce similar vibrations when they are not in breeding condition. These vibrations may sometimes be vocalized (Bogert, 1960).

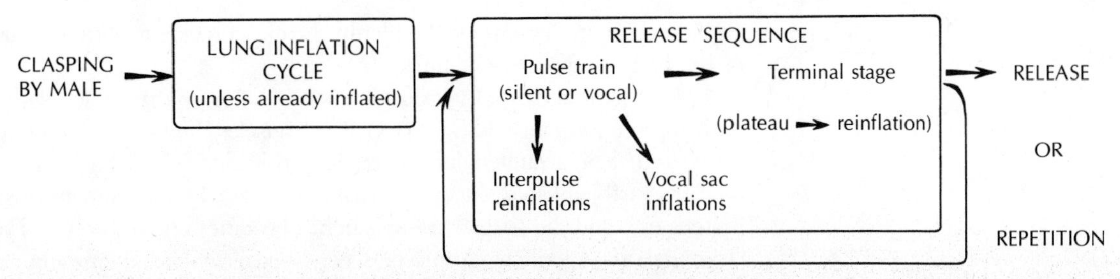

Figure 5-29 Diagram showing the sequence of events that occurs when a male toad is clasped by another male. Unless the release sequence causes the clasping male to release, the behavior will be repeated.

Some frogs also produce a variety of rain calls, as well as territorial calls, sometimes during showers or on overcast days and sometimes at night and far beyond the restricted mating season. The function of these calls has not been as clearly documented as for the release and mating calls. They may serve to define territory or to communicate some general level of information about spacing to other members of the group. This may be advantageous when the frogs are spread out in their feeding rather than concentrated in their breeding territories. Finally, there are various warning and distress calls or screams. These are generally produced with the mouth wide open and emitted when the animal is suddenly frightened or when it has been seized by a predator. Here again there is no experimental information on positive selective advantages to the emitter or to other individuals of the species.

All of these vocalizations use air and hence show some relation to breathing. The release and mating calls are obviously most closely associated with breeding behavior and are therefore seasonal vocalizations. The release sequence was chosen for our studies because it was the easiest to elicit, and we used specimens of the Gulf Coast toad (*Bufo valliceps*), rather than a frog, because they could be obtained at the time from breeding choruses in the most nearly natural condition (Martin and Gans, 1972).

The release sequence of *Bufo valliceps* includes a series of major variables (Figure 5-29). When a toad is first clasped, it inflates itself in a sequence reminiscent of the inflation cycle. A series of pulsatile movements then shakes the entire body. The pulses of the train can be felt most strongly in the axilla, namely the place where the forelimbs of a clasping male would usually rest. By the third pulse there will ordinarily be some associated vocalization, although silent trains also occur. Vocalized pulse trains always involve closed nostrils, a swelling of the floor of the buccal cavity, and a decrease in the body diameter. A few pulses later one may see stepwise inflations of the vocal sac. These tend to coincide with sharp rises of the radiated amplitude of the sound. The release sequence stops after a few seconds. After a variable interval the

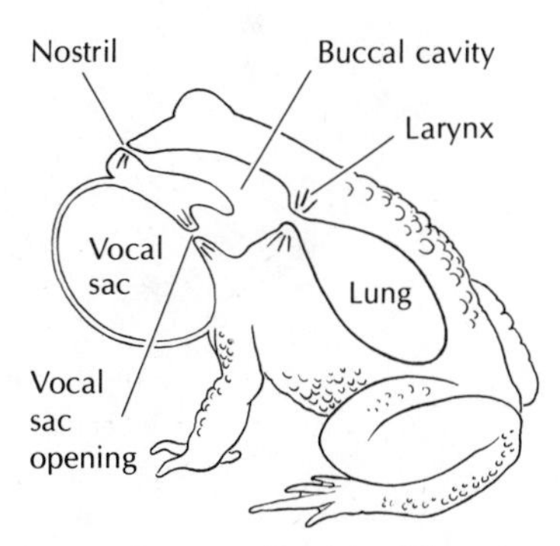

Figure 5-30 Vocalization differs from ventilation in involving three rather than two chambers and three rather than two valves associated with the pulmonary system.

swollen vocal and buccal sacs suddenly shrink, coincident with a return of the body to its previous inflation level.

The anatomical basis of vocalization is equivalent to that of ventilation with three important additions: (1) The muscles of the body wall are active and their arrangement has to be considered. (2) The vocal sac may be partly inflated during the call; the state of the sac's intrinsic musculature and slip-shaped closures hence becomes important. (3) The larynx serves not only as closure between buccal cavity and lungs, but as the source and initial modifier of the vocalizations. It proves impossible to describe the basic mechanism of vocalization in terms of two or three groups of muscles acting upon the gaseous contents of one chamber. Each of the three distinct chambers—buccal cavity, lung, and vocal sac

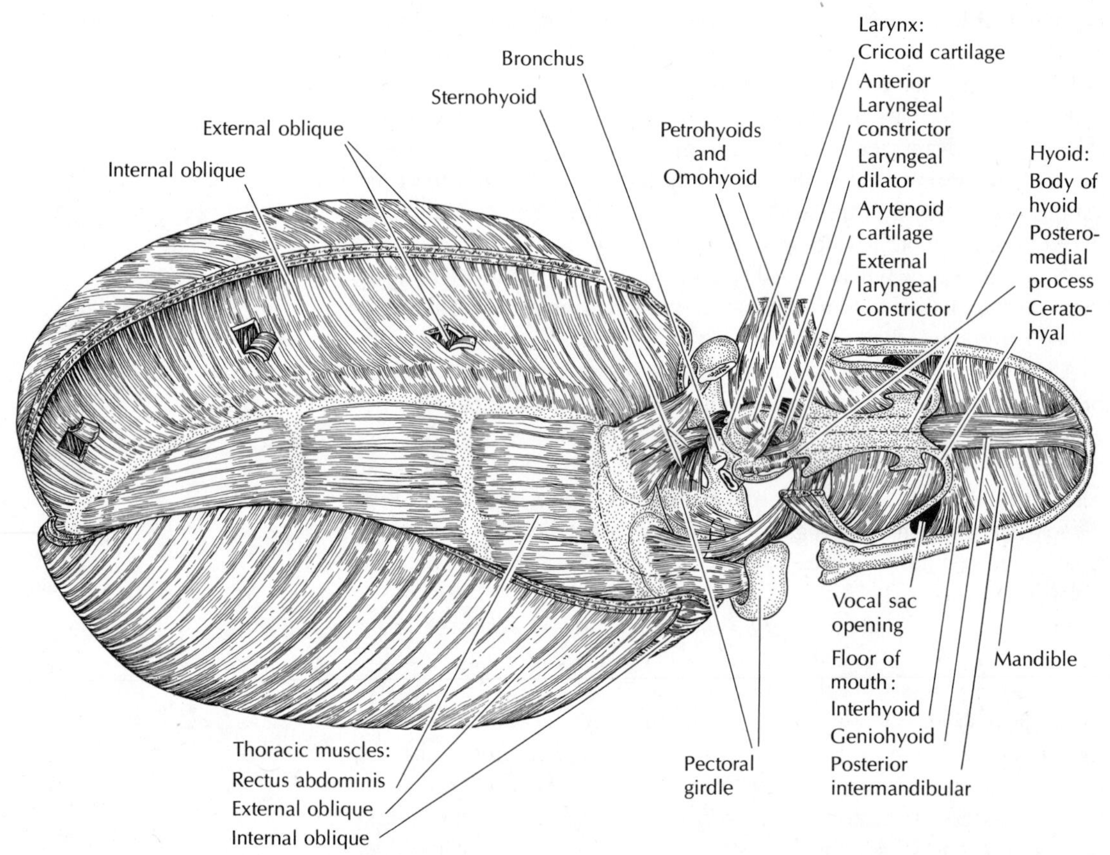

Figure 5-31 Multiple groups of muscles participate in the vocalization of toads. Those of the floor of the mouth and the petrohyoid grouping serve mainly in inflation of the lungs, in the reverse pulsing of air, and in control of the vocal sac. The muscles surrounding the larynx control the valves and sound production. Vocalization differs from ventilation in that the sheetlike muscles of the body wall contract in synchrony to pressurize the lung further and force its compressed contents across the laryngeal vibrators. (After Martin and Gans, 1972)

(Figure 5-31)—has its intrinsic muscular pump, and the state of each of the three associated valves needs to be examined.

Mechanical Components of Vocalization

The electromyograms obtained during the process of vocalization indicate that the pulmonary inflation cycle that precedes the toad's release sequence proceeds as already described for the bullfrog (p. 220).

The pulse train of the release sequence is here characterized by grouping the multiple muscles surrounding and separating the three chambers into mechanical components defined by their structural association. This preliminary decision does not imply that all elements of a component act in synchrony; it does provide a first-level description which may then be refined.

The pulse train involves a series of pressure pulses within the lung and the visceral cavity. The electromyograms indicate that these pressure pulses are coincident with bursts of activity in the rectus and oblique muscles of the body wall (Figure 5-32). Because of the regular association of these two events and since the magnitude of these electromyograms (when integrated) has proven predictive for the magnitude of concomitant pressure changes, it becomes clear that contraction of the muscles of the body wall is producing the pulmonary pressure peaks.

As indicated earlier (p. 219), these pulses must act upon the visceral cavity as a whole; the digestive and urogenital systems undergo the pulses in parallel with the lungs. Simultaneous electromyographical records from multiple sites of the body wall indicate that activity starts in the posterior muscle fibers and that a wave of contractions then passes anteriorly. The pulmonary pressure continues to rise briefly after cessation of such electrical activity, and does not ordinarily return to baseline between pulses. This collectively suggests that the contractions accelerate the visceral mass in a forward direction. Its inertia continues to build up or maintain some level of compression of the pulmonary contents after the muscles cease firing. In general, the peak pulmonary pressure during a release sequence will be six to ten times the peak pressure seen during ventilation and even several times the highest pressure seen in an ordinary inflation cycle.

The first one or two (nonvocal) pulses of a release train may not be reflected in the buccal pressure, and there are entire trains of silent pulses that apparently proceed without opening the larynx. On the other hand, there are nonvocal pulse trains that show regular increases in the buccal pressure. The interpulse drop in pulmonary pressure is always more sharply marked than is that between pulses in the buccal cavity. The terminal shrinkage of the buccal cavity is coincident with a protracted peak of buccal pressure during which the pulmonary pressure

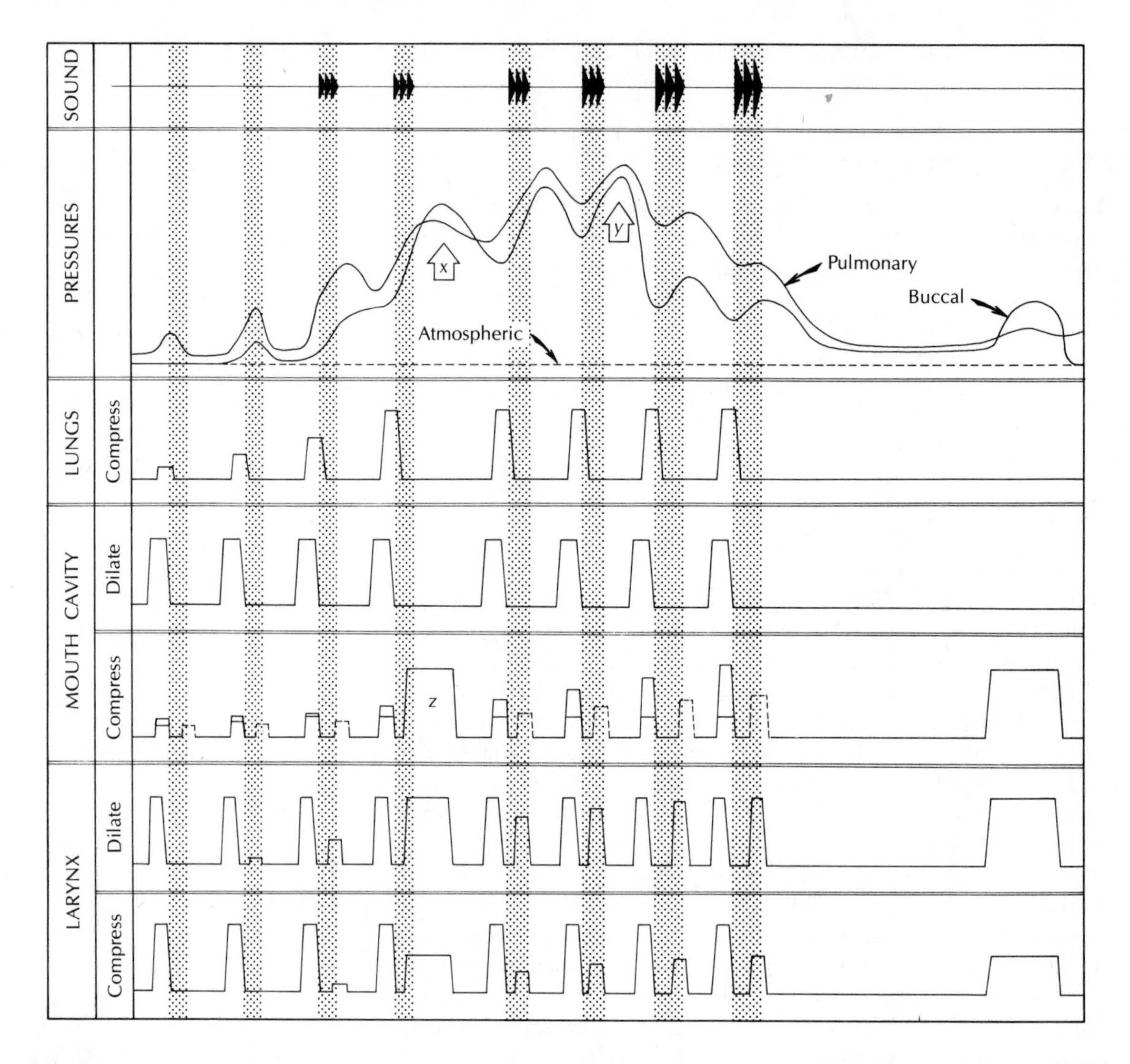

Figure 5-32 This sketch diagrams the major events that may occur during a short, vocalized release sequence. The sound track shows that the initial pressure pulses tend to be silent. Sound amplitude immediately reflects pulmonary reinflation (at *x*, resulting from activity of the buccal compressors at *z*) and the distention of the vocal sac (at *y*). Except for reinflation pulses, maximum activity of the laryngeal muscles occurs before rather than after each vocalization pulse. Cessation of vocalization is followed by terminal reinflation after a variable interval. (After Martin and Gans, 1972)

rises as well. The nares close for the entire release sequence. The stepped rise in buccal pressure with each pulse would seem to reflect the shift of slugs of air from lung to mouth, and the terminal pressure peak reflects a return of air to the lung (reinflation).

Two important pressure changes may occasionally be noted in the buccal cavity. Inflation of the vocal sac is always coincident with a drop in buccal pressure. Alternatively, the buccal pressure will occasionally show a sudden rise between pulses of a train, briefly exceeding the pulmonary pressure. During the next pulse, the buccal pressure drops back to a level lower than that observed before the rise; however, the pulmonary pressure will show a rise after each such buccal pressure peak. These then are the mechanical events that need to be explained.

The anterior intermandibular muscle starts to fire at the beginning of a release sequence and shows increasing activity until the end of the terminal reinflation which returns the body to its starting state. The muscle obviously acts, as in the frog, to maintain closure of the nares (pp. 222–223).

In the toad, as in the bullfrog, the sternohyoids are again the only dilators of the buccal cavity. They fire strongly and evenly just before and then synchronously with each firing of the muscles of the body wall. These muscles apparently distend the posterior aspect of the buccal cavity at this time, thus increasing the pressure differential (that required for laryngeal opening or activation) across the larynx. The diverse compressors of the buccal cavity differ in activity depending on the place of the pulse in the series, whether it is vocalized, and whether the vocal sac is inflated. In general, these buccal compressor muscles fire at low amplitude more or less coincident with the firing of the sternohyoid; they also fire variably, but much more strongly, at the end of sound emission. These postsound contractions are particularly strong during the sudden increases in buccal pressure between the sound pulses of a train. This makes it clear that we are dealing with pulmonary reinflations within the pulse train. Finally, all buccal compressors fire in a prolonged bout during the terminal reinflation. This shifts the air back from vocal sac and buccal cavity to lung.

The pressures and muscular activities of the vocal sac have, as yet, been inadequately studied. The available evidence does indicate that rising buccal pressure does not, by itself, inflate the sacs. The great compliance of the sac, furthermore, keeps its pressure significantly less than the buccal one. Inflation of the vocal sac results in distension and, like that of the buccal cavity, is stepwise, increasing very significantly apparently by repeated openings of the valvular slits. The interhyoid normally fires with the other compressors of the buccal cavity but ceases its activity when the vocal sac inflates. As this muscle is supposed to have given rise to the compressor of the vocal sac and forms the anterior edge of the slit-shaped connection between vocal sac and buccal cavity, it certainly deserves further study.

The preceding analysis suggests that the basic pressure pulses driving the release sequence are established by the contraction of the trunk

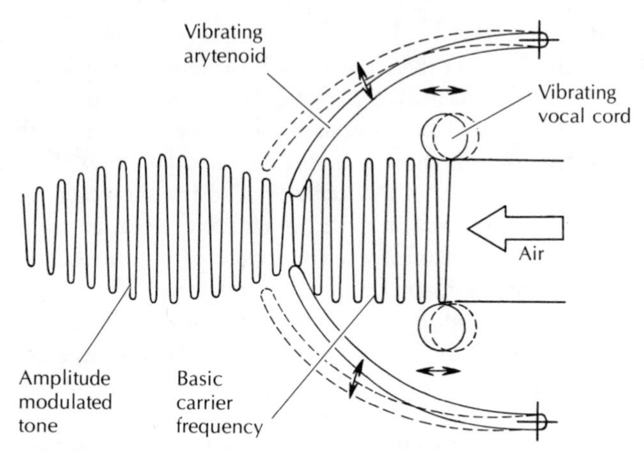

Figure 5-33 Simplified sketch to show how the vibrating vocal cords establish the basic carrier frequency and how the vibrating arytenoids impose amplitude modulation upon this.

musculature. During vocalized and some silent pulses, the larynx seems to open briefly, letting slugs of air escape from the lungs. As the nares are closed, this air distends the buccal cavity starting with the posterior portion, which undergoes some mechanical dilation by contraction of the sternohyoid coincident with the beginning of each pulse. At an undetermined pressure level the slits to the vocal sac are opened actively, and the buccal gas enters and distends the vocal sac. This leads to a great increase in the level of the radiated sound. The buccal compressors may fire one or more times during a pulse train, driving the buccal pressure above pulmonary and reinflating the lung. Such reinflation increases the amount of available air that can be passed anteriorly across the larynx and thus increases the duration of the pulse train. This, then, explains the basic directions and power sequence of the air movement; it explains neither laryngeal control nor the resultant vocalization.

Sound Production

Sound production apparently involves both passive and active components (Martin, 1971). The passive components do not involve muscular control and can be produced by blowing air through the relaxed laryngeal slit, indeed through an isolated larynx. Martin (1971) showed that the larynx serves as a self-exciting vibrator, producing a pulsating signal from a continuously pressurized energy source (the compressed air of the lungs). The air stream apparently causes the vocal cords to vibrate at a basic carrier frequency. In *Bufo valliceps* these cords are attached to the arytenoids without independent tensor muscles, hence the fre-

Figure 5-34 The laryngeal apparatus of the toad shows double suspension. The arytenoid shells are free to rotate on the surrounding ring-shaped cricoid cartilage and this, in turn, pivots on the posterior hyoid horns (*a*). Since the three pairs of muscles attach to the hyoid rather than the cricoid, they only control the cricoid movements indirectly. Contraction of the laryngeal dilator pulls the shells toward the hyoid, rotating them about the pivot on the cricoid (*b*). Contraction of the anterior laryngeal constriction not only pulls the shells together, but its shortening may be seen to rotate the cricoid around its pivot on the posterior hyoid horn (*c*). When both constrictors are active, they depress the system. The sliding movement of the constrictors over the edge of the arytenoid is shown in Figure 5-35.

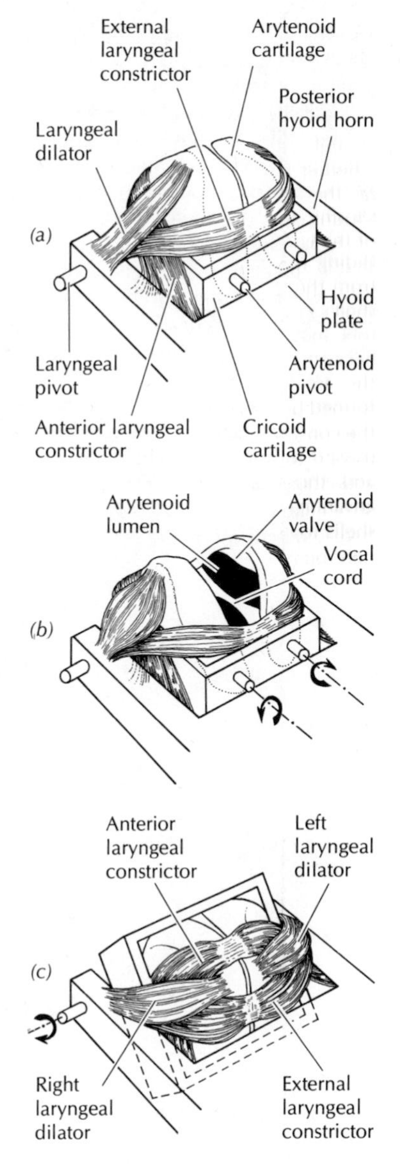

quency of their vibrations is not under direct muscular control but varies mainly with the pressure drop across the larynx. The air stream also induces a vibration of lower frequency on the arytenoid cartilages. Their movements thus affect or modulate the amplitude of the basic carrier frequency. Consequently, the resonance of the mechanical vibrator and its basic carrier frequency and the nature of the amplitude modulation are passive components of vocalization in *Bufo valliceps*.

What active components were discovered in the myographical studies? The first and most curious observation was that the two major constrictors and the dilatator of the larynx always fired simultaneously just before the occurrence of the sound pulse (or vocalization) of a train. This synchronous firing differed from the preconceived notion that the opening of the larynx might be induced by contraction of the dilatator muscle alone. Furthermore, the dilatator generally fired much more strongly at the *end*, rather than at the *beginning*, of the sound pulse. It must be admitted that it was some time before we could believe that these results were real.

The significance of this coincident firing pattern is probably to be found in the structural arrangement of the larynx (Figure 5-34). The dilatators each attach to the sides of an arytenoid shell. The constrictors attach neither to a median raphe nor to the arytenoids, but each attaches to the one of the opposite side. When they contract they may then slip freely over the shell-shaped arytenoids. Since the forces produced by these constrictors pass between the roughly circular curve formed by the arytenoid edge and its center, both constrictors will, in shortening, move from top and bottom toward the middle of the arytenoid edge (this part is closer to their origins than are the extremes). This pushes them toward a position close to the laryngeal slit and thus the position of the greatest mechanical advantage for closure. Neither constrictors nor dilatators attach primarily to the cricoid cartilage, but to the hyoid. Contraction of these laryngeal muscles, therefore, does not only induce opening and closing forces on the arytenoids but also depresses the entire larynx between the posterior horns of the hyoid plate.

Figure 5-35 The edges of the arytenoid shells are smoothly curved up to the central opening. Their separation, whether by internal pressure or action of the dilators, poses no problems. Closing involves not only contraction of the constrictor muscles but also their sliding from the edges of the shell, far from the laryngeal opening toward the shell's center. The contracting constrictors inevitably slide in this direction because their lines of action lie between the arc and the center of the curve formed by the arytenoid edges. Thus, as the constrictors shorten, they are forced toward the location of the opening (*b*) and thus into a mechanically more advantageous position for forcing the shells together. (The cricoid cartilage is here omitted to simplify the figure.)

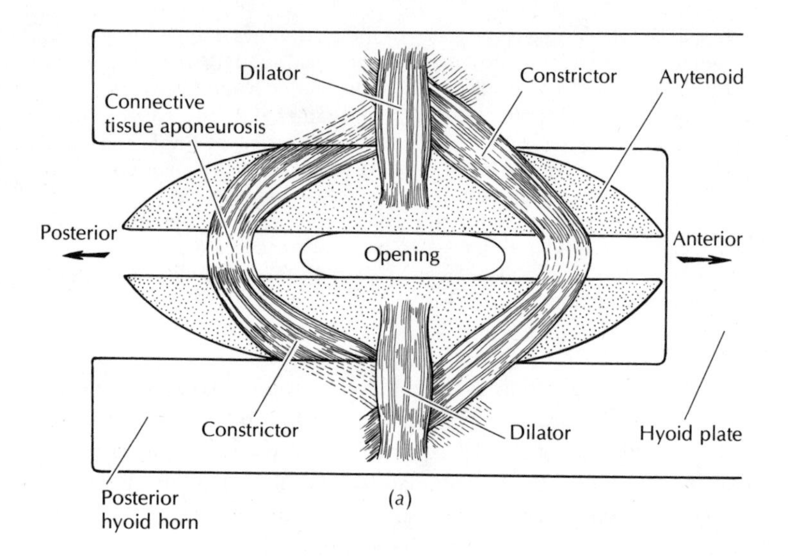

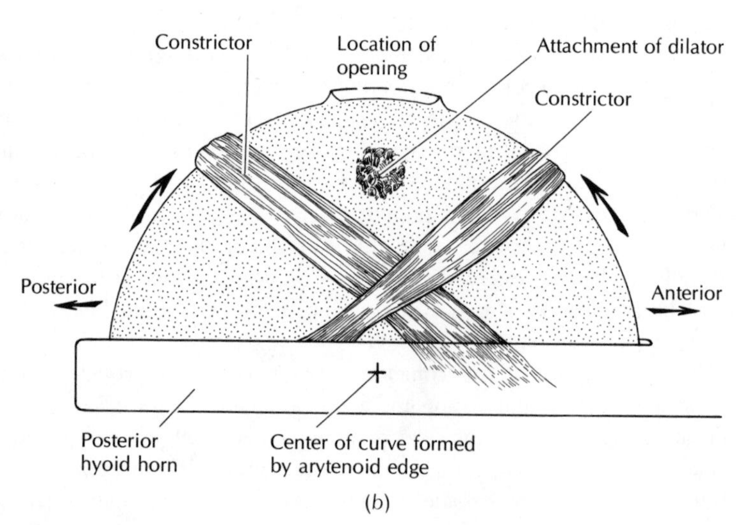

Since each sound pulse ensues immediately upon cessation of activity in the laryngeal muscles, it follows that the previously closed arytenoids are separated by the pulmonary pressure pulse after this muscular relaxation. Air may then flow through the larynx. As it flows it sets the vocal cords and arytenoids into vibration and, with this, starts the sound pulse. The dilatators must later act to pull the arytenoids and their mechanically coupled vocal cords farther apart and stabilize them out of the air stream. Therefore the wide opening of the larynx, rather than its closure, terminates the sound pulse. It also facilitates sudden reverse flow from buccal cavity to lungs when the buccal compressors fire to produce interpulse reinflations.

The laryngeal mechanism of *Bufo valliceps* allows the toad to radiate a signal of specified and constant frequency for a controllable period. Muscular control does not markedly affect this frequency.[10] The sound pulse is started by allowing the pulmonary pressure to open the larynx. If the arytenoids were to be separated by contraction of the dilatators, the vibratory characteristics of the system would change, because the mass of the arytenoids would include that of the tensed muscles. The increased inertia would modify the fundamental frequency of vibration. The present system gives a sharp onset of an unmodified signal.

The sharp termination of the sound pulse is due to the dilatator muscles. These pull the arytenoids out of the air stream and may simultaneously affect the mechanical coupling of the vocal cords. The cutoff occurs most quickly as the movement occurs against minimal resisting force; the pressure of the escaping gas acts to enhance the lateral movement of the arytenoids. The toad would have to impose far greater force if it were to terminate the sound pulse by pushing the slit closed against the pressure of escaping gas. Such closure of the slit would also induce a gradual change in the inertia of the vibrator and in the velocity of air across the larynx, and with this the activation pressure. Both the inertial and the velocity change would affect the fundamental frequency emitted at the very end of the sound pulse. *Bufo valliceps* does not thus modify the pulse.

The significant firing of the buccal compressors just after the sound pulse induces a slight rise in buccal pressure when the larynx is wide open. This reduces the pressure differential while the elastic closure mechanism (p. 218) shuts the larynx. The interpulse reinflations are produced simply by an increase in the magnitude of the buccal pulse; otherwise the firing sequence stays the same as that during laryngeal closure.

The described release sequence varies least among individuals taken from breeding choruses, and most in animals kept in the laboratory for some months or in those obtained outside of the breeding season. Responsiveness increases but variability is only insignificantly reduced in nonbreeding toads injected with physiological doses of the hormone Antuitrin-S, known to stimulate sexual activity. Unfortunately, we have no evidence to indicate that the least variable release sequence is the most effective in promoting release.

This somewhat abbreviated description characterizes the basic mechanism of the release sequence. Preliminary experiments indicate that the sequence of muscular activity is similar in mating calls. It is most

[10]Martin (1971, 1972) has documented a general shift in the tones emitted between an excised larynx and by the organ *in vivo*. This shift presumably affects the damping properties of the adjacent tissues in the intact animal rather than synchronized muscular activity.

interesting to note the extent to which ventilation and vocalization involve the same muscles and, indeed, similar motor sequences. This lets one speculate about reasons why frogs and toads have apparently been unable to modify their breathing from pulse pumping to aspiration. Any major shift would obviously induce simultaneous modifications in the sound system. If, as is likely, such change would result in reduced efficiency in mate finding or in reproductive isolation, it would incur immediate selective disadvantage.

The inherent inefficiency of the pulse-pumping pattern was apparently outweighed by the advantages of the associated vocalization mechanism. Here again a selective compromise would seem to have locked the species into a structural pattern that limited the kinds of niches occupied. The jet-stream mechanism provided a respiratory specialization that decreased the energy cost per unit gas exchange after the pattern was determined. Limits on the body size and temperature levels tolerated by frogs could then be relaxed, but not eliminated.

Limits and Questions

Rather than relying merely on dissection and observation, the approach in this chapter has emphasized the measurement and recording of events on freely moving animals. The actions of individual muscles and of their major components were recorded; movements and pressures were also determined or plotted from motion picture, television, or cinefluoroscope records. It was then possible to make quantitative statements regarding the mechanical events during the breathing and calling of frogs and toads. The degree of certainty is hence greater than in the studies discussed in earlier chapters, where complex and very rapidly occurring events often had to be deduced or extrapolated from surface indications.

This approach has provided some quantitative basis to biomechanical theory by showing how the same muscle grouping will act when performing different functions. In so doing it furnished further examples of the perfection by selection of fundamentally inefficient mechanisms that utilize the same structures, indeed the same subroutines, as utilized by other mechanisms of more critical selective importance. Yet all of the present cases are as yet documented in only a most preliminary fashion.

The breathing mechanism of frogs described above has been confirmed thus far in only a few species of *Rana* and of *Bufo*. Except in a few species of *Bufo*, the active components of vocalization have not yet been analyzed in detail; somewhat less complex analyses give some information about the patterns of vocalization in other species of *Bufo*, and in *Rana* and *Scaphiopus* (Schmidt, 1972). Yet two of three suborders and at least ten families of frogs have never even been sampled. For that matter, we completely lack similar studies of the breathing mechanics of salamanders. Thus many of the sweeping generalizations here offered about evolutionary sequences in the development of air breathing may well be premature; they must be treated as hypotheses, subject to additional test, rather than as didactic statements to be perpetuated in textbooks.

The biomechanical methods applied to the measurement of frog breathing and vocalization have wide generality. Each of the cases dealt with in earlier chapters, indeed any analysis of motion in multicellular animals, may well benefit from such approaches. Prey capture or mastication, locomotion or warning display, all are now far more open to detailed characterization and analysis.

The methods now available, furthermore, have particular advantages when applied to the sound-communicating systems of frogs. Investigators have begun studies of the sound receptor and its central connections (Capranica, 1965; Capranica et al., 1973). Some central neural mechanisms of sound production have been shown to be capable of analysis (Schmidt, 1965, 1972, 1973). It is now possible to perform more detailed analyses on the mechanism of such animal vocalization.

One critical evolutionary question has always concerned the nature of changes in incipient species, in particular the nature and origin of isolating mechanisms. Frogs furnish an admirable system in which to study their development. Which isolating mechanisms develop by modifying the neural response to different signals? Which isolating mechanisms develop by changing the mechanical-acoustic properties of the passive components? Which isolating mechanisms develop by modification of the properties of Sherrington's "final common path"? Similar questions have already been asked about insects (cf. Bentley and Hoy, 1972). We now have the opportunity of asking them about vertebrates. With this we may take the biomechanical approach further toward the ultimate issue of nervous control.

An ability to quantify the nature and magnitude of signals on the final common path lets us approach truly basic concepts of functional change. Is structural protoadaptation (p. 14) indeed utilized to permit more varied behavior (read more varied motor firing sequences)? Does protoadaptation indeed establish a selective advantage for further structural change? The working hypotheses alluded to in the first chapter (which may now be reread) may finally be tested.

On all of these questions biomechanical analyses have provided some preliminary answers; they have let us develop new hypotheses yet to be tested. What is still more important is that such analyses direct us to a new start by forcing us to refine our questions.

REFERENCES

Alexander, R. McN. (1967). Functional design in fishes. Hutchinson Univ. Library, London, 160pp.

Aronson, L. R. (1944). The mating pattern of *Bufo americanus*, *Bufo fowleri* and *Bufo terrestris*. Amer. Mus. Novitates (1250):1–15.

Ballintijn, C. M., and G. M. Hughes (1965). The muscular basis of the respiratory pumps in the trout. J. Exp. Biol., 43:349–362.

Basmajian, J. V. (1967). Muscles alive—their function revealed by electromyography, 2nd ed. Williams and Wilkins Co., Baltimore, xi+421pp.

———, and G. Stecko (1962). A new bipolar electrode for electromyography. J. Appl. Physiol., 17:849.

Bentley, D. R., and R. R. Hoy (1972). Genetic control of the neuronal network generating cricket (*Telegryllus gryllus*) song patterns. Anim. Behav., 20(3):478–492.

Bogert, C. M. (1958). "Sounds of North American frogs." Folkways Records, N.Y. FX 6166.

——— (1960). The influence of sound on the behavior of amphibians and reptiles. In Animal sounds and communication (W. N. Tavolga, ed.). AIBS, Washington, D. C., pp. 137–320.

Buller, A. J. (1965). Mammalian fast and slow skeletal muscle. The scientific basis of medicine. Annual Reviews, 1965(11):186–201.

Capranica, R. R. (1965). The evoked vocal response of the bullfrog. Res. Monograph 33. M.I.T. Press, Cambridge, Mass., 110pp.

———, L. S. Frishkopf, and E. Nevo (1973). Encoding of geographic dialects in the auditory system of the cricket frog. Science, 182(4118): 1272–1275.

Cowles, R. B., and R. L. Phelan (1958). Olfaction in rattlesnakes. Copeia, 1958:77–83.

de Jongh, H. J., and C. Gans (1969). On the mechanism of respiration in the bullfrog, *Rana catesbeiana*: A reassessment. J. Morph., 127(3):259–290.

Denison, R. H. (1941). The soft anatomy of *Bothriolepis*. J. Paleont., 15:553–561.

Gans, C. (1962). The tongue protrusion mechanism in *Rana catesbeiana*. Amer. Zool., 2(4):524.

——— (1966). An inexpensive arrangement of movie camera and electronic flash in the study of animal behavior. Anim. Behav., 14(1):11–12.

——— (1970). Respiration in early tetrapods—the frog is a red herring. Evolution, 24(3):740–751.

*——— (1971). Strategy and sequence in the evolution of the external gas exchangers of ectothermal vertebrates. Forma et Functio, 3:61–104.

——— (1973). Sound production in the Salientia: Mechanism and evolution of the emitter. Amer. Zool., 13(4):1179–1194.

———, H. J. de Jongh, and J. Farber (1969). Bullfrog (*Rana catesbeiana*) ventilation. How does the frog breathe? Science, 163:1223–1225.

Goslow, G. E., E. K. Stauffer, W. C. Nemeth, and D. G. Stuart (1972). Digit flexor muscles in the cat: Their action and motor units. J. Morph., 137(3):335–352.

Gutmann, W. F. (1969). Zu Bau und Leistung von Tierkonstructionen 9. Die Entstehung der Wirbeltiere. Natur und Museum, 99(2):45–55.
*Hill, A. V. (1970). First and last experiments in muscle mechanics. Cambridge Univ. Press, Cambridge, xv + 141pp.
Jonsson, B., and S. Reichmann (1968). Reproducibility in kinesiologic EMG-investigations with intramuscular electrodes. Acta Morph. Neerl.-Scand., 7:73–90.
Kallen, F. C., and C. Gans (1972). Mastication in the little brown bat, *Myotis lucifugus*. J. Morph., 136(4):385–420.
*Katz, B. (1966). Nerve, muscle, and synapse. McGraw-Hill Book Co., New York, xi + 193pp.
Martin, W. F. (1971). Mechanics of sound production in toads of the genus *Bufo*: Passive elements. J. Exper. Zool., 176:273–294.
*———— (1972). Evolution of vocalization in the genus *Bufo*. In Evolution in the genus *Bufo* (W. F. Blair, ed.). Univ. of Texas Press, Austin and London, pp. 279–309.
————, and C. Gans (1972). Muscular control of the vocal tract during release signalling in the toad *Bufo valliceps*. J. Morph., 137(1):1–27.
McMahon, B. R. (1969). A functional analysis of the aquatic and aerial respiratory movements of an African lungfish *Protopterus aethiopicus*, with reference to the evolution of the lung-ventilation mechanism in vertebrates. J. Exper. Biol., 51:407–430.
Osse, J. (1969). Functional morphology of the head of the perch (*Perca fluviatilis* L.): An electromyographic study. Netherlands J. Zool., 19(3):289–393.
———— (1972). Respiratory and feeding mechanisms of *Amia calva* with notes about the role of the branchiostegal apparatus. Abst. Amer. Soc. Ichthyol. Herpetol., 52nd Meet., Boston, p. 76.
Rahn, H., K. B. Rahn, B. J. Howell, C. Gans, and S. M. Tenney (1971). Air breathing of the garfish (*Lepisosteus osseus*). Respir. Physiol., 11:285–307.
Schmidt, R. S. (1965). Central mechanisms of frog calling. Behavior, 26:251–285.
———— (1972). Release calling and inflating movements in anurans. Copeia, (2):240–245.
———— (1973). Central mechanisms of frog calling. Amer. Zool., 13:1169–1177.
Sherrington, C. S. (1904). Correlation of reflexes and the principle of the common path. Proc. Brit. Assoc. Adv. Sci. (Cambridge), 728–741.
Spallanzani, Abbe (1784). Dissertation relative to the natural history of animals and vegetables. J. Murray, London, vol. 2; translation of 1780. Dissertazioni di fisica animale e vegetabile. Modena.
*Wilkie, D. R. (1968). Muscle. Studies in biology, no. 11. St. Martin's Press, New York, 64pp.

INDEX

For the sake of brevity, some terms have been combined; for instance, the English forms of scientific names frequently appear with the Latin term. Entries for muscles and bones refer to those pages where they are defined and illustrated; only some of the incidental citations are indexed.

All entries are indexed by page numbers. Italicized page numbers refer to *figures* and their captions, numbers followed by a *b* to *boxes,* and those followed by an *f* to *footnotes.*